FOOD SAFETY

Dr. Renu Agrawal is the Ex. Chief Scientist at CSIR-CFTRI, Mysuru, India and rural program coordinator for outreach activities at CFTRI. She has over 20 patents to her name, with 65+ publications in International and National Journals. She has presented more than 160 papers in various national and international conferences and delivered more than 600 talks. She has guided students for M.Sc, M.Phil and PhD in Biotechnology and Microbiology. She has represented India as a team leader taking scientific delegation for the Asia meet at Indonesia, Bali nominated by DST, Government of India. She had her PhD from University of Rajasthan, Jaipur. Her areas of specialization include biotransformation and probiotics. Her research involves lactic acid bacteria, yeasts and Bacillus as probiotic properties inhibiting harmful diseases and reduce the use of antibiotics improving health in the society. Her group has developed greener probiotic technologies used by the common man. As CSIR-800 Rural development coordinator she led CSIR for CSIR 800 to give a better quality of life to 800 million people in India through scientific and technological intervention. Many awareness camps, demonstrations of technologies were conducted.

She has been awarded at multiple scientific associations. Some honourable mentions are Scientific fellow awards as FAMI, FNABS, FSAB, FISBT. She has been awarded Karamveer Chakra by REX-UN, Best woman scientist NABS, Woman achiever from International Guild of woman achievers, Reflections Excellence Award from Rotary Innerwheel, Kissan Ratna puraskar, Abdul Kalam Lifetime achievement Award, Nari Shakti Award to name a few. She has been selected as the "Best woman scientist" by National Academy of Biological Sciences, felicitated by Rotary Inner wheel for scientific excellence with "Suguna Award" and felicitated at various scientific and societal forums. In Rotary Innerwheel she was awarded the Best secretary for two times and also Best ISO twice in the district 318. She is the editor of six Research Journals and has authored scientific books on probiotics and Utilization of agricultural wastes. She has written many book chapters with Marcell Dekkar, USA and Taylor and Francis, UK. Her work has won many best paper awards at the national and international conferences. She has taught as faculty of M.Sc Food Technology at CFTRI and to ACSIR students. She has served as the president of 'Association of Microbiologists of India", Mysore chapter. She has been on the board of many international Technical Advisory committees at Dubai, Bulgaria, Argentina and Indonesia. She has been on the National committees of NCERT of National Science Talent Search, scientist selection boards of UPSC, DST and DBT and many International technical committees. She is an Executive committee board member of "National Academy of Biological Sciences".

She regularly writes for City Today and BiostandUps a science magazine. She has been nominated as an Expert Committee member on Scheme for Young Scientists and Technologists (SYST), Department of Science and Technology, Govt. of India, nominated as a member of National Experts Advisory Committee (NEAC) from Government of India, Ministry of Science and Technology for evaluating and monitoring various programmes involving children, Children science congress, innovative research and outreach. Other committees of DST, Gov. of India where she is a member are TIASN (technological intervention for addressing societal needs, ASAR , CODER, projects to be taken in cold and Desert areas, "Pandit Deen Dayal Upadhyay Gram Sankul Pariyojna", AWSAR for augmenting writing skills and 'KIRAN" which involves projects for women. She has been nominated to the Consultative Group made by the Principal Scientific Advisor, Dr. Vijay Raghavan, cabinet rank, who reports to the prime minister of India for inputs on science and technological developments in the country. She has been honoured as "SHERO" by Biostandups, an online science magazine and is on their editorial board. She is also involved in the upliftment of rural education of Karnataka through "Pratham" in teachers training workshops and in translating course material to be made into short videos to make science easy and fun for the children. For 35 years, Dr. Renu has been leading innovations and change in organisations. Her programs and interviews are telecasted on Doordarshan (DD science).

Currently she is a science communicator, reviewer, author , visiting professor in many universities and colleges and a National Expert Advisor in the ministry of science and technology. Her commitment to work can be seen as she was taken as Project Advisor by CFTRI even after her retirement. She has been serving the society by her work. She served as the secretary of "CSIR Pensioners Welfare Association, Mysuru centre. She is a philanthropist and has been running many fellowships and national level competitions for poor meritorious students, constructed pre fixed auto stand in Mysore at Hardinge circle for public use. She has donated a serum analyser for rapid detection of HIV to Asha Kirana hospital at Mysuru.

FOOD SAFETY
Making Foods Safe and Free From Pathogens

Renu Agrawal
FAMI, FNABS, FSAB, FISBT
Ex. Chief Scientist
Food Microbiology and Fermentation
Central Food Technological Research Institute
Mysuru, Karnataka

CRC Press
Taylor & Francis Group
Boca Raton London New York

CRC Press is an imprint of the
Taylor & Francis Group, an **informa** business

NEW INDIA PUBLISHING AGENCY
New Delhi – 110 034

First published 2021
by CRC Press
2 Park Square, Milton Park, Abingdon, Oxon, OX14 4RN

and by CRC Press
6000 Broken Sound Parkway NW, Suite 300, Boca Raton, FL 33487-2742

© 2021, Renu Agrawal

CRC Press is an imprint of Informa UK Limited

The right of Renu Agrawal to be identified as the author of this work has been asserted by her in accordance with sections 77 and 78 of the Copyright, Designs and Patents Act 1988.

Print edition not for sale in South Asia (India, Sri Lanka, Nepal, Bangladesh, Pakistan or Bhutan).

British Library Cataloguing-in-Publication Data
A catalogue record for this book is available from the British Library

Library of Congress Cataloging-in-Publication Data
A catalog record has been requested

ISBN: 978-0-367-70199-4 (hbk)

Centre for Natural Biological Resources and Community Development
(Recognized as SIRO by DSIR, GOI)

Bangalore
26-08-2019

FOREWORD

Human beings have been concerned about the concept of food safety since time immemorial. Around the globe there has been major improvement in food safety management in the last twenty years. The book "Making Foods Safe and Free From Pathogens" authored by Dr. Renu Agrawal aims to create awareness in food safety management. Looking into the progress made on the ways to make food safe and how to make them free from pathogens, this book, will explain food hygiene, the journey of research been taken in food safety till date and the challenges that we are going to face in future to ensure food safety and its wholesomeness. It also includes the role and responsibilities of the various sectors of society, namely governments, food industry, consumers and academia. It is important that for the empowerment of the public they understand how to go about it. It has been found that the problems with Asian food safety involves primary producers who work under difficult environmental conditions, where the raw materials has to be transported to long distances and complex supply chains. The final destination to the consumer is the retail foods store, a restaurant, which are usually presumed to be safe, nutritious and tasty. However the consumer is not aware of the safety of the food or adulteration if any. This book will help the community at large to know and be aware of food safety. It will also help the consumers in keeping the environment green, clean and healthy and ultimately improving human health. The book also provides a science and risked-based intervention approach to the food industry for the control of food safety hazards. The book has been divided into 15 chapters. Chapter1 deals with the introduction to Food safety. Chapter 2 deals with What is food Safety? wherein it describes the handling, preparation and storage of food that can prevent food-borne diseases and about Food Recall..Chapter 3 deals with How can we make the food safe? How important it is the way we cook our food. It also deals with while cooking how can there be cross contamination starting from raw foods. Chapter 4 deals with How good are

41, RBI Colony, Anand Nagar, Bengaluru – 560024
Email: djbagyaraj@gmail.com; jemravi@gmail.com; Landline: 080 – 23335368

Centre for Natural Biological Resources and Community Development
(Recognized as SIRO by DSIR, GOI)

organic foods? It has been observed that in recent times organic and local foods are preferred as they use less pesticides and/or fertilizers. The author has brought out the details of these foods, like others, can be exposed to harmful bacteria during the growing and harvesting process and how they can be protected. Chapter 5 deals with the Role of Packaging material in Food and the systems where the chemicals do not leach into the food during cooking. Chapter 6 deals with the Food adulteration and the harmful side effects as it is an unethical and criminal malpractice but is very common in countries of the South-East Asia region. Chapter 7 deals with the Food Pathogens. Chapter 8 deals with genetically modified Foods and their safety. Chapter 9 deals with the Role of WHO in food safety and legislations. Chapter 10 deals with HACCP and GMP practices. Chapter 11 deals with the Role of Bacteriocins in Food Safety. Chapter 12 deals with the safety of Space foods and what food should be carried in a spacecraft. Chapter 13 deals with Food security bill. Chapter 14 deals with Challenges in food safety and Chapter 15 deals with Conclusions and future perspectives of food safety. Main features of the book includes: the three food safety hazards: microbial, chemical and physical, with a special emphasis on microbial hazards and food recalls are explained. It also explains a practical guidance on how to control food safety hazards. The information in the book is presented in a straight forward, instructive manner.

This book provides a very comprehensive review starting from the introduction of food safety to the future implications. The book provides the most recent updates in the area with a holistic approach. The author has been working in the field of food safety and has contributed towards novel bacteriocin production by lactic acid bacteria. I am certain that the book will be useful to students, teachers and researchers interested in this topic.

D. J. Bagyaraj, Ph.D., FNA, FNASc, FNAAS, FNBS, FPSI, FAMI, FSAB
INSA Honorary Scientist & Chairman
Centre for Natural Biological Resources and Community Development
Bangalore-560024

41, RBI Colony, Anand Nagar, Bengaluru – 560024
Email: djbagyaraj@gmail.com; jemravi@gmail.com; Landline: 080 – 23335368

Preface

The food we eat has become the main focus in these days. Everybody is conscious to eat safe food. The public is concerned about the concept of food safety and is keen to know methods to go about it. This book will enrich the readers on the major improvement been made in food safety management in the last twenty years. This book, will explain food hygiene, the journey of research been taken in food safety till date and the challenges that we are going to face in future to ensure food safety and its wholesomeness. It also includes the role and responsibilities of the various sectors of society, namely governments, food industry, consumers and academia. It also deals with HACCP, GMP practices and Food laws. It is very important to empower the public with all these aspects which will help them to understand and also allow them towards safe food. This book is unique as it has included the causes of food allergies, adulteration, genetically modified seeds and crops, GM fruits and vegetables and the effect on human body. It has also discussed the difference between traditional and organic farming. The book will be helpful to know the foods to be used in space shuttle. The book has also discussed the role of FDA and WHO in food safety which is a very important aspect in food safety. The role of bacteriocins obtained from bacteria of GRAS status; as natural preservative is very important. The author has discussed this aspect in detail. The role of packaging in food is another very important aspect in keeping the shelf storage of food. This has also been discussed in details. At the end, the author has looked into the future challenges in food safety. It will also help the consumers in keeping the environment green, clean and healthy and ultimately improving human health. This book gives a very comprehensive review starting from the introduction of food safety to the future implications. The book provides the most recent updates in the area with a holistic approach.

Renu Agrawal

Contents

1

Introduction

Human beings have been concerned about the concept of food safety since time immorial. Around the globe there has been major improvement in food safety management in the last twenty years. Looking into the progress made on the ways to make food safe and how to make them free from pathogens, this book, will explain food hygiene, the journey of research been taken in food safety till date and the challenges that we are going to face in future to ensure food safety and its wholesomeness. It also includes the role and responsibilities of the various sectors of society, namely governments, food industry, consumers and academia.

In the beginning of the twentieth century many modern food safety policies have been established. In the packing of meat and in food processing industries a lot of work has been done (Sinclair 1906). Another second generation policy includes the outbreak in the United States in early 1990s' of pathogen *E. coli* O157:H7 and the BSE scandal of the late 1990s in the UK Lopman et al., (2003) have discussed on viral outbreak in europe. Currently, the crisis has come due to the economic and technological transformations which have taken place in food and the food supply system. As these changes have many health risks the public is very concerned. Therefore, the control forms of safety regulation and food safety laws have come into the picture. These policies include continuous line inspection, visual product inspection and detailed specification of approved hygiene practices. There is a need of a public health-focused policy which works towards the preventive integrated management of foodborne hazards from farm to table. This includes the use of modern science based risk management tools which will increase the efficiency accurately towards public health. A lot needs to be worked out for better implications. Current efforts to modernize food safety policy around the world are being worked out by the use of risk analysis and cost-benefit analysis in public administration and also the quality management in industry. The role of public economics plays an important role in food safety policies. For the social benefits the policy makers need to take decision on the costs and strengthening of the risk assessment. Today there are globalized markets and this can influence the health globally. Emerging economies like China are industrializing and urbanizing rapidly like that of Europe and North America.

Adrianna and Gay (2016) have written a review on the impact of preparation and cooking on nutritional quality of vegetables and legumes. According to US Pharmacopeia (2010) and Gale and Hu (2009) the adulteration of pet food, milk and toothpaste has demonstrated the need for an institutional capacity of industry and government. The levels of diseases presently reflect not only the past but also the current investments in controlling foodborne hazards. In case there is a failure to maintain these controls it might lead to the reemergence of disease problems. Barker et al., (1999) found that *E. Coli* survives in soil protozoa.Sudershan et al., (2014) have discussed food borne infection in Hyderabad. Sobal and Lachaussee (1993) have looked into the social science perspective of the consumers. In 1999, Belgian animal feed was unintentionally contaminated with dioxin in polychlorinated biphenyls (PCBs) and distributed to approximately 2,500 farms. In the winter of 2008–2009 due to the neglect of a leaky roof, more than 700 people became sick by *Salmonella* infected peanut products in the U. S. (U. S. CDC 2009). Therefore, there is a possibility of failure in private management and public enforcement (Covaci et al. 2008). It has been observed that without continuous control of diseases like Trichinosis in U. S. it can reemerge. Anonymous et al., (2008 b) have studied the effect of climate change on food safety.

For food safety it needs each human being to be responsible. It has been noted that many a times most of the human beings would have encountered with a foodborne illness. Bloomfield et al., (2012) have studied the infection in home food. This could happen due to the consumption of food contaminated by microbial pathogens, toxic chemicals or radioactive materials. Some foods cause food allergy which needs to be addressed as it is an emerging problem. Many foodborne diseases cause serious ill effects and sometimes even death. Bennet et al., (1999) have studied the integrated approach to food safety. Today, food safety has become extremely important due to the changing food habits, popularization of mass catering establishments and the globalization of food supply. WHO has taken a pivotal role in order to promote efforts for improving food safety from farm to plate. The World Health Day is celebrated on 7th April. In 2015, the slogan given by The World Health Day was "From farm to plate, make food safe". In most countries of the South-East Asia Region the food production and marketing system is being carried out informally and because of this, there are, many challenges to enforce food safety regulations. This allows a lot of contamination. Traditionally, the need is for hot and well-cooked food and milk to be boiled (pasteurized). This has been the practice mainly to prevent foodborne infections. In many countries street food is very common.

These need good hygienic conditions and drinking water. Hygienic conditions are improving, provided that potable water supply and clean facilities are available. (WHO, 2002) have looked into the ways to reduce the risks. The introduction of bottled drinking water has contributed to prevent many waterborne diseases especially diarrhoea. However, there is an urdent need of political awareness and consumer education on food safety. This will allow strengthening in enforcing the food standards, improving hygienic practices and in preventing foodborne illnesses. The "WHO Five keys to safer food" is Ln8 after safer food:(WHO, 2019) the basis for educational programmes to train people handling food and also in educating the consumers. These will help in preventing foodborne illnesses. The Five keys are to keep the food surfaces clean; wash all utensils, plates, platters, and cutlery as soon as used; separate raw food from cooked food; cook food thoroughly; to the appropriate temperature, keep food at safe temperatures; both for serving and storage and use safe water and raw materials.

The food safety must start at the farm level itself. The more usage of agro-chemicals, pesticides, growth hormones and veterinary drugs may have harmful effects on human health. The microbial and chemical risk factors should be checked at the farm-level (using water contaminated by industrial waste or poultry farm waste for irrigation of crops). Good agricultural practices should be used in order to reduce the harmful microbes and chemicals. Today, a lot of emphasis is being given to Organic farming (without the use of pesticides). Organic products are expensive as compared to commonly available food items. However, there is a tendency among health- conscious consumers to buy organic foods. Louise et al., (2012) have studied the consumer perception on food safety in Asian and Mexican restaurants.

Attention to food safety should be given during harvest, transport, processing, storage and finally during food preparation and storage by the consumers. In recent years the food habits are changing and therefore much attention is required towards processed, frozen or ready-to-eat foods. These foods are gaining popularity due to product diversification, busy lifestyle and mass production practices. It has been found that in the cities where both husband and wife work outside home usually there is a tendency to buy meat, milk and vegetables and fruits on the weekend and store them in the freezer or in the refrigerator. Microwave ovens are often used for reheating of food. Using a refrigerator and microwave are part of daily life in urban settings, most users do not know how to safely store and reheat food. Michael et al., (2015) have brought out the role of food industries to prevent illnesses in US. Looking into this and considering that there is a knowledge gap among the consumers and food handlers, specific focus has been given on these concerns in this book.

2

What is Food Safety

It is a discipline of science that describes handling, preparation and storage of food that can prevent food-borne diseases. When there are two or more cases of a similar illness from the ingestion of the same common food it is known as a food-borne disease outbreak (FSN, 2018). In order to avoid potential health hazards, food safety is very important. This needs safety between industry, market and the consumer. For food safety the labelling on food material is vital. This includes the origin of food, steps taken for food hygiene, food additives, pesticide residues, policies on biotechnology, food guidelines for the management of governmental import, export inspection and certification systems for foods. The food should be safe in the market with safe preparation and delivery of the food for the consumer. Foegeding et al., (1996) have done an assessment of risk factors with food borne pathogens. Warnock et al., (2007) have reported intervention to food safety.

When pathogens are in the food they may be transmitted to the consumer. It may result in an illness or death of the person or other animals. The pathogens include bacteria, viruses, molds and fungus. These multiply in the food medium. Joon et al., (2012) have found pathogenic bacteria in meat products of Korea. The developed countries have high intricate standards for the preparation of food; however, in the developing countries there are less standards and enforcement of the standards which are followed. Apart from food another major source of infection is water, which is very critical in spreading many diseases. It is possible to prevent food poisoning if good water is used for all purposes. However, it becomes difficult as in the food supply chain a large number of persons are involved apart from the fact that pathogens can be introduced into foods no matter how much precautions are taken. Cowden (2000) has studied the notification on food poisoning to improve health. Frenkel et al., (2015) studied the amount of wastes from food processing industries.

Food quality and food safety
Safety measures must be taken from the farmers/suppliers to the consumer level. This will ensure safety, nutritional quality and acceptability of the delivered

foods. The safe food has more acceptability by the consumer. While purchasing any food the consumer looks for two things, the food safety and its quality. While buying juice the consumer looks for quality as presence/absence of heavy metals and pesticide residue which is food safety along with the acidity in the juice concentration which becomes the quality issue. This needs to be monitored as Critical Control Points (CCP). There should be assurance that food will not cause harm to the consumer when it is prepared and/or eaten. Only then the food can be considered as safe. Food contaminants can be of various types which include physical agents. These include non living physical things which we can see with our eyes like bone pieces, metal pieces, hairs, feathers, glass, stones and dirt. The biological agents include the living things present in foods like bacteria, viruses and parasites. The major food contaminants as bacteria include; *Salmonella, Staphylococcus aureus* and *E. coli.* There can be viruses like hepatitis A, rotavirus, or other viruses. There can be parasites like *Trichinella spiralis* and *Cryptosporidium parvum.* Bennet et al., (1999) have studied an integrated approach to food safety. Hald et al., (2013) have looked into the advances in microbial food safety. Hall et al., (2013) have associated health with the natural outbreaks. Health (2015) has studied the partnership for food safety education. Randell et al., (1997) have given the CODEX for international trade.

There can be chemical agents present in the food as harmful chemicals. Among the toxins there are naturally occurring mycotoxins (aflatoxins), added intentionally. These include pesticides, hormones, antibiotics and food additives. There can be chemicals added non-intentionally which include cleaning and sanitizing agents. Most of the time the body shows harmful effect which may happen after eating food. These include headache, nausea, vomiting, diarrhea, fever or more severe symptoms. Food quality is a complex concept that includes appearance, flavour, texture and nutrient content. It is an expression of degree of excellence of food. It is the quality that influences a product's value to the consumer. Good quality may be the origin, colour, flavour, texture and processing method of the food. Bad quality of food shows visible spoilage, contamination with filth, discolouration or off odours or tastes. Sometimes the unsafe food may appear to be of good quality such as tainted meat. It influences the nature and content of the food control system. With lot of awareness in general public there is a vital need to maintain food security. The quality of food which comes to the market should be of good quality and must be safe for consumption. There should not be any source of disease and infection. Therefore, securing food safety and quality has become a matter of international significance and a responsibility of food producers and governments. In the supply chain, the food products go through various stages like different processes which describe the travel of food from a farm to the consumers¼ table. The various steps from food production to the consumer are very time and space consuming. There are great chances of food getting contaminated during production, transport and food storage or food preparation. In order to enable food quality and sanitary safety of food products, companies have to follow legislations, standards and norms at every stage of the supply chain. FDA (2009) has given a report of the risk factors involved in food borne illnesses.

How can the food quality be evaluated?

In order to routinely monitor the food quality and to ensure that the foods being produced in the food industry are acceptable by the consumers. It is important to evaluate both sensory and objective evaluation of the food. Traditionally, qualities of foods are evaluated by our sensory organs (eyes, nose, mouth or recent instruments). By food regulatory authorities the sensory evaluation is made by constituting a panel of judges. It is a scientific discipline which analyses and measures the human responses to the composition of food and drink. It includes appearance, touch, odour, texture, temperature and taste. It is not only the methods to analyse foods for pesticides, veterinary drug residues and adulteration. Martin et al., (1971) have studied the different aspects of food quality.SABS ISO 9000(2000) explains the quality management systems and fundamentals. Van der Spiegel (2005) developed an instrument to measure food quality management. Buluswar et al., (2018) have looked into the urban food quality initiatives.Under the objective evaluation the product is checked by chemical, physiochemical, microbial and physical method of analysis. There are chemical methods which can determine the nutritive value of foods before and after cooking. They can also detect the products obtained after decomposition and adulteration in foods. However, usually the measurement of physical properties is made by instruments like the appearance and volume of foods. For the routine quality control of foods and food products an objective evaluation of food is required. This is done by instrumentation. Today, the sensory properties in any product are very vital as the consumer is able to check it immediately. In recent times, to understand both the product and the consumers together newer methods and tools are required. Objective measurements are very fast, reliable and repeatable. The subjective measurements are sensory analysis which can be modulated by the use of different psychophysical model and understanding of fundamental physical behaviour of food in the mouth. Many statistical tools and mathematical modelling techniques are useful in correlating the instrument measurement with the panel measurements.

Martin et al., (2012) have written a good article on the food quality evaluation of learning techniques. According to them the food safety includes all hazards that may be chronic or acute. It makes the food injurious to the health of the consumer. It includes all the steps starting from production, handling, storage and preparation of the food to prevent infections in the food production chain. It must also ensure the quality of food which should be maintained to promote good health. It also helps to make sure that our food contains enough nutrients for a healthy diet. Unsafe food and water means that it has been exposed to dirt and germs or is rotten. This causes infections or diseases such as diarrhoea, meningitis and many others. These infections make the people sick and cause different diseases. As the body immunity goes down the absorption of nutrients becomes difficult. In case there is deterioration in the quality and in the nutrients of food, malnutrition can take place.

Rama et al., (2009) have studied the effect of water hygiene to reduce diseases. There should be separate water sources for livestock and humans. Contamination of drinking water needs to be prevented, especially from human and animal faeces. It is important to dispose off the household waste in bags and bins. This helps in avoiding the decaying of food and infection of flies which carry disease causing microorganisms. Many diseases can be prevented by using pit latrines and washing the udders of livestock before milking, keeping milking pails clean and properly stored. While purchasing foods it is important to check the label details. The food while preparation should be cooked properly as it destroys the germs. In order to get rid of germs it is important to wash hands with soap and clean water after going to the toilet, changing babies' nappies and working with animals. Washing hands before working and eating with food, cleaning of fruits and vegetables in clean water before eating and cooking reduces infection. Washing dishes and utensils immediately after using and storing them in a clean place is a step forward towards food safety. While purchasing fruits and vegetables it is good to purchase fresh ones. Boiling water and milk for at least 5 minutes is found good. Cracked eggs should not be used and before using, it is advisable, to wipe eggs with a clean cloth. Food should be always cooked or reheated properly to avoid overcooking as it can destroy the nutrients. Cooking food in too much water is not advisable as the nutrients get lost and they go into the water. While keeping raw and cooked food, care must be taken, to keep them separately when preparing or storing them. This will reduce contamination of pathogens. In case of keeping the food at room temperature the cooked food should not be kept for a longer time. Or otherwise it is better to consume right away. At refrigerated temperatures not many pathogens can grow and it keeps the food safe. On raw meat the pathogenic microorganisms grow very rapidly and therefore, it should be kept in the freezer. Covered and clean containers protect food from dust, insects or rats and coolness helps food to last longer. Metal tins can get rusted. Keeping, in the food grade clean plastic containers with lids, has been found useful. Incorporation of human and animal faeces into water should be completely prevented and therefore pure water should be used for cooking rather than from rivers and lakes. Keeping Chickens as pets is very risky as their faeces carry germs and attracts flies, rats and other organisms which spread disease. Scallan et al., (2011), Scharff et al., (2010), Roberts et al., (2007) have estimated the societal costs of US food borne illnesses. Saramento et al., (2004) have given a road map for food industries in California.

What is a Food Recall and how it is done?

Many a times some foods may be a concern to the consumers. At that time the food manufacturer /distributor initiates to take foods off the market. This procedure is known as Food recall. Sometimes the government agencies (USDA or FDA) take action in this regard. This protects the public from products that may cause health problems or may be fatal. It also happens when the products may be adulterated or misbranded. In a product recall a manufacturer usually requests the consumer to return the product after the product is identified to have defects that might have a danger to the consumer or put the maker/seller at risk of legal action. Under Food recall it is necessary to know who will decide, how unsafe is the product and what are the results of the preliminary investigation. The Food Safety and Inspection Service (FSIS) must notify the public, during a recall, the FSIS' must ensure how much is a recall effective, keeping documentation on the recalls undertaken for a particular product, the ways to identify the recalled products and documentation of recall information's. In USA the FSIS inspects and regulates meat, poultry and processed egg products which are produced in federally inspected plants. It is the responsibility of FSIS to ensure that these products are safe, wholesome and labelled with all details. Other food products are regulated by the Department of Health and Human Services' Food and Drug Administration (FDA). FDA (2009) has given a report for risk factors in food borne illnesses. The recalls are voluntary. In case of a problem and if the company refuses to recall its products, then FSIS has the legal authority to detain and seize those products in commerce. Unsafe foods can be known by four primary means as the company that manufactured or distributed the food, information to the FSIS of the potential hazard, the results received by FSIS could indicate that the products are adulterated or misbranded. Investigation by FSIS field inspectors, and epidemiological data submitted by the State or local public health departments or other Federal agencies like the Food and Drug Administration (FDA) are very vital. It is important to report unsafe, unwholesome or inaccurately labeled food. FSIS has given guidelines for the safety of meat and poultry products. If FSIS is informed about a potentially unsafe or mislabeled meat or poultry product in commerce, then the agency usually conducts a preliminary investigation to determine whether there is a need for a recall or not. The preliminary investigation may include contacting the manufacturer of the food for more information; interviewing any consumers who allegedly became ill or injured from eating the suspect food; collecting and analyzing food samples; collecting and verifying information about the suspected food; discussions with FSIS field inspection and compliance personnel; contacting state and local health departments and documenting a chronology of events. In usual practice FSIS notifies the public through a Recall Release for Class I and Class II recalls. It issues a Recall Notification Report (RNR) for Class III recall issues. (The RNR provides substantially the same information as the Recall Release. The Recall Release is made public to the media outlets where the product had been given. These are posted on the FSIS Website and distributed to FSIS email subscribers. Whenever possible, FSIS includes pictures of the recalled product labels as part of the FSIS online Recall Release posting. For every Class I recall, FSIS develops a list of retail consignees that have, or had, the recalled products in their possession. There is a list of retail consignees prepared which includes the name, street address, city and state of each retail consignee. This is posted within 3 to 10 days from the date of the recall. Thereafter, the retail consignee list is updated periodically as additional retail consignee information becomes available. Usually, the recall press releases and RNRs are posted on the FSIS Recalls area of the Web site, at www. fsis. usda. gov/Fsis_Recalls/Open_Federal_Cases/index. asp. There is a facility to the public for a request to receive FSIS press releases and recall announcements by subscribing to the Agency's email subscription service. Further information can be obtained from www. fsis. usda. gov/News_&_Events/ Email_Subscription/index. asp. FSIS' newsletters, including the Constituent Update, are also available via email at www. fsis. usda. gov/News_&_Events/ Newsletters/index. asp. In case, the recalled products are purchased by USDA and distributed through a food distribution program, (such as the National School Lunch Program) in that case the FSIS notifies the Federal agency responsible for the food program, and that agency will hold the product. If the adulterated or misbranded products have entered commerce, the FSIS Recall Management Division convenes the Recall Committee meeting, a standing committee within FSIS. This Committee consists of FSIS scientists, technical experts, field inspection managers, enforcement personnels and communication specialists. They evaluate all the available information and then make recommendations to the company about the need for a recall. If the Recall Committee recommends a recall, the Committee classifies the recall on the relative health risk basis.

These include: Class I - which involves a health hazard situation in which there is a reasonable probability that eating the food will cause health problems or death. Class II- recall involves a potential health hazard situation in which there is a remote possibility of adverse health consequences from eating the food.

Class III- recall involves a situation in which eating the food will not cause adverse health consequences. The Recall Committee also verifies all details. They advise the company and provide an opportunity for the firm to offer any information.

FSIS conducts effectiveness checks throughout the distribution chain in order to ensure that the recalling firm makes all reasonable efforts. If FSIS determines that the recalling firm has been successful in contacting its consignees and has made all reasonable efforts to retrieve and control products; it notifies the firm that the recall is complete and no further action is expected. There is a Recall Management Division that maintains comprehensive case files for all recalls which is coordinated by FSIS. Information on open and closed federal cases can be found on the FSIS Web site at www. fsis. usda. gov/Fsis_Recalls/index. asp.

The labelling on all the containers of meat, poultry and egg products should be labeled with a USDA mark of inspection and establishment (EST) number assigned to the company where the product was produced. Poultry Product "EST" Number Establishment numbers for poultry plants can be identified with the prefix "P" for "Plant" prior to the number. The plant number is also permitted to appear on the exterior of the container which includes either on a metal clip to close casings or on aluminum trays placed within containers. A statement of the location should be printed near or connected to the official inspection legend, such as "P. No. on Metal Clip" or "P No. on Pan. " The number may not be applied over any required labeling information. EGG PRODUCTS "EST" Number Establishment numbers for processed egg products are found within the egg products shield or on the principal display panel prefaced with the term "Plant" or the prefix "P. " If any additional information on recalls of food and other products are required by the consumers the information can be obtained from The USDA Meat and Poultry Hotline at 1-888-MP Hotline (1-888-674-6854) weekdays from 10 a. m. to 4 p. m. ET (English or Spanish).

Recall of food product has interest for all. The industry, the government and the consumer. A recall is defined as an action to remove from sale, distribution and consumption of any food which may pose a safety hazard to consumers. The food traders voluntarily recall the food item. However, under the Public Health & Municipal Services Ordinance and food on sale for human consumption must be wholesome, unadulterated and uncontaminated. It should be properly labelled and fit for human consumption. Violations of any kind can lead to the prosecution against the manufacturer, the importer or the distributor of the food. Specifically, under Section 59 of the Ordinance an authorised officer may examine or seize any food which is considered to be unfit for human consumption.

The Food Safety Ordinance empowers the Director of Food and Environmental Hygiene ("the Authority") to make a food safety order in relation to food that will protect the public health under section 30. This order usually prohibits the import of any food, prohibits the supply of any food, directs for any food supplied to be recalled, directed that any food be impounded, isolated, destroyed or otherwise disposed of. It may also prohibit the carrying of an activity in relation to any food or permit the carrying of such activity in accordance with conditions. In a voluntary recall the government's main role is to monitor the progress of the recall and assess the adequacy of actions taken by the food trader concerned. It is the Food and Environmental Hygiene Department that makes sure that the product is destroyed or suitably improved after a recall is completed. The publicity about a recall is made by the Food and Environmental Hygiene Department at the time when public needs to be alerted about a health hazard or that clarification of the situation needs to be made. In cases of public health emergencies, it alerts the public before a decision on recall is reached. They can also come out with a Food Safety Ordinance. Role of any food industry is to carry the prime responsibility of implementing the recall and ensuring compliance with the recall procedure at its various stages including the follow-up checks. This ensures that recalls are successful and that subsequent batches of the food products are safe for human consumption. A recall should be undertaken in consultation with the Food and Environmental Hygiene Department and preferably with prior agreement on the recall strategy. During the recall process, company personnel must inform all the relevant parties of the latest developments. If the recall involves products exported overseas, the company concerned should notify even to the overseas recipients of the recalled stocks. A recall can be initiated arising from the reports/complaints referred to the food trader from a variety of sources. It could be anyone from the manufacturers, wholesalers, retailers, medical practitioners, government agencies and consumers. A recall of food manufactured overseas may also be initiated by reports appearing in overseas bulletins and similar publications by the health authorities or from the information received directly from such authorities. Recalls are carried out in the shortest possible time. Traders are allowed to develop their own recall procedure which helps to respond immediately in any emergency. This would efficiently retrieve any product which is potentially unsafe from the market. Food traders should immediately notify the Food and Environmental Hygiene Department on the cause of recall along with the risk to public health and safety. The information should be provided to the Department by fax or email. It can also be in the form of a press release, letter to the concerned parties, paid advertisement in the media, making public announcement and putting up posters in stores for receiving the food concerned from customers. Enough of telephone hotline services need to be available for any enquiries. The adequacy of the traders' action is assessed by the Food and Environmental Hygiene Department for prompt announcement of recall/ prohibition of supply or import through the media, setting up a customer enquiry service with an agreement amongst the importers, distributors and retailers on recall/ prohibition of supply or import arrangement. The food concerned should be removed promptly from the shelves at retail end. A proper investigation should be taken into the cause of the defect and the remedial action taken (investigation report with improvement measures to be submitted to Food and Environmental Hygiene Department). The manufacturers need to be very careful in storing such food separately and should not make them available for use. The retailers should keep a proper record of the quantity of the food withdrawn to be returned to the distributor. Products may be recovered from the supermarkets, return via distribution chains / direct return from consumers to a central site/recovery sites. There must be accurate records of the amount of recovered product and the batch codes of the product recovered. After recovery, products may be corrected or reprocessed before release to the market or if it is fit for human consumption. Otherwise the product is to be destroyed. At regular intervals the trader should provide a progress report at the request of the Department and later submit a final report within a specified time as decided by the Department. The reports should include the important information which includes, the circumstances leading to the recall; the action taken by the trader including details of any publicity; the extent of distribution of the relevant batch; the result of the recall (quantity of stock returned, outstanding, etc.); the method of disposal or otherwise record of destruction for returned food and the investigation report on cause of defects and the action proposed to be implemented so that in future recurrence of the problem can be prevented. Without the satisfactory reports the Department can take further action. These can be stepped-up inspection, extension of recall action or making a food safety order.

To be effective, recall notification must reach as far as the product has been distributed. The effectiveness of the recall is assessed on the amount of product returned as a percentage of the amount of product which left the manufacturer while taking into account the retail turnover of that product. Around the world it is the cooperation between the food traders and the regulatory authorities has proven over the years to be the most reliable method to remove potentially dangerous products from the market. These guidelines enhance the efficiency and transparency in the recall of food products. The implementation of such guidelines will hopefully minimize the loss inflicted on the traders and the community at large. Food recall is defined as an action taken to remove foods which may pose any safety risk from the sale, distribution or consumption to consumers. Food recall is a fundamental tool in the management of risks in response to food safety events and emergencies. A food recall may be initiated as a result of a report or complaint obtained from many sources that include manufacturers, wholesalers, retailers, government agencies and consumers. Recalls are conducted by food businesses to ensure that potentially hazardous or unsafe foods are not consumed. Primary cause of food recall is due to microbiological contamination, chemical or radionuclear contamination in food. It is mandatory for consumer protection in industrialized countries where food labelling is compulsory for prepackaged food and commonly practised by multinational food industries. In 2014, the latest year on record, 8,061 food products were recalled by the Food and Drug Administration, while there were an additional 94 recalls by the U. S. Department of Agriculture. Of those USDA recalls, two-thirds were "Class 1. This meant that the food would cause illness or death. Food Safety Modernization Act became a law in 2011. This (FSMA) granted the FDA an authority to force the companies to recall foods. The responsibility for recalls is given to the USDA, FDA, Food Safety Inspection Service and the companies manufacturing and selling these products. The USDA primarily handles meat, poultry and egg products. Companies are also required to follow a HACCP plan (1997 guidelines) that identifies possible hazards in the food chain and identifies ways to check and control for them. The recalls that make it into the headlines are even worse still. In 2014, nearly nine million pounds of beef were recalled after the Food Safety Inspection Service discovered it came from "diseased and unsound animals" that had not been properly inspected. It was the Peanut Corporation of America which had sent out salmonella-tainted peanut butter which resulted in the deaths of nine people and over 700 illnesses. An investigation resulted in the recall of two years' worth of peanut butter produced by the company, and eventually, jail time for the company's CEO.

Business Standard staff (2017) studied the guidelines for a product recall is the process of retrieving and replacing defective goods for consumers. When a company issues a recall, the company or manufacturer absorbs the cost of replacing and fixing defective products. Many a times for bigger companies, the costs of repairing faulty merchandise can accumulate to multi-billion dollar losses. Recently, car manufacturers Toyota (TM), General Motors (GM), and Honda (HMC) have suffered the embarrassing consequences of product recalls. Not only affecting automobiles, product recalls have occurred in the food, medicine and consumer electronics industries. Smaller companies with less cash flow find it very difficult to sustain the financial losses. However, large enterprises are able to withstand the financial consequences.

Some Notable Historical Recalls

Public that purchases goods needs confidence correctly and safely on consumerism especially in America. It is the responsibility of a number of government agencies to test and recognize faulty products. These agencies include the Consumer Product Safety Commission (CPSC), Food and Drug Administration (FDA) and National Highway Traffic Safety Administration (NHTSA), to name a few. If the food is unsafe or defective and is given to the public a recall is then issued by the supplier. In the early 2000's Ford (F) issued a recall of 6. 5 million vehicles with Firestone tyres. The defective tyres resulted in 1,400 complaints, 240 injures and 90 deaths in the U. S. Likewise, Toyota has issued a number of massive recalls beginning in 2009, ultimately recalling over 10 million vehicles due to numerous issues including gas pedals that stuck and faulty airbags. The drug industry has also suffered from devastating recalls. In the early 2000s, the drug manufacturer, Merck (MRK) recalled arthritis medication Vioxx, which increased the risk of heart attacks. The drug cost Merck $4. 85 billion in settled claims and lawsuits. Recently Keurig, a coffee machine manufacturer, recalled 7. 2 million single-service brewing machines due to claims of overheating. Regardless of the industry in which the recall occurs, it is evident that large companies are able to withstand both financial and reputation costs. According to the consumer protection laws, manufacturers and suppliers have to bear the costs of a product recall. Though insurance usually covers a minimal amount to replace defective products, however, many a times the product recalls result in lawsuits. A significant recall can cost multi-billion dollars. The lost confidence of the customers brings a long-term loss.

There has been a lot of transformation due to the quick and more efficient means of transportation in the global supply chain. In order to remain competitive, companies have increased global supply chains, offshoring and outsourcing at the cost of product reliability. For example, Apple (AAPL) iPhones can be broken down to hardware, casing and assembly from Mongolia, China, Korea and Europe. In such cases, the final product should comply with regulations in the country in which it is sold. This often results in recalls. Taking responsibility and quick action is the safest ways to save brand recognition from product recalls. While settlement claims and repair expenses can be robust, a decrease in stock prices will have longer lasting effects. To be competitive, the companies incorporate manufactured parts from around the world at the cost of reliability. In the food industry the recalls create a huge threat to the businesses and can result in catastrophic outcome both for the company and consumer. Foods and the ingredients in food products are increasingly grown, processed and consumed in different locations around the globe. The FAO/WHO International Food Safety Authorities Network (INFOSAN) coordinates communication regarding the contaminated food which has been internationally distributed to allow for food recalls in importing countries. Alerts are communicated to national INFOSAN Emergency Contact Points in each country for their attention and action.

Food safety is everybody's concern, and it is difficult to find anyone who has not encountered an unpleasant moment of foodborne illness at least once in the past year. Foodborne illnesses may result from the consumption of food contaminated by microbial pathogens, toxic chemicals or radioactive materials. Food allergy is another emerging problem. While many foodborne diseases may be self-limiting, some can be very serious and even result in death. Ensuring food safety is becoming increasingly important in the context of changing food habits, popularization of mass catering establishments and the globalization of our food supply.

3

How Can We Make The Food Safe

According to Futter (2001) it is very important the way we cook our food. If the cooking is inadequate it causes food poisoning. While cooking there can be cross contamination starting from raw foods. It includes hands or utensils which can cause food poisoning. Most foods, especially meat, poultry, fish and eggs should be cooked thoroughly to kill most types of food poisoning bacteria. Generally the food should be cooked to a temperature of 75 °C or hotter. When food is cooked, it should be eaten promptly, kept hotter than 60 °C or cooled, covered and stored in the fridge or freezer. Some people are more at risk from food poisoning than others. In order to buy safe food in the market like buying pre-cut produce as a half a watermelon or cut veggies from a salad bar, it is important to choose items that are refrigerated or surrounded by ice. They must be kept in separate bags. FDA (2019) has studied the contamination in lettuce. Shopping at a farmer's market is a great way to get locally-grown, fresh fruit, vegetables, and other foods. However, safety of these produce needs to be checked. Many markets have their own food safety rules and vendors must comply with them along with the government regulations. Fruits and vegetables must be washed thoroughly under running water just before eating, cutting or cooking. Fruits and vegetables must never be washed with soap or detergent or using commercial produce washes. Even if it has to be peeled before eating, it is still important to wash it first. Any bacteria present on the outside of items like melons can be transferred to the inside when it is cut or peeled. They should be refrigerated within two hours after preparation. The juice or cider must be treated (pasteurized) to kill harmful bacteria. Pregnant women, children, older adults and people with weak immune systems should drink only pasteurized or treated juice. Raw milk can harbor dangerous microorganisms like *Salmonella, E. coli,* and *Listeria.* World India (2017) has looked into the food regulations. These can pose serious health risks. Pregnant women, older adults and people with weakened immune systems are at higher risk for illness caused by *Listeria.* One source for these bacteria is soft cheese made from unpasteurized milk. It is vital to check the label to make sure that it is made from pasteurized or treated milk. Make sure that eggs are properly chilled at the market. FDA requires that untreated shell eggs must be stored and displayed at 45°F. While buying eggs it needs to be ensured that the eggs are clean and the shells are not cracked. Meat should be properly chilled even in the market and kept in closed coolers with adequate amounts of ice to maintain cool temperatures. If it is brought in an insulated bag or cooler during transit it is very beneficial to keep the quality. It should be always kept separately from your other purchases, so that the juices from raw meat (which may contain harmful bacteria) do not come in contact with produce and other foods. Essays (2018) have written a very comprehensive overview of food safety.

Pregnant women, young children, the elderly and anybody with a suppressed immune system come under vulnerable groups. Therefore, it is important to take care while preparing, cooking, serving and storing food for these groups. Food safety should be kept in mind while cooking high-risk foods as food poisoning bacteria grow more easily on some foods than others. Commonly the high-risk foods include; raw and cooked meat, including poultry as chicken and turkey and utensils in which food is kept like casseroles. High risk foods should be kept out of the 'temperature danger zone' of between 5 °C and 60 °C. In case they have been kept in these conditions for up to two hours the food should be reheated, refrigerated or consumed. If high-risk foods have been left in the temperature danger zone for longer than two hours, but less than four hours, they should be consumed immediately. Any high-risk foods that have been left in the temperature danger zone for more than four hours must be thrown. Zulfiqar et al., (2017) have found the prevalence of *Staphylococcus* in commercially processed food products in Karachi. Xevier (2007) has associated health with inadequate housing.

Food Preservation

According to Zhong and Wang (2019) Much care needs to be taken while preparation of food at home and reheating. The source of water is essential in food preparation as it is used for washing the food before cooking, used as a cooking medium, used to clean containers of food before and after preparation and used as the most important beverage. Therefore, it is important that water used for drinking and cooking purposes must be without harmful pathogenic bacteria. Water free from pathogenic bacteria and which is palatable is known as potable water. Potable water must be used for making ice as it is used in cold foods and in cold drinks. Another factor in the preparation of any food item is the oil. It has been observed that the cooking oil is reused for frying many times. Lot of precautions should be taken while frying, as oil, is an essential part of cooking. Oil should be poured at the beginning into the pan and not later when the pan is very hot. Each batch of food should be prepared in fresh oil.

However, it is not found practical and there is a tendency to reuse it many times. The reuse of oil gives rise to "free radicals" which can cause many illnesses. These free radicals can be carcinogenic and atherosclerosis which can lead to increase in bad cholesterol levels, blocking the arteries.

In order to minimize the health hazard of reusing oil it is advisable to use sunflower, soybean, mustard and canola oil which have a high smoke point. They do not break down at high temperatures and therefore become suitable to be used for frying. Oils which do not have a high smoke point such as olive oil should only be used for sautéing. The leftover oil after cooking or frying should be cooled down and then transferred into an airtight container through a strainer. In case when the reused cooking oil becomes dark in colour or becomes greasy and is also sticky than it is it is better to change the oil. The vessels of frying should be cleaned between uses. Food safety and illness (2010) and Food code (1993) have provided means for better health.

Many vegetables or fruits contain pesticide residues without any noticeable smell, taste or visual defect. Therefore, precautions must be taken to reduce dietary exposure to pesticide residue if agricultural products are sold in common market. The fruits and vegetables must be thoroughly rinsed and scrubbed. Many times they need peeling. Outer leaves of leafy vegetables like cauliflower and cabbage could be removed. Fat should be trimmed from meat, poultry and fish. Oils and fats in broths and drippings must be discarded. The little fish takes up fewer pesticides and other harmful residues and should be used rather than the bigger ones. To kill the bacteria that can cause foodborne illness it is important to cook meat properly. Temperature and cooking times varies depending on the type and size of meat and cooking method. Visual check of meat is also important. No pink meat should be left. Meat changes colour when it is cooked. The juices should run clear on piercing the thickest part of the meat. It should also be checked inside by cutting the meat open with a clean knife which should be piping hot and should be steaming. This type of meat product is good to eat. However, in chicken it is well cooked if it is white and there is no pink flesh. Fish is done when it is opaque and flakes easily with a fork. In recent times, in a modern kitchen a meat thermometer is usually used which measures the internal temperature of cooked meat and poultry or any other dishes. This assures that a safe temperature has been reached and that harmful bacteria have been destroyed. Many times there is cross-contamination when different foods are kept in one place. It is the transfer of harmful microorganisms from one item of food to another via a non-food surface such as human hand, equipment or utensil. There can be a direct transfer from a raw to a cooked food item. Contaminated water may be potential source of food contamination. FSMA (2011) has written to amend the federal food, drug and cosmetic act.

Recently, microwave ovens are being utilized for reheating, cooking and defrosting. Special glass cookwares which are microwave safe should be used. Many plastics are not safe in microwaves, where they can melt, form cracks that can cause food leakage or dissolve and can have many side effects. Most metals, including metal paints on ceramics or glass, are usually not safe either because of the increased risk of sparks and fire. The dish should be covered with a lid or plastic wrap approved for microwave use. Enough space must be kept in between the food and the top of the dish not allowing the plastic wrap to touch the food. Loosen the lid or wrap so that the steam can pass out. The steam helps in destroying the harmful bacteria. Cooking bags are available which can be used that provides safe and even cooking. They help in stirring, rotating or turning the food upside down during cooking in the microwave. Many microwave ovens have a turntable that stir and turn the food from top to bottom. After removing food from the microwave it should be left for at least 3 minutes. It completes the cooking process. The food can cook less evenly than in a conventional oven. For uniform cooking the food items must be arranged evenly in a covered dish. In a microwave oven, the air in the oven is close to the room temperature so the temperature of the food surface is often cooler than food in a conventional oven where the food is heated by hot air. While thawing a food item and defrosting in a microwave the packaging should be removed. Foam trays and plastic wraps should not be used as they are not stable to heat and also to the high temperatures. Melting or warping of the trays or wraps from the heat of the food may cause harmful chemicals to leach. Many microwaves have special settings for defrosting. Microwaves are absorbed by water than ice. Therefore, turning the food during the defrosting process improves the evenness of the heat. Cooking of poultry in a microwave oven, whole or stuffed form is not recommended. This is due to the cooking on the outside of the food than inside. Due to this the desired temperature does not reach inside. This is required for destroying harmful bacteria. Microwaves cause water, fat, and sugar molecules to vibrate 2. 5 million times per second which produces a lot of heat. After the oven is off or food is removed from the oven, the molecules continue to generate heat. The additional cooking after microwaving stops is called "carry-over cooking time," "resting time," or "standing time. " It is found to be more and for a longer time in dense foods like whole chicken roast than in foods which are less dense that include breads, small vegetables and fruits. The temperature of a food increases to several degrees during this time. Additionally, microwave heating is often uneven with focal spots of intense heat may be near areas of cold food and therefore, standing time allows for more even heat dispersion. Pasteurisation is the most popular method of heat treatment. It involves heating milk to high temperatures to kill harmful bacteria that causes illness. Milk is heated to a high temperature and then rapidly cooled to kill the heat stable and heat labile pathogens. It does not kill all bacteria. Pasteurization alone does not give a safe shelf-stable product without proper storage. Pasteurized milk must be kept at low temperature probably in a refrigerator. It can also be kept for longer period in the freezer chamber. Commercially, the milk, fruit juice or cream are pasteurized at a higher temperature (132°C) for a shorter time which is known as ultra-high temperature (UHT) food processing. The UHT milk or juice passes through heating and cooling stages in quick succession and is then immediately put into a sterile Tetra pack shelf-safe carton. By this procedure the product can be kept upto six months without refrigeration or preservatives. Tetra pack products are sterile and safe to consume until opened. After opening it should be stored in the refrigerator. These products are labelled as ultra-pasteurized. Usually it is believed that food items can be kept in a freezer for an indefinite period. However, if a refrigerator is set below 5°C most of the foods are protected. Over time, the chilled foods also get spoilt. The low refrigerated temperatures slow down the bacterial growth but the growth is not inhibited completely. Some bacteria like *Listeria* grow well at low temperature and cross-contamination may occur if raw and cooked food or salads are open and kept together.

Refrigerators should be kept clean and dry at all times. In order to ensure food safety, wholesomeness and quality of food materials is important and the refrigerator should never be overstocked. Overfilling reduces the circulation of cool air and makes it difficult for proper cooling or chilling of food materials and beverages. The refrigerator/freezer doors should not be opened frequently as the temperatures rise and pathogens may grow. At the same time it also saves energy. It is vital to keep raw meat, poultry and seafood in sealed containers or they can be also wrapped in order to prevent raw juices from contaminating other foods. In case of large quantities of food item being prepared it can be divided into small portions which could be put in shallow containers before being refrigerated. The foods can be covered or wrapped with paper which will help in retaining moisture and preventing them from picking up odors from other foods. The perishable foods can be thrown out every week. Food preservation is designed to prevent spoilage and decay of seasonal and perishable foods by increasing the shelf-life of food thus increasing the supply and availability throughout the year. Technological advancements such as canning, vacuum packing, refrigeration, freezing and tetra packing have increased the shelf-life of food and have improved the quality of food material. The principle of all preservation techniques is to restrict food spoilage so that food can be used safely in a palatable form at a later time. Food preservation inhibits the growth of harmful microorganisms and arrests the biochemical breakdown of tissues and does not allow the transformation of the cell contents. The various methods used are heat, cold, drying, fermentation, radiation and chemicals. Man has been using ice or snow to preserve foods from time immorial. Fermentation of agricultural produce also was known from early times which are still practised for the preservation of vegetables such as kimchi, sinki, gundruk and pickles. Other popular domestic methods used for food preservation used earlier were smoking, drying and salting. Smoking of meat and fish for preservation and flavouring is an old practice.

Rodgers et al., (2016) have looked into the technicalities of minimally processed functional foods. Looking into the microbiological aspects two methods of traditional food preservation are used. These include altering the environmental conditions in order to see that spoilage microorganisms do not grow in the food. These include dehydration, pickling, salting, smoking and freezing. Removal of water, use of acid, use of oil and spices, use of chemical preservatives and use of low temperatures can help. Removal of microorganisms and enzymes in which most of the microorganisms present in the food are killed and enzymes are inactivated include canning, cooking, irradiation and blanching. Heating food at high temperature helps in the preservation through coagulation of proteins, inactivation of their metabolic enzymes and destruction of microorganisms. Foods with high fat are difficult to be stored in the frozen state. Oily fish are the most suitable for canning. Canning retains the natural flavour of the fish. *C. Botulinum* spores are not always killed by the high heat of the canning process, but the poison that these bacteria make does not occur in acidic conditions. Canning or pickling of vegetables or meat in airtight containers with enough acid to prevent toxin formation is found to be good for preservation of food material. Cans that appear to have swelled or bulged or unopened jars with pressure pop-up centres that have popped up are contaminated and should be thrown away.

All foods should be cooked at 75 °C. The procedure to cook food is very important. Different foods need a different approach: Aim for an internal temperature of 75 °C or hotter when you cook food. This helps to kill most of the food-poisoning bacteria. A thermometer can be used to check the internal temperature of foods during the cooking process. Mince, sausages, whole chickens or stuffed meats should be cooked right through till the centre. There should not be any pink meat and the juices should be clear. Steak, chops and whole cuts of red meat can be cooked as per ones preference as food poisoning bacteria are mostly on the surface. Fish should be cooked until the flakes can be easily taken with a fork. Foods made from eggs such as omelettes and baked egg custards should be cooked thoroughly. Extra care should be taken while preparation of foods that contain raw egg, such as homemade mayonnaise, tiramisu and egg nog. Bacteria present on egg shells and inside the egg can contaminate these types of food and cause food poisoning. Foods with raw eggs should not be given to pregnant women, young children, elderly people and people with a suppressed immune system.

Microwaves are a quick and convenient way to cook food. However, if they are not used correctly, they can cook food unevenly. This may leave food partially cooked or not reaching a uniform temperature of 75 °C. While cooking food in the microwave it should be cut into even sized pieces with thicker items to be towards the outside edge of the dish. While storage of food one must wait until the steam stops rising. After which, it can be covered and kept in the fridge. In this manner the food is kept at appropriate temperature. Ideally the cooked food can be stored in the fridge for a few days. In case of storing cooked food for a longer time it can be frozen immediately after cooling in the fridge. Cooked food must be kept separately from the raw food. These include raw meats, poultry and fish. Raw meats and poultry must be at the bottom of the fridge to avoid raw juices dripping onto other food. All foods should be covered or sealed while storing in fridge. While reheating the food it should be taken to steaming temperature above 75 °C or could be boiled. When reheating food in a microwave oven same procedure should be followed. The food supply in the United States is among the safest in the world. However, when certain disease-causing bacteria or pathogens contaminate the food, they can cause foodborne illness which is known as food poisoning. There is an estimation of about 48 million cases of foodborne illnesses annually which result in an around 128,000 hospitalizations and 3,000 deaths. When these dangerous foodborne bacteria are consumed they cause illness within 1 to 3 days after consumption. However, sometimes it may take up to 6 weeks. Usually the symptoms are vomiting, diarrhea and abdominal pain. Sometimes there are flu-like symptoms like fever, headache and body ache. Mostly the healthy people recover within a short period of time. However, in chronic cases they can cause life-threatening health problems. Following the four simple steps helps in protection. These include cleanliness in the complete food chain. Washing hands with warm water and soap for at least 20 seconds before and after handling food and also after using the bathroom, changing diapers, and handling pets is important. The cutting boards, dishes, utensils, and counter tops must be washed with hot soapy water after preparing each food item. Paper towels for cleaning up kitchen surfaces is a good option as they can be thrown. In case of cloth towels, they should be treated in the hot cycle. The fresh fruits and vegetables must be washed under running tap water, including those with skins and rinds that are not eaten. Scrubbing of firm produce with a clean produce brush is helpful. Canned foods should be cleaned especially near the lids before opening. Foods like raw meat, poultry, seafood and eggs must be separated from other foods in the grocery shopping cart, grocery bags and refrigerator. The cutting boards for fresh produce should be different from that of raw meat, poultry, and seafood. Cooked food should never be placed on a plate that previously held raw meat, poultry, seafood, or eggs unless the plate has been washed in hot, soapy water. It has been found that color and textures are unreliable indicators of food safety. Using a food thermometer is the only way to ensure the safety of meat, poultry, seafood, and egg products for all cooking methods. These foods must be cooked to a safe minimum internal temperature to destroy any harmful bacteria. Refrigeration or freezing of meat, poultry, eggs, seafood, and other perishables within 2 hours of cooking or purchasing has been found useful. The food items should be refrigerated within 1 hour if the temperature outside is above 90° F. The food should never be thawed at room temperature. Defrosting of food must be done in the refrigerator, in cold water or in the microwave. Food thawed in cold water or in the microwave should be cooked immediately.

According to Zhong and Wang (2019) bacteria, viruses, parasites, chemicals, other contaminants, additives and adulterants can cause over 200 diseases ranging from diarrhoea to even cancer. Every year diarrhoea caused by contaminated food and water kills 2. 2 million people, including 1. 9 million children, globally. Unsafe food and water kills an estimated 700 000 children in WHO's South-East Asia Region every year. Access to safe food remains a challenge in the Region. Whether as individuals, families, farmers, contributors to and handlers of the food chain or policy makers, food safety should be made a priority. Safety in food is very critical for public health as foodborne diseases affect people's health and well-being. In case the food is unsafe it creates a vicious cycle of diseases and malnutrition, particularly affecting infants, young children, the elderly and people with low immunity. It reduces the socio-economic development as there is a strain on the health care systems. It also reduces the national economies, tourism and trade. As the food passes through multiple hands from the farm to fork, it is important to ensure food safety in many ways. Prevention of food diseases can improve food safety and quality through application of good farming practices by using agro chemicals or veterinary drugs only in the prescribed amount. Good storage, transportation, retail and restaurant practices are equally important to make food safe.

Street foods have become common these days as they are economical. However, they are emerging as a source of foodborne diseases where focus should be kept. Another major item is the impact of climate change on food production, distribution and consumption. Due to emerging biological and environmental contamination in the food chain newer pathogens are emerging. They are antimicrobial resistant and pose a lot of challenge in the safety of our food. It is vital that countries have a comprehensive food safety policy, legislation and national food safety programmes for all sectors and aspects of food safety. Most of the countries around the globe have food safety policies but following them is a big challenge. For export of food commodities this is usually followed but not always enforced for food in the domestic market. Food safety and quality needs stringent control and inspection mechanisms for both. Another major problem is the adulteration of food as informal food production and distribution systems are deeply entrenched at the community level. Contamination of mustard oil is usually done with argemone oil which started in 1998 and contamination of imported milk and infant formula with melamine in 2008. These cases raised the awareness of food safety among consumers and policy makers around the world. According to FDA it has been found that some U.S. hogs, poultry and farmed fish ate animal feeds which contained Chinese ingredients along with an industrial chemical which is called as melamine. However, the FDA reported that people who ate meat from those animals did not have much risk of melamine-related health problems. There was an outbreak of *Salmonella* in peanut butter and *E. coli* in fresh spinach in USA. Fracis et al., (2014) have studied the hybrid food preservation program.

In the traditional way the quality of food is evaluated by our sensory organs like eyes, nose or mouth. Recently scientists have shown that it can be done by the use of instruments. Sensory evaluation is commonly practised by food regulatory authorities who consist of judging the quality of food by a panel of judges. Parameters for the evaluation include measuring, analyzing and interpreting the qualities of food as perceived by the senses of sight, taste, touch and hearing. Sampling of the food in a careful manner is necessary for sensory evaluation. However, combinations of experts are needed to measure the contamination of food by pesticides, veterinary drug residues and adulteration. In the objective evaluation chemical, physiochemical, microbial and physical parameters are analysed. In chemical methods nutritive value of foods before and after cooking and products of decomposition and adulterants in foods are analysed. The most widely employed objective evaluation is the measurement of physical properties by the use of instruments. Measurements of the appearance and volume of foods are also important. Food safety means to limit the presence of hazards which are chronic or acute and make the food injurious to the health of the consumer. It also includes production process, handling, storing and cooking food in a way so as to prevent any infection and contamination in the food chain. This ensures to maintain the food quality and wholesomeness promoting good health. EFSA (2008) has come out with the maintenance of list of QPS microorganisms to food and feeds. Drewnowski et al., (1997) have studied the new trends in global diet.

4

How Good are Organic Foods

Today, organic and local foods are preferred as they use less pesticides or fertilizers. However, these foods, like others, can be exposed to harmful bacteria during the growing and harvesting process. The farmers and distributors must use good sanitary practices to minimize food contamination. Consumers should cook food properly. The street food vendors are a traditional and indigenous fast food approaching in most countries of the South-East Asia Region. As these are economical they cater to millions of consumers. Among safe street foods there are many contributing factors that are associated with the safety of street food. WHO had conducted a survey in 1993 covering 100 street foods in 100 countries around the world. It revealed that the major health threat comes from raw and undercooked food, infected food handlers and inadequate hygiene practices in processing and storing such food. If the street food is cooked well and is served hot, then the chances of food poisoning is less. The water quality, hygienic conditions and the level of cleanliness especially in the summer season are the major factors for food poisoning. However improvements in physical infrastructure and hygienic conditions are growing as consumers are also demanding quality food. Michael et al., (2017) have looked into organic versus conventional foods.

Recently, the sale of organic foods has hit $28. 4 billion. According to the Nutrition Business Journal, the organic food products will reach much higher in the days to come. Despite the popularity the public is very confused as per a study published in the International Food and Agribusiness Management Review. A team of researchers had surveyed consumers across the U. S. and Canada. They discovered that 11% of the people believed that foods labelled "organic" were also grown locally, 20% believe that local produce is grown organically, 40% of consumers think "organic" food is more nutritious than conventional food and 29% believe that "local" products are more nutritious than their imported equivalents. The organic and conventional farming systems both have the potential to produce food that is safe for human consumption. The organic foods do not have any chemical pesticides, however, a study has shown that some organic pesticides can be dangerous. Organic foods may have higher nutritional value than conventional food, according to some research due to the absence of pesticides and fertilizers the yield of phytochemicals (vitamins and antioxidants) are increased that strengthens their resistance to pathogenic microorganisms and weeds. It might have some residue due to the pesticide treatment given by the approved ones for organic farming or because of airborne pesticides from conventional farms. However, the health outcomes are ambiguous because of safety regulations for maximum levels of residue allowed on conventional produce. The companies use very confusing terminologies as "all natural" / "free range". Sometimes the health foods have no legal definition enforceable by the Food and Drug Administration or the Federal Trade Commission. From a legal perspective word "Organic" is quite clear. To be labelled organic, a producer must abide by a stringent set of government standards. The USDA qualifies produce as organic if no synthetic pesticides, chemical fertilizers or genetically modified organisms (GMO) are used. Controlling the pests and crop nutrients it is a must to be managed by natural physical, mechanical and biological control methods. In case of producing organic meat, eggs and dairy, the farmers, must provide non-GMO livestock with year-round outdoor access. Usage of growth hormones / antibiotics is prohibited when farming is done by organic means. According to a 2010 report from the U. S. Department of Agriculture although **local** has a geographic connotation, there is no consensus on a definition in terms of the distance between production and consumption. Farm Act, of 2008 has stated that any food which is labeled "local" must be produced in the "locality or region in which the final product is marketed, so that the total distance that the product is transported is less than 400 miles from the origin of the product. Looking into the health part the organic foods give a good picture. A 2012 study conducted by Stanford's Center for Health Policy concluded that organic produce is not more nutrition-dense than its generic counterparts. The belief of the consumers (29%) on the local foods to be more nutritious may be true as , it has been found out that when a produce is picked the nutrients begin to degrade. Many studies have shown that a peach or berry if plucked at nearly ripe stage are more nutritious than an organic fruit or if picked before or after its peak of ripeness.

The USDA defines organic agriculture as "products using methods that preserve the environment and avoids most synthetic materials like pesticides and antibiotics. " Yet the term itself, "organic," has become a buzzword for health. This health relates to and encompasses the wellness of consumers, producers, farmland, general environment, local communities and local economies.

It is important to know the reasons for using organic or local produce. These include greater nutritional value in produce, when grown locally; crops are picked at their peak of ripeness. However, conventionally, they are harvested early for shipping and to be distributed to the retailers. The crop's nutrient value is kept for organic crops as the time between harvest and distribution is very less. Phytochemicals like antioxidants, vitamins and minerals may be lost during the processing and transport of conventionally grown crops. Moreover, local, organic food usually stays in its natural, pure state as there are no artificial chemical additives or presence of any other chemicals. There is a lot of support for the amount spent by a farmer which can be reinvested with other businesses and services within the community. In such cases the money stays with them rather than going out to large corporate retail stores. This also gives jobs to the local population. According to the United Nations Food and Agriculture Organization; more than 75 percent of agricultural genetic diversity had been lost in the 20th century. It has been observed that in conventional farm practices mono-cropping with limited plant varieties is done which brings in a need for small, biodiverse farms to preserve food heritage. As the community needs to be strengthened it is advised to choose local and organic food. The consumer has a direct interaction while buying from local sources. The organic produce has less environmental contaminants and are preferred. However, the produce with chemical fertilizers and pesticides may have harmful effects to the environment including soil destruction and erosion, chemical run-off into water systems and development of weeds and bugs which are resistant to herbicides and pesticides. Organic farming does not use synthetic or petroleum-based pesticides or fertilizers resulting in less water and soil contamination. Well-managed and maintained healthy farms conserve fertile soil and leave clean water to the communities. Such farms may also serve as a habitat for certain wildlife. Another major problem is of pollution and this reduces the carbon levels. It also improves the food security. Presently there is a growing world food crisis due to the continuous population growth and increased use and consumption of biofuels. Economically poor communities are the most vulnerable to rising food prices and shortages. These create food deserts. The development of local organic farms can enhance food security by providing communities and neighborhoods with fresh produce. It helps in food supply and there is a lowered risk of food contamination. There is safety needed during harvesting, washing, shipping and distribution of conventionally grown food. According to the U. S. Bureau of Labor Statistics, careers in conventional farming are decreasing. The initial investment is less, as, huge amounts of expensive chemical fertilizers, pesticides and genetically-modified seed stocks are not required.

In USA there is a Slow Food organization with a mission of good, clean and fair food for all. This nourishes a healthful lifestyle. The food is produced in ways that preserve biodiversity, sustainability and ensures animal welfare without harming human health. It deals with better, cleaner and fairer world by emphasizing that daily food choices determine the future of the world's environment, economy and society. In the United States, Slow Food has 170 local chapters and 40 university campus chapters that coordinate local activities, projects and events. According to Dr. Edward group(2013) organic typically means "Certified Organic". Growers have to go through a lengthy and expensive certification process overseen by the USDA or other credible certifiers such as Oregon Tilth to be "certified organic". They should not be genetically modified. For plants, organic also means that farmers don't irradiate their crops and for animals it means that they have been given only organic feed for a year or more with no use of antibiotics and no growth hormones. Sometimes the farmers may be farming organically, but they may not be certified because there is a requirement of time commitment certification. Stringent supporters of local business have pointed out that, organic or not, buying local is better for the environment because sometimes organic foods have to be transported to long distances. This uses oil and fuel which increases the emissions of CO_2 and increasing their carbon foot print. Sometimes, all of these factors, outweigh the benefits of buying organic products. Lee et al., (2016) have studied the beneficial effects of chestnut shell against Campylobacter jejuni in cheken meat.

On the other side of the table, local foods that are not organically grown have all the problems of many other commercially produced foods. They are less nutritious and have fewer flavors. Most non-organic foods contain high levels of pesticide residues, heavy metals and other harmful contaminants. It has been found that even after washing they persist in the produce and goes into the human body when consumed. Non organic foods have a detrimental effect on the environment because the herbicides and pesticides used leach into the water supply and spoil the local ecosystems and are also absorbed by the roots. Pesticides are poisonous and adversely cause cancer. Mainly they are present in fruits like apples, grapes, nectarines, peaches and pears, legumes like peas, beans and peanuts; Berries like strawberries, raspberries, blueberries and goji berries; vegetables like bell peppers, celery, peas and spinach. A thorough washing of all of these is very important before consumption. Many foods that have a thick skin or husk that are peeled off prior to eating need not be purchased organically produced unless a strict organic food diet is prescribed. These include bananas, pineapple, corn and avocados, tropical fruits like kiwi, mangoes and papaya; vegetables like asparagus, broccoli and cauliflower. Most cities around the world have farmer's markets that offer a variety of locally produced fruits, vegetables, sauces and meats. Another product is local honey which is very beneficial as it is resistant to allergens. There is a website Localharvest. org that helps people find farmers markets and other local, sustainable grown foods. Sometimes insisting the provision store persons to keep farm produce will help the community. Whenever possible, it is healthier to buy local and organic products.

Having a kitchen garden at home for fruit and veggies helps in many ways. It is easy to grow seasonal fruits and veggies in barrels or flower boxes. They do not require much maintenance and grow with minimal care. What can be grown in a particular season, and when should it be grown, depends on the place of residing and accordingly the produce can be grown. With the advent of internet many online books are available in the market. These give a helping hand to the people who have just started this line. Generally the Garden nurseries or centers provide the brochures for methods of organic farming. They also outline the plants and their time of growing best during the year. These provide information on local farming. Recently, the area of farmers' markets has become a growing trend. It has been found out that with 64 percent, of the more than 8,400 farmers' markets, surveyed by USDA, there was an increase in the number of people purchasing from them. This shows that the consumers are more interested in buying local food. Buying locally helps in supporting the local economy and it could also reduce the amount of energy required to ship food from other areas. In some cases, fruits or vegetables may be fresher in case they are purchased locally. Looking into the safety aspects of farmer's market produce, the studies show, that the microbial contamination is very low. At present, the small farmers are exempted from most food safety regulations, as that of Food Safety Modernization Act. However, the U. S. Food and Drug Administration (FDA) released a final rule on the food safety act in November 2015 to prevent food problems before they occur. Enforcing the same regulation on the small farmers may be harsh as they will go out of business, due to lack of available resources. There will also be an impact on local economies. According to Diez-Gonzalez, the Minnesota Department of Health, promoted the adoption of Good Agricultural Practices (GAP) among the small farmers. However, the state does not have the resources to enforce, so the GAPs are recommendations rather than regulations. Therefore, it is important for the consumers to look for best practices while selecting their produce. Consumers can check the produce is clean and stands out for cleanliness which includes gloves, clean utensils for food handling, covered garbage cans, coolers for perishables and use of clean bags. Reusable groceries are very popular these days as it is an eco-friendly way to transport food. However, there is a separate tote for the use of raw meat, poultry and ready-to-eat foods like fresh produce and breads. In a survey done by the Centers for Disease Control and Prevention (CDC) the sprouts, leafy greens and tomatoes present a higher risk than some other crops. Produce like onions, carrots and peppers have very little food safety problems.

According to Diez-Gonzalez there are mixed opinions on some of the disinfecting products. However, washing with water is advisable to reduce any contamination on the surface of the produce. As of today "The perception right now is that local is good and everything organic or natural is good," says Diez-Gonzalez. Organic farming takes more time to produce crops because they refrain from using the chemicals and growth hormones that are commonly used by the conventional farmers. Also the production-oriented government subsidies reduce the overall cost of crops. Since organic food production strictly avoids the use of all synthetic chemicals, it does not pose any risk to the soil and underground water contamination. However, in conventional farming in which tons of artificial fertilizers and pesticides are used but it helps to conserve biodiversity. The stated principles of organic agriculture are health, ecology, fairness and care (Greger, 2017).

5

Role of Packaging Material in Food

In this century the usage of plastics had become very common. However, it is important to know as to what kinds of plastics are used for food handling and storage and the health hazards caused by using it. It has been observed that plastic packaging plays a significant role for the enhanced shelf life and in the ease of storage and cooking for many foods. Some kinds of plastic materials which are widely used for handling and storage of food and water include polyethylene terephthalate (PET). It is used in soft drinks, water, sport drinks, ketchups, salad dressing bottles, peanut butter, pickle, jelly and jam jars. This plastic is strong, heat resistant, resistant to gases and acidic foods. It can be transparent or opaque. It does not leach any chemicals that are suspected of causing cancer or disrupting hormones and can also be recycled.

Usage of High density polyethylene (HDPE) has been found to be common in milk, water, juice bottles, yogurt, margarine tubs, grocery, trash and retail bags. High-density polyethylene is stiff and strong but is not heat stable (it melts at a low temperature). It does not leach any chemicals that have a role in causing cancer or disrupting the hormones. HDPE can be recycled. Low-density polyethylene (LDPE) is used into making films of various types, some bread and frozen food bags and also in the making of squeezable bottles. Low-density polyethylene is relatively transparent. Many of the films are not heat stable either and may melt the food in case it is touching it. However, polypropylene (PP) is more heat resistant, harder, denser and more transparent than polyethylene and is therefore used for heat-resistant microwavable packaging and sauce or salad dressing bottles. Polycarbonate is clear and heat resistant. It is also durable and often used to make refillable water bottles and sterilisable baby bottles, microwave ovenware, eating utensils, plastic coating for metal cans. Tiny amounts of bisphenol A are formed when polycarbonate bottles are washed with harsh detergents or bleach (sodium hypochlorite). At high levels of exposure, bisphenol A is potentially hazardous as it mimics the female hormone estrogen. In addition, polystyrene (PS) and polyvinyl chloride (PVC) are also used during food material transportation and handling in supermarkets. Modern food safe plastic bags are plasticizer-free which do not leach harmful chemicals into the food during cooking.

The different types of plastics are made up with chemicals which may be toxic to health. If plastic packaging is used properly then it lowers the migration of chemicals. While using household plastics like cling films and bags it is important to take care. Instructions should be followed while using cleaning products on containers, bottles and lids. Therefore, using the correct type of plastic for each role should be taken. Microwave-safe plastics should be used in the microwave. During microwave cooking cling film must not touch the food as it melts at a low temperature. In case the food is stored in cling film it should be removed before cooking in a microwave. Before keeping in the refrigerator a corner of the dish uncovered must be left for the steam to escape. It avoids the film to be blown off and settling on to the food. One can freeze food in ice-cream containers but should not be heated in the microwave. Food additives are substances not normally consumed as a food by itself and are not normally used as a typical ingredient of the food, whether or not it has nutritive value. The addition of additives in foods is for a technological purpose in the manufacturing process. They can be added during the preparation, treatment, packing, packaging and transport of food. The term does not include contaminants or substances added to food for maintaining or improving nutritional qualities. During fortification one or more essential nutrients are added to a food, whether or not it is contained in the food. This helps in preventing or correcting a demonstrated deficiency of one or more nutrients in the population or specific population groups and is able to deliver nutrients to large segments of the population without requiring radical changes in food consumption patterns. These days the times are changing rapidly and so are the food habits. With changing lifestyles more highly processed foods have become common. In order to ensure nutritional adequacy of the diet fortification has become common. Supplementation/Fortification with micronutrients into food has become a valid technology for reducing micronutrient malnutrition. Food technology has been developed for fortification of vitamins and micronutrients in cereals and cereal based products, milk and milk products, fats and oils. According to the World Health Report (2000) fortification of iodine, iron, vitamin A and zinc is important due to these deficiencies being among the world's most serious health risk factors. Salt iodization to control iodine deficiency disorders has been revolutionary in public health nutrition over the last 30 years. The international community for promotion of food fortification has mainly focused on the three most prevalent deficiencies: vitamin A, vitamin B complex, iodine and iron. Another major problem is of food adulteration. It is mainly found in the informal food production and marketing services and also in places where the enforcement of food regulation is weak. Food adulteration is an unethical and often criminal malpractice which is commonly found in the South-East Asia region.

According to Marsh (1993) to keep the food safe in the supply chain many advances have been made in food processing and food packaging. Good packaging maintains the benefits of food processing after it is complete and enables foods to safely transport for long distances. Food packaging technology should balance food protection with other issues, including energy and material costs, social and environment and regulations on pollutants in disposing municipal solid waste.

Packages, food scraps, yard trimmings and durable items such as refrigerators and computers are some of the items which are included in Municipal solid waste (MSW). Efforts in the legislation and regulation to control packaging are based on the mistaken perception that packaging is the major burden of MSW. In a survey, the U. S. Environmental Protection Agency (EPA) has found that approximately 31% of the MSW which was generated in 2005 was from packaging-related materials, including glass, metal, plastic, paper, and paperboard. This has been almost constant since the 1990s despite an increase in the total amount of MSW. Nonpackaging sources such as newsprint, telephone books and office communication generate more than twice as much MSW (EPA, 2006a). It is only'the commodity called as food which is consumed 3 times per day by every person. According to Hunt et al., (1990) it is the food packaging that forms almost two-thirds of total packaging waste by volume. Approximately 50% (by weight) consists of food packaging of the total food packet. During the century the packaging issues have not been properly understood. Discarding the packaging materials takes a lot of complicated efforts especially for the environment. Many scientists have done a lot of work to look into this complicated problem and the ways to solve them.

If the role of packaging is seen, according to Coles (2003) the principal role of any food packaging is to protect food products from outside influences and damage, to contain the food and to provide consumers with ingredient and nutritional information. Convenient foods have taken a top position and the demand is increasing day by day. The goal of food packaging is to contain food in a cost-effective manner that would satisfy the industry requirements along with the desires of the consumers, maintaining food safety and minimizing environmental impact.

Protection/preservation

Food packaging can protect the products from spoilage retaining the beneficial effects of processing. It also extends the shelf-life and maintains/ increases the quality and safety of food. Packaging provides protection from major external influences like chemical, biological and physical.

Environmental influences such as exposure to gases (typically oxygen), moisture (gain or loss) and light (visible, infrared or ultraviolet) can be protected. Many different packaging materials can provide a chemical barrier. Metals and glass are a good barrier to chemicals and other environmental agents. However, few packages are purely glass or metal since closure devices are added to facilitate both filling and emptying. Many times the closure devices contain materials that prevent permeability. These can be found in plastic caps and gasket materials. In the metal cans the lids are sealed after filling. Plastic packaging offers a large range of barrier properties however it is found to be more permeable than glass or metal.

Barrier to microorganisms (pathogens and spoiling agents), insects, rodents and other animals is best provided by biological protection. This prevents diseases and spoilage. Also it maintains the conditions to control senescence (ripening and aging). It functions through a multiplicity of mechanisms, including preventing access to the product, preventing odor transmission and maintaining the internal environment of the package.

Protection of food physically is very helpful as it shields the food from mechanical damage and includes cushioning against the shock and vibration encountered during distribution. These are usually made out of paperboard and corrugated materials. These help to resist various impacts, abrasions and crushing damage. These are widely used as shipping containers and as packaging for delicate foods such as eggs and fresh fruits. It is an important aspect of physical packaging which protects consumers from various hazards. It has been found that the child-resistant closures are beneficial. Plastic packaging for products like shampoo to soda bottles has reduced the danger of glass containers. Food packaging has given a positive impact to the environment and has reduced food waste through out the supply chain. Many countries have reported food wastage in the range of 25% for food grain to 50% for fruits and vegetables (FAO, 1989). This usually happens due to inadequate preservation/protection, storage and transportation. As appropriate packaging extends the shelf-life of foods it prolongs the usability by reducing the waste. Rathje et al., (1985) found that the per capita waste generated in Mexico City contained less packaging, more food waste and one-third more total waste than generated in comparable U. S. cities. In addition, Rathje et al., (1985) observed that packaged foods result in 2. 5% total waste as compared to 50% for fresh foods in part because agricultural by-products collected at the processing plant are used for other purposes while those generated at home are typically discarded. Therefore, packaging may contribute to the reduction of total solid waste.

Prior to purchase the consumers are able to see the package of any product. Any distinctive or innovative packaging can increase the sale due to competition in the market. Packaging and labels also provide information to the consumer. Package labeling must also look into the legislation which would satisfy the legal requirements for product identification, nutritional value, ingredient declaration, net weight and manufacturer's information. The package label also conveys important information about the product like cooking instructions, brand identification and also pricing in different states in any country. All of these enhancements reduce the waste disposal. The Codex Alimentarius Commission defines traceability as "the ability to follow the movement of a food through specified stage(s) of production, processing and distribution" (Codex Alimentarius Commission, 2004). The objectives include improving supply management, to facilitate trace-back for food safety and quality purposes, differentiate and market the foods with subtle or undetectable quality attributes (Golan et al.,2004). Food manufacturing companies incorporate unique codes onto the package labels of their products. This helps to track the products throughout the distribution process. Codes are available in various formats (printed barcodes or electronic radio frequency identification (RFID) and can be read manually and/or by a machine. With increasing revolution in this century there are many convenience features like ease of access, handling, disposal, product visibility, resealability and microwavability. It is, therefore, that packaging is playing a very important role in minimizing the efforts necessary to prepare and serve foods. Today, food packaging has brought forward oven-safe trays, boil-in bags, and microwavable packaging which enables the consumers to cook a complete meal without any preparation. There are many new closure design supplies available which have allowed the ease of opening, resealability and different features for dispensing. Innovative flexible bags are available with a scored section that provides access to the food product. A membrane with a peelable seal cover and again reclosing it after opening is very helpful. These convenience features add value and competitive advantages to products. However, the waste disposal requirements needs to be seen. Special packaging are designed to reduce the risk of tampering and adulteration. Although any package can be breeched but tamper evident features cannot easily be replaced. These are like banding, special membranes, break away closures and special printing on bottle liners or composite cans such as graphics or text that irreversibly change upon opening. There are other measures like special printing with holograms that cannot be duplicated easily. Tamper-evident packaging usually requires additional packaging materials, which exacerbates disposal issues, but the benefits are high. An example of a tamper-evident feature that requires no additional packaging materials is a heat seal used on medical packaging that is chemically formulated and changes the color when opened. Packaging is also very important while giving a gift, additional product, or containers for household use. If the packaging is reused many times it is possible to delay the entry into the waste stream.

Role in determining the shelf life of a food product depends on the package design and construction. The right selection of packaging material and technologies help in maintaining the product quality and freshness during distribution and storage. The most common materials used traditionally in food packaging are glass, metals which include aluminum, foils, laminates, tin plate and tin-free steel, paper / paperboards and plastics. Moreover, a wider variety of plastics have been introduced in both rigid and flexible forms. With innovation the scientists have combined different materials to get the combined benefits. As research to improve food packaging continues, advances in the field may affect the environmental impact of packaging.

The U. S. Food and Drug Administration (FDA) regulate the packaging materials under section 409 of the federal Food, Drug and Cosmetic Act. The primary method of regulation is through the food contact notification process that requires what manufacturers notify to FDA, prior (120d) to marketing a food contact substance (FCS) for a new use. An FCS is "any substance intended for use as a component of materials used in manufacturing, packing, packaging, transporting or holding of food if the use is not intended to have a technical effect in such food" (21 USC §348(h)). All FCSs that may reasonably migrate to food under conditions of intended use are identified and regulated as food additives unless classified as generally recognized as safe (GRAS) substances. If the history of food packaging is seen glass was most common. Initially it was around 3000 BC (Sacharow and Griffin, 1980) that food packaging started with glass. The production of glass containers involves a long procedure. This involves heating a mixture of silica (the glass former), sodium carbonate (the melting agent), lime stone/calcium carbonate and alumina (stabilizers) to high temperatures until the materials melt into a thick liquid mass which is put into the moulds. Broken glass which is known as cullet can be recycled and can be used in the manufacture of glass. This accounts for 60% of all raw materials of glass utilized. Glass containers used in food packaging are usually surface-coated that provides lubrication in the production line and eliminates scratching and line jams. A Glass coating usually preserves the strength of the bottle to reduce breakage (McKown, 2000). Being odorless and chemically inert with almost all food products, glass has several advantages for food-packaging applications. It is not only impermeable to gases and vapors, it also maintains, the freshness of the products for a long period of time with keeping quality of taste or flavor. It can also withstand the high processing temperatures of both low-acid and high-acid foods. It also gives good insulation and can be molded into different shapes. As it is transparent, the consumers can see the product. Different colours of glass can also protect light-sensitive products. However, the heavy weight increases the transportation costs; it is brittle and could break in case of pressure, impact, or thermal shock. Looking into these problems metal seems to be versatile in all packaging forms as it gives protection, has barrier properties, formability, recyclability and consumer acceptance. The most common are aluminum and steel. These are utilized in the making of cans, foils and laminated paper / plastic packaging. Aluminum which is derived from bauxite ore is a lightweight, silvery white metal. In order to improve the properties of its strength magnesium and manganese are added. It is resistant to corrosion, its natural coating of aluminum oxide is an effective barrier to air, temperature, moisture and chemical attack. It has flexibility and surface resilience, malleability, formability and embossing potential. It protects from moisture, air, bad odors, light and microorganisms. Pure metal is utilized in the packing of soft-drink cans, pet food, seafood and prethreaded closures. The disadvantage is the high cost as compared to other metals (steel) and that it can not be welded. Foils of aluminum are made by rolling pure aluminum metal into very thin sheets, followed by annealing which makes it possible to be folded tightly. The thinner foils are used to wrap food and thicker foils used for trays. These give protection against moisture, air, odors, light and microorganisms. As it is inert to acidic foods it is a preferred metal for packaging. Aluminum foils cannot be made from recycled aluminum without pinhole formation in the thin sheets. For lamination in aluminum packaging the binding of aluminum foil to paper or plastic film is done to improve its barrier properties. Thin gauges of this metal helps in the application of various kinds. It is also laminated to plastic which helps in heat sealability, but it does not completely prevent moisture and air. It is used in packing high value foods which include dried soups, herbs and spices. A less expensive packaging is metallized film. Metallized films are plastics containing a thin layer of aluminum metal (Fellows and Axtell, 2002). These films have improved barrier properties to moisture, oils, air, odors and due to the highly reflective surface of the aluminum it is liked by the consumers. Metallized films are specifically used in packing snacks. The individual components of laminates and metallized films are technically recyclable. However, sorting and separating is economically expensive. Another useful metal is that of tin plate. It is produced from low-carbon steel. It is made by coating both sides of black plate with thin layers of tin. The processing is done by dipping sheets of steel in molten tin or by the electro-deposition of tin on the steel sheet. These containers are usually lacquered to provide an inert barrier between the metal and the food product. Lacquers which are utilized are epoxy phenolic and oleoresinous groups or vinyl resins. It protects from gases, water vapor, light and odors. They can be heat-treated and sealed, making it suitable for sterile products. It has good ductility and formability and can therefore be used for containers of many different shapes. They are usually used in making cans for drinks, processed foods and aerosols. Powdered foods and sugar or flour based confections can also be kept in tin containers. It is a good substrate for modern metal coating and lithoprinting technology that enables it for graphical decorations. It is low weight along with high mechanical strength and can be easily shipped and stored. It can be also recycled many times without loss of quality and is significantly cheaper than aluminum.

Electrolytic chromium or chrome oxide coated steel also known as tin-free steel requires a coating of organic material to be corrosion resistant. The chrome/ chrome oxide makes tin-free steel not suitable for welding. However, the adhesion of coatings as that of paints, lacquers and inks is extremely good. This tin-free steel has good formability and strength. It is also cheaper than tinplate. It is mainly used in food cans, can ends, trays, bottle caps and closure cans. In addition, it can also be used to make large containers like drums for bulk sale and bulk storage of ingredients or finished goods (Fellows and Axtell ,2002). Plastics are made by the condensation (polycondensation) or addition polymerization (polyaddition) of monomeric units. During polycondensation, the polymer chain grows by condensation reactions between molecules and is accompanied by formation of low molecular weight byproducts such as water and methanol. It involves monomers of minimum of two functional groups like alcohol, amine or carboxylic groups. In polyaddition, polymer chains grow by addition reactions, in which two or more molecules combine to form a larger molecule without the formation of any by-products. In polyaddition the unsaturated monomers are linked together to form a monomer chain by breaking the double or triple bonds. It is beneficial in food packaging. Liquid and moldable plastics can be made into various different sheets, shapes and structures which offers good design flexibility. They are preferred as they are resistant to chemicals, lightweight and are economical with a wide range of physical and optical properties. Most of the plastics are heat sealable, easy to print and can be integrated into production processes, filled and sealed in the same production line. The only problem is their variable permeability to light, gases, vapors and low molecular weight molecules. They can be divided into two categories as thermosets and thermoplastics (EPA, 2006 a). Thermosets are polymers that solidify or set irreversibly when heated and cannot be remolded. As they are strong and durable, they are used in automobiles and in construction applications like adhesives and coatings but cannot be used in food packaging applications. Thermoplastics are polymers that soften upon exposure to heat and return to their original condition at room temperature. Thermoplastics can easily be shaped and molded into various products like bottles, jugs and plastic films. Therefore, they are ideal for food packaging. Moreover, virtually all thermoplastics are recyclable (melted and reused as raw materials for production of new products) although separation poses some practical limitations for certain products. The recycling process requires separation by resin type as identified by the American Plastics Council. Usage of residual monomer and components in plastics, including stabilizers, plasticizers and condensation components such as bisphenol A has side effects on the health. Some of these concerns are based on scientific studies when very high levels are used. For public safety, FDA carefully reviews and regulates substances which are used in making plastics and other packaging materials. Any substance that can reasonably be expected to migrate into food is classified as an indirect food additive subject to FDA regulations. A threshold of regulation is defined as a specific level of dietary exposure that typically induces toxic effects and therefore poses negligible safety concerns (21 CFR §170. 39). It may exempt substances used in food contact materials from regulation as food additives. FDA revisits the threshold level if new scientific information raises concerns. Furthermore, FDA advises consumers to use plastics for intended purposes in accordance with the manufacturer's directions to avoid unintentional safety concerns. Even after these concerns the use of plastics in food packaging has continued to increase due to the cheap cost of materials and functional advantages like that of thermosealability, microwavability, optical properties and unlimited sizes and shapes over that of traditional materials such as glass and tinplate (Lopez-Rubio et al., 2004). Many different types of plastics are being used as materials for packaging food, including polyolefin, polyester, polyvinyl chloride, polyvinylidene chloride, polystyrene, polyamide and ethylene vinyl alcohol. There are more than 30 types of plastics have been used as packaging materials (Lau and Wong, 2000) but only polyolefins and polyesters are very common. Polyolefin is a collective term for polyethylene and polypropylene, the two most widely used plastics in food packaging and other is less popular olefin polymers. Polyethylene and polypropylene both possess a successful combination of properties. These are flexibility, strength, lightness, stability, moisture, chemical resistance and easy processability. All these properties are well suited for recycling and reuse.

Polyethylene is the simplest and most inexpensive plastic made by polymerization of ethylene. There are two types of polyethylene: high density and low density. HDP (High-density polyethylene) is very stiff, strong, tough and resistant to chemicals and moisture, gas permeable, easy to process and easy to form. Therefore, it is used to make milk, juice packets and water bottles. It is also used in cereal box liners; margarine tubs, grocery, trash and retail bags. LDP (low-density polyethylene) is flexible, strong, tough, easy to seal and resistant to moisture. It is little transparent and is predominately used in film applications and in applications where heat sealing is necessary. Some of the examples of low-density polyethylene are bread and frozen food bags, flexible lids and squeezable food bottles. Polyethylene bags are sometimes reused (grocery and nongrocery retail). HDP containers, especially milk bottles are the most recycled among the plastic packages. Polypropylene has good resistance to chemicals and is effective at barring water vapour which is harder, denser and more transparent than polyethylene. As the melting point is high (160 °C) it is suitable for applications where thermal resistance is required, like hot-filled and microwavable packaging. Popular uses include yogurt containers and margarine tubs. When used in combination with an oxygen barrier such as ethylene vinyl alcohol or polyvinylidene chloride, polypropylene provides the strength and is a moisture barrier for ketchup and salad dressing bottles. Polyethylene terephthalate (PET or PETE) is a polycarbonate and polyethylene naphthalate (PEN) are polyesters. These are condensation polymers which are formed from ester monomers that result from the reaction between carboxylic acid and alcohol. The most commonly used polyester in food packaging is PETE. It is formed when terephthalic acid reacts with ethylene glycol. It is a good barrier to gases (oxygen and carbon dioxide) and moisture. It is resistant to heat, mineral oils, solvents and acids but not to bases. It is PETE used as the packaging material for many food products, like beverages and mineral waters. The use of PETE is also common to make plastic bottles for carbonated drinks (van Willige et al., 2002). It is because it is transparent like glass and has quality of gas barrier for carbonation retention and is also light in weight. Major applications of PETE are in the packaging of containers (bottles, jars and tubs), semi rigid sheets for thermoforming (trays and blisters) and thin oriented films (bags and snack food wrappers). PETE is found as amorphous (transparent) and also as semicrystalline (opaque and white) thermoplastic material. Amorphous PETE has better ductility, it is not stiff and has good strength, ductility and is hard. Recycled PETE from soda bottles is used as fibers, for insulation and in nonfood packaging applications. Another packaging material is polycarbonate which is formed by the polymerization of a sodium salt of bisphenol acid with carbonyl dichloride (phosgene). It is clear, heat resistant and durable. It is mainly used in items such as large returnable/refillable water bottles and sterilizable baby bottles. Some care is required while cleaning polycarbonate because using harsh detergents such as sodium hypochlorite is not recommended as they catalyze the release of bisphenol A, a potential health hazard. An extensive literature analysis by vom Saal and Hughes (2005) suggests the need for a new risk assessment for the low-dose effects of this compound. PEN (Polyethylene naphthalate) is a condensation polymer of dimethyl naphthalene dicarboxylate and ethylene glycol. It has high glass transition temperature. As it does not allow carbon dioxide, oxygen and water vapor it is better than PETE and PEN. It provides better performance at high temperatures, allowing hot refills, rewashing and reuse. However, PEN is three times the cost than PETE. Because

PEN provides protection against transfer of flavors and odors, it is well suited for manufacturing bottles for beverages such as beer.

An addition polymer of vinyl chloride called as Polyvinyl chloride (PVC) is quite heavy, stiff, ductile, strong, amorphous and a transparent material. It is resistant to chemicals (acids and bases), grease and oil and has good flow characteristics. It is mainly utilized in medical and non food applications. Among food it is used in bottles and packaging films. Because it is easily thermoformed, PVC sheets are widely used for blister packs such as those for meat products and in pharmaceutical packaging. Due to the flexibility quality it can be transformed into materials with the addition of plasticizers such as phthalates, adipates, citrates and phosphates. Non food packaging applications of phthalates are in cosmetics, toys and medical devices. Safety concerns have emerged over the use of phthalates in certain products, such as toys (FDA, 2001; Shea, 2003; European Union, 2005). Phthalates are not used in food packaging materials in the United States (HHS , 2005) due to safety concerns but nonphthalate plasticizers like adipates are used. DEHA (di-(2-ethylhexyl) adipate) is used in the manufacture of plastic cling wraps. These alternative plasticizers have the potential to leach into food but the levels are very low as compared to phthalates. Low levels of DEHA have shown no toxicity in animals. PVC is difficult to recycle as it is used for such a variety of products, which makes it difficult to identify and separate. In addition, incineration of PVC presents environmental problems because of its chlorine content. Polyvinylidene chloride (PVdC) is an addition polymer of vinylidene chloride. It is heat sealable and has barrier properties to water vapor, gases and fatty and oily products. It is used in flexible packaging as a monolayer film, a coating, or part of a co-extruded product. Major applications include packaging of poultry, cured meats, cheese, snack foods, tea, coffee and confectionary. It can be utilized for hot filling, retorting, low-temperature storage and modified atmosphere packaging. PVDC contains twice the amount of chlorine as PVC and therefore also presents problems with incineration. Polystyrene is an additional polymer of styrene which is clear, hard and brittle with a low melting point. It can be mono-extruded, co-extruded along with other plastics. It can be injection molded or foamed to produce a range of products. Foaming produces an opaque, rigid, light weight material with impact protection and thermal insulation properties. It is used in the protective packaging as that of egg cartons, containers, disposable plastic silverware, lids, cups, plates, bottles and food trays. In expanded form, polystyrene is used for nonfood packaging and cushioning. It can be recycled or incinerated. Other material polyamides which are commonly known as nylon (a brand name for a range of products produced by DuPont) is used in many products. Polyamides were initially used in textiles. It is formed by the condensation of diamine and diacid; polyamides are polymers in which the repeating units are held together by amide links. Different types of polyamides can be differentiated by a number of carbons in the originating monomer. Nylon-6 has 6 carbons and is used in packaging. It has mechanical and thermal properties similar to PETE, has similar usefulness like boil-in bag packaging. Nylon also offers good chemical resistance, toughness and low gas permeability. Ethylene vinyl alcohol (EVOH) is a copolymer of ethylene and vinyl alcohol. It has barrier to oil, fat and oxygen. It is moisture sensitive and is therefore used in multilayered co-extruded films in situation where it is not in direct contact with liquids. Plastic materials can be manufactured either as a single film or as a combination of more. There are two ways of combining plastics. It can be either laminated or co-extrusion can be done. In lamination two or more plastics are bonded together. Plastic can be bonded to another material like paper or aluminum. Bonding is commonly achieved by use of water, solvent or solid based adhesives. After the adhesives are applied to films then both the films are passed between the rollers after applying a pressure which binds them together. Laser can be used for lamination rather than adhesives as for thermoplastics (Kirwan and Strawbridge, 2003). Lamination enables reverse printing, in which the printing is buried between layers and thus not subject to abrasion and can add or enhance heat sealability. Co-extrusion can be done where different layers of molten plastics are combined during the film manufacture. This process is more rapid but requires materials that have thermal characteristics that allow co-extrusion. As in co-extrusion and lamination of multiple materials are combined, it makes the recycling complicated. However, combining materials results in the additive advantage of properties from each individual material and often reduces the total amount of packaging material required. Therefore, co-extrusion and lamination can be sources of packaging reduction. The use of paper and paperboards for food packaging has been from time immemorial. It started in the 17th century with accelerated usage in the later part of the 19th century (Kirwan, 2003). These are sheet materials which are made from an interlaced network of cellulose fibers derived from wood by using sulfate and sulfite. These fibers are pulped and/or bleached and after treatment with chemicals like slimicides and other strengthening agents it produces the paper product. FDA regulates the additives used in paper and paperboard food packaging. Paper and paperboards are commonly used in corrugated boxes, milk cartons, folding cartons, bags and sacks and wrapping paper. It is also used in making tissue papers, paper plates and cups. In primary packaging of food the paper is treated, coated, laminated or impregnated with materials like waxes, resins or lacquers which improve its functional and protective properties. There are many different types of papers that can be used in food packaging. These include the following:

- Kraft paper is produced after a sulfate treatment process. It is available in several forms and colors like natural brown, unbleached, heavy duty and bleached white. The natural kraft is the strongest of all papers and is commonly used for bags and wrappings. It is also used to pack flour, sugar and dried fruits and vegetables.

- Sulfite paper is lighter and weaker than kraft paper, it is glazed to improve the appearance and to increase its strength and also oil resistance. It can be coated for higher print quality and is also used in laminates with plastic or foil. It is used to make small bags or wrappers for packaging biscuits and confectionary.

- Grease proof paper is made by a process known as beating. The cellulose fibers are given a hydration period that breaks the fibers and makes them gelatinous. The fine fibers are packed densely to provide a surface which is resistant to oils but not to wet agents. It is used to wrap snack foods, cookies, candy bars and other oily foods. It is being replaced by plastic films.

- Glassine is a grease proof paper taken to extreme hydration which produces a very dense sheet with a highly smooth and glossy finish. It is used as a liner for biscuits, cooking fats, fast foods and baked goods.

- Parchment paper is made from an acid-treated pulp (passed through a sulfuric acid bath). This makes it smoother and also impervious to water and oil enhancing the wet strength. However, it does not provide a good barrier to air and moisture. It is not heat sealable and is used to pack fats like butter and lard.

Paperboard is thicker than paper with a higher weight per unit area and is made in multiple layers. Mainly containers for shipping are made by it. These include boxes, cartons and trays. According to Soroka, (1999) the various types of paperboard are:

- White board, which is made from many layers of thin bleached chemical pulp. It is typically used as the inner layer of a carton. It is coated with wax or laminated with polyethylene for heat sealability. It is the only form of paperboard recommended for direct food contact.

- Solid board, has a lot of strength and is very durable. It has multiple layers of bleached sulfate board. It can be laminated with polyethylene to be used in the making of liquid cartons (known as milk board). Solid board is also used to pack fruit juices and soft drinks.

- Chipboard, is made from recycled paper but it contains blemishes and impurities from the original paper. This makes it unsuitable for direct contact with food, printing and folding. It can be lined with white board to improve the appearance and strength. It is the cheapest form of paperboard and can be used to make the outer layers for the tea cartons and cereals packets.

- Fiberboard, can be solid or corrugated. The solid type has an inner white board layer and outer kraft layer and it provides good protection against impact and compression. When laminated with plastics or aluminum, solid fiber board can improve the barrier properties. It is used to pack dry products like coffee and milk powder. The corrugated type is also known as corrugated board which is made with two layers of kraft paper with a central corrugating material. The resistance to impact abrasion and crushing damage makes it widely used for shipping bulk food and case packing of retail food products.

Paper laminates are coated or uncoated papers based on kraft and sulfite pulp. They can be laminated with plastic or aluminum to improve various properties. Along with polyethylene it becomes heat sealable and improves the gas and moisture barrier properties. However, lamination increases the cost of the paper. Laminated paper is used to pack dried products such as soups, herbs, and spices.

In order to protect human health, to preserve natural resources and the environment a proper waste management system is important. EPA strives to motivate behavioral change in solid waste management through nonregulatory approaches which includes pay-as-you-throw and Waste Wise. In pay-as-you-throw systems, residents can be charged for MSW services on the basis of the amount of trash they discard. This will create an incentive to generate less trash and increase material recovery through recycling and composting. These types of programmes may reduce the wastage to at least 14% to 27% per year. Waste Wise (launched in 1994) is a voluntary partnership between EPA and U. S. businesses, institutions, nonprofit organizations and government agencies to prevent waste, promote recycling and buy recycled content products. In a program on Environmentally Preferable Purchasing program in the year 2005 more than 1800 organizations had taken part in the WasteWise program (EPA, 2005). It helps federal agencies and other organizations to purchase products with lesser or reduced effects on human health and the environment as compared to other products. Pollution prevention is the primary focus, with a broader environmental scope than just waste reduction.

From a regulatory point of view, EPA guidelines for solid waste management emphasizes the use of a hierarchical, integrated management approach (EPA,1989) like the reduction at source, recycling, composting, combustion and landfilling. Waste disposal methods, combustion and landfilling are governed by regulations which are issued under subtitle D of the Resource Conservation and Recovery Act (40 CFR Parts 239-259). Source reduction (waste prevention) is the reduction in the amount and/or toxicity of the waste that is generated by changing the design, manufacture, purchase or use of the original materials and products. EPA considers source reduction as the best way to reduce the impact of solid waste on the environment as it reduces the waste generation. Source reduction includes using less packaging, designing products to last longer and reusing the products and materials (EPA, 2002). To achieve this, it is recommended to use light weight packaging materials, purchasing durable goods with large sizes (which use less packaging per unit volume). They can also be refillable containers and toxic-free products must be selected. It has many environmental benefits which include conservation of resources, protection of the environment and prevention of greenhouse gas formation. One way to achieve source reduction is through light weight, which is by using thinner gauges of packaging materials. This can be done neither by reducing the amount used or by using alternate materials. Girling (2003) has reported that the average weight of glass containers decreased by nearly 50% from 1992 to 2002. Similarly, aluminum cans were 26% lighter in 2005 than in 1975, with approximately 34 cans being made from 1 pound of aluminum, up from 27 cans in 1975 (Aluminum Assn., 2006). According to EPA (2004), Anheuser-Busch Companies Inc. light weighted their 24-ounce aluminum cans in 2003, which resulted in reducing the use of aluminum by 5. 1 million pounds. The amount of aluminum used in foil laminates has also been reduced. Moreover, steel cans have been lightweighted, with cans now at least 40% lighter than those of 1970. The amount of tin has been drastically reduced from pre-World War II levels of 50 pounds of tin per ton of tinplate steel to a current average of 6 pounds per ton (Miller, 1993). With the new packaging materials, plastic containers have reduced the weight. The weight of 2-L PETE soft drink bottles has decreased by 25% (from 68 to 51 g) since 1977, resulting in a savings of more than 206 million pounds of plastic packaging each year (American Plastics Council, 2006 a,b). Similarly, the 1-gallon plastic milk jug has undergone a weight reduction of 30% in the last 20 years. Reducing the weight has been achieved in the paperboard industry by using thinner gauge materials. For example, Anheuser-Busch saved 7. 5 million pounds of paperboard by decreasing the thickness of its 12-pack bottle packaging (EPA, 2004).

Another way to achieve source reduction is by reusing it. It has been found out that some glass containers, especially bottles are reused many times after washing with powerful detergents. Plastic refillable containers are made from PETE, PEN, or high-density polyethylene and polycarbonate are declining due to logistical difficulties. Today the manufacturers are also offering refill products to achieve source reduction. This is found especially with nonfood items like household cleaners. Refillable glass containers for beverage use have been replaced by thinner one-way glass or plastic containers to reduce transportation and cleaning costs. PETE containers have been depolymerized and repolymerized to avoid any contamination problems. However, it has not been found to be feasible.

Recycling diverts materials from the waste stream to material recovery. It involves reprocessing the material into new products. A typical recycling program entails collection, sorting, processing, manufacturing and sale of recycled materials and products. However, to make recycling economically feasible such products and materials must have a market. Marsh et al., (1991) have come out with an effective management of food package.

The purchase of products made with recycled materials is promoted by EPA's Comprehensive Procurement Guidelines (CPG). EPA designates products that can be made with recovered materials and recommends practices for buying these products. Later procuring agencies are required to purchase the product with the highest recovered material content level possible. EPA has selected more than 60 recycled products under the CPG program and has also proposed many other products. Almost all packaging materials (glass, metal, thermoplastic, paper and paperboards) are recyclable. Various factors play into any economic assessment of recycling, including costs for collection, separation, cleaning or reprocessing and transportation (energy). However, market and application for recycled products is a must. For instance, materials reclaimed through metal and glasses recycling are considered safe for food contact containers as the heat used to melt and form the material is enough to kill the microorganisms and pyrolyze organic contaminants. Reprocessing of plastic takes a large amount of heat to destroy microorganisms and is still not sufficient to pyrolyze all organic contaminants. Another problem is that the postconsumer recycled plastics are not generally used in food contact applications.

In general, recycling rates have been on the rise (EPA, 2006a). A total of 30 million tons of containers and packaging were recycled in 2005 (40% of amount generated). Because of increased collection and demand for recycled glass the process has grown in recent years. About 90% of recycled crushed glass (cullet) is used as raw material to make new containers. Recycling of aluminum cans has risen. According to the Aluminum Assn. (2006) it has increased to 52% in

2005 as compared to 50% in 2003. Rates of plastic recycling especially that of PETE and high-density polyethylene bottles have increased significantly since the 1990's (American Plastics Council, 2004). EPA considers composting to be a form of recycling. Composting is the controlled aerobic or biological degradation of organic materials like food and yard wastes. It involves arranging organic materials into piles and providing sufficient moisture for aerobic decomposition by microorganisms. The piles should be turned periodically to promote aeration and to prevent anaerobic conditions. The resulting humus, a soil-like material, is used as a natural fertilizer. This reduces the need for chemical fertilizers. Organic materials continue to be a large component of total MSW (about 25% for food scraps and yard trimmings, which makes composting a valuable alternative to waste disposal. It has been worked out how different food stuffs work (https://health.howstuffworks.com)

Combustion/incineration

The controlled burning of waste when done in a designated facility is known as combustion. It is an alternative in cases where the wastes cannot be recycled or composted. This helps in bringing down the MSW volume by almost 70% to 90%. The incinerators can be made to produce steam that can provide heat or could also generate electricity (waste-to-energy combustors or WTE facilities). Many plastics are obtained from petroleum feedstocks. These possess a high heat content which is advantageous for waste-to-energy incineration. In 2004, the United States had 94 combustion facilities of which 89 were WTE facilities. It had a processing capacity of around 95000 tons per day or about 13% of MSW (Kiser and Zannes, 2004). Presently, there are three types of incinerators and they are also known as municipal waste combustors (MWC's): mass-burn incinerators, refuse-derived fuel incinerators and modular combustors. In Mass-burn incinerators all the types of wastes can be used unless the item is very big and is unable to go through the feed system. Integrated waste is put on a grate which moves through the combustor. The air is forced into the system, above and below the grate, to promote complete combustion. As these incinerators burn the waste in a single stationary chamber and are typically constructed on site they are different from other MWC's. In order to recover the combustion heat for energy production they are installed with boilers. In 2004, 65 of the total 89 WTE facilities (77%) in the United States mass-burn technology was employed to process approximately 22 million tons of MSW. Refuse-derived fuel (RDF) incinerators are utilized in the use of preprocessed waste and can remove noncombustibles and recyclables. The combustibles are shredded into a uniform fuel that has a higher heating value. Kiser and Zannes (2004) have written a good article on the facilities available in USA. In case of mass-burn incinerators all waste can be used without preprocessing however, lesser than in mass burn. Out of the total U. S. MWC units in 2004 these are 10% (9 out of the total 89).

Landfilling

Landfills are the safest disposal of MSW and the residues of recycling and combustion operations can also be utilized. It is also environment safe. Most of the times it is the federal and state regulations that govern the location and operation of the landfills. These are carefully designed structures in which waste is isolated from the surrounding environment and also from the groundwater. If the landfill is designed properly it is able to manage the leachates and collects the landfill gases (methane and others) for potential use as an energy source. Having passed through or emerged from landfill waste, leachate contains soluble, suspended or miscible materials from the waste. Recently, there have been modifications in landfill design which is known as a bioreactor that can enhance aerobic and/or anaerobic degradation of leachate and organic waste (EPA, 2006c).

Due to the many problems of the environment, slow degradation in landfills along with the use of synthetic packaging materials in large amounts a lot of recent work has taken place in the fields. Recently, advanced landfill technology has been developed, more stringent environmental regulations are made for the landfills and biodegradable packaging materials has been developed. Now the landfills are engineered in a way that environmental contamination is prevented and managed to ensure compliance with federal regulations (40 CFR Part 258) or equivalent state regulations. A landfill reclamation approach has been set up that allows the expansion of existing MSW landfill capacity and preclusion of land acquisition for new landfills (EPA,1997). EPA also runs the landfill methane outreach program, promoting the use of landfill gas as a renewable energy source. The bioreactor developed by EPA for the landfills are designed in a way that they rapidly degrade organic waste with the addition of liquid or air to speed up microbial degradation. There are 3 types of bioreactors namely; aerobic, anaerobic and hybrid. EPA is also taking an initiative to identify the standards of bioreactor and recommend any operating parameters. Mainly the degradation in landfills is anaerobic degradation. During the process the microorganisms slowly break down the solid wastes. These may be primarily organic-based materials like wood and paper (in the absence of oxygen) which gets converted to carbon dioxide, methane and ammonia. It is through the compacted solid waste that the leachate is pumped. This can accelerate the process of degradation. In order to prevent ground water contamination the leachate should be contained in a system like a combination of liners and storage systems. At the end the leachate is processed into a stable residue that can be disposed off safely. Anaerobic degradation is mostly used to treat biosolids (sewage sludge) and organic waste contaminants. More research is underway. Biodegradable polymers can be obtained from replenishable agricultural feedstocks, animal sources, marine food processing industry wastes or microbial sources. According to Tharanathan (2003) the biodegradable materials break down to produce environmentally friendly products such as carbon dioxide, water and quality compost.

These polymers from cellulose and starch have been quite common (Miles and Briston, 1965). Cellophane is the most common cellulose-based biopolymer. There are many starch-based polymers, which swell and deform after getting exposed to moisture. These are amylose, hydroxylpropylated starch and dextrin. Other starch-based polymers are polylactide, polyhydroxyalkanoate (PHA), polyhydroxybuterate (PHB) and a copolymer of PHB and valeric acid (PHB/ V). A starch derivative formed from the microbial fermentation of polylactide does not degrade on exposure to moisture (Auras et al., 2004). PHA, PHB, and PHB/V are also formed by bacterial action on starches (Krochta and DeMulder-Johnston,1997). Biodegradable films are also produced from chitosan, which is a derivative from chitin of crustacean and insect exoskeletons. Chitin is a biopolymer with a chemical structure similar to cellulose.

Another form of biodegradable polymer are edible films or thin layers of edible materials which are applied on the food as a coating or placed on/ or between food components. These inhibit the migration of moisture, gases and aromas and improve the mechanical integrity or handling characteristics of the food (Institute of Food Technologists,1997). Films that are edible can be obtained from plant and animal sources such as zein (corn protein), whey (milk protein), collagen (constituent of skin, tendon and connective tissue) and gelatin (product of partial hydrolysis of collagen).

Synthetic polymers can be partially degraded by blending them with biopolymers, incorporating biodegradable components (such as starches) or by adding bioactive compounds. The biocomponents are degraded to break the polymer into smaller components that work through various mechanisms. They can be mixed with swelling agents which expand the molecular structure of the plastic upon exposure to moisture to allow the bioactive compounds to break down the plastic. The development of biodegradable polymers has taken acceptance by the public. It solves the problems of solid waste disposal and litter to substituting renewable resources (plant origin) for nonrenewable resources (oil, coal and natural gas) as raw materials. However, it is not a solution for all solid waste management problems. Change from synthetic polymers to biopolymers will not have much effect on the source reduction and incineration. However, recycling is complicated by the existence of blended or modified polymers unless they are separated from the recycling stream. In a landfill, plastics, which are biodegradable do not have benefit. This happens as the landfills exclude oxygen and moisture which are required for biodegradation. If biopolymers are used extensively than there should be sufficient plant materials for sufficient quantities of packaging polymers to be made. Presently, it has been found that economically, bioplastics are very expensive than other petroleum based polymers, so substitution would likely result in increased packaging cost.

Sometimes the biodegradable packaging is not practical even after so many advantages. Biodegradable plastics are better as they break down even after being littered. Especially in the marine environment the litter is very hazardous to the marine life. In military applications it is useful where they do not have traditional disposal. Specific but minor functions for biodegradable polymers include limiting moisture, aroma and migration of lipids in between the food components. Recently, many companies are making it. These are namely, Nature Works, LLC (a stand-alone company wholly owned by Cargill Inc.) manufactures polylactide from natural products (corn sugar). After use the polymer can be hydrolyzed to recover lactic acid. Wal-Mart Inc. is utilizing biopolymers polylactide for packing fresh cut produce (Bastioli, 2005).

Recycling of plastics is not allowed to easily identify polymers and systems. This is very stringent in high-density polyethylene milk bottles and PETE soda bottles. Some polymers can be taken into thermoplastic resins. This is utilized in the making of park benches and playground equipment. Beverage containers have been a visible component in litter. It is the most improper disposal of solid waste. These include bottles of beer, soft drink and other beverage containers. For reusing an economic incentive is being provided to ensure that the used bottles are returned. Beverage containers made from metal, glass and plastics can be recycled. However, it is possible that with biodegradable containers these problems could be prevented. There has been an increase of approximately 37% over the 179. 6 million tons generated in 1988 (EPA, 1990; IFT, 1991) and a decrease of 1. 6 million tons from 2004 (EPA, 2006a). The decrease in waste generation is due to the decrease in each individual waste generation rate. EPA analyzes MSW in 2 ways: By materials (paper and paperboards, glass, metals, plastics, rubber, leather, textiles, wood, food scraps and yard trimmings). By major product categories (containers and packages, nondurable goods, durable goods and other wastes). Containers and packaging waste products are mostly from food packets like soft drink cans, milk cartons and cardboard boxes. Apart from this there are newspapers, magazines, books, office papers, tissue papers, paper plates and cups. A lot of waste also comes from clothing and footwear. Other products producing a lot of wastes are household appliances, furniture, furnishings, carpets, rugs, rubber tyres, batteries and electronics. In a survey it has been found that waste percentage among materials was; paper and paperboard 34.2% (84 million tons) while food scraps were 11.9% (29.2 million tons) of the total MSW. Glass, aluminum and plastics contributed 5.2% (12.8 million tons), 1.3% (3.2 million tons) and 11. 8% (28.9 million tons),respectively. It was found that containers and packaging formed the highest portion of the total solid waste generated at 31. 2% (76.7 million tons) followed by nondurable goods at 25.9% (63.7 million tons). Along with the waste generation the recovery through recycling has also increased. In 2005, a total of 79 million tons (32.1%) of MSW were recovered through recycling and composting. However, around 58.4 million tons were recovered by recycling and the rest (20.6 million tons) by composting (EPA, 2006). It was in 2005 that the net per capita recovery was highest (1.5 pounds per person per day). The most recovered were (39. 9% of amount generated) followed by nondurable goods (31%). Recovery of yard trimmings was the highest at 62% (20 million tons), followed by paper and paperboard at 50% (42 million tons) and metal at 37% (7 million tons). The MSW has increased in quantities many folds. In 2005 approximately 168 million tons (68%) of MSW were discarded into the municipal waste. From these 33.4 million tons (20%) were burnt before disposing it (EPA, 2005) and 133.3 million tons were directly discarded in landfills. Proper waste management faces many limitations. It requires careful planning, financing, collection along with good transportation system. Along with an increase in the population the solid waste generation has increased and this has brought many challenges, as these wastes if not disposed cause a number of diseases.

It has been found out that the opposing pressures in food packaging like source reduction and convenience are very common. In case of unit packages or dispensability and microwavability additional packaging is required. It does not help in source reduction efforts. Recently, there has been a lot of demand by the consumers. However, they must evaluate if the convenience is adding safety in any way or not. The price at source reduction can be accelerated if consumers are willing to accept the loss of convenience and modify the purchasing. Refillable plastic containers have been developed as a strategy for source reduction but their use has declined in favor of non returnable containers.

Factors to reduce cost at source reduction are towards more economical bulk packs that needs less packaging material per unit of the product. In case of ratio of package dimensions remaining constant, increased size will increase the enclosure dimensions as a square function and increase the volume as a cube function. As the volume increases it may need less packaging per unit volume. These are good for warehouses and clubs reducing the cost. However, large portion sizes for small families leads to a lot of food waste increasing the waste. When the materials have to be reused and recycledit needs to be cleaned in order to remove any safety hazard posed by contaminants. In order to do this the materials are usually washed with powerful (usually caustic) detergents. This creates liquid waste which should be properly treated. Moreover, transportation costs can be high, depending on the proximity of each plant (Stilwell et al, 1991). Another major problem is that shipment charges of reusable or recyclable containers over distances requires a lot of energy as compared by refilling. Recycling of crushed glass (cullet) requires it to be transported to a limited number of glass manufacturing companies. With increase in oil prices the transportation is a big challenge. Studies can determine the environmental impact and resource demands of different waste management.

Reuse of bottles also brings in a lot of contamination. Many times it has been found that the bottles which were used for garden chemicals, gasoline transfer, or any other nonfoods pose a lot of hazard when returned to a food establishment. Also in the times when the bottles are not cleaned properly before recycling, the contamination gets transferred into the new pack which causes a health risk. Therefore, it is important that they are rinsed properly to remove food residues. Used soft drink bottles also attract insects and pests into a food establishment giving place for the growth of toxic microorganisms. According to Carolina Recycling Assn. (2002) this is a major concern among the food establishments.

In order to avoid any contamination in landfills towards air and ground water, proper design, construction and management are very important. It also helps in protecting the environment. In earlier times when the environmental impact to toxic matter was not considered the landfills were made on the non-expensive lands like wetlands, marshy lands, quarries, spent mines and gravel pits. It is covered with soil to reduce odors, litter and attack by the rodents. It was in1991, that it was noticed that landfills in wetland areas created groundwater contamination. This brought in the promulgation of MSW Landfills Criteria (40 CFR Part 258). It looks into the location restrictions, operating practices and requirements for composite liners, leachate and groundwater monitoring. If the landfills are not designed properly they could contaminate the groundwater. It can also happen when the water from rain or the waste itself permeates into the landfill and dissolves toxic substances. Certain substances can enhance the extraction of any conditions which may be acidic or alkaline. Under the standard conditions the composite liners prevent leachate from reaching ground water and help its collection and treatment before disposal. The standard methods have shown to minimize groundwater contamination which limits the air and water permeation of waste. It protects degradation of the organic material within landfills. Research work by the EPA's into bioreactor's supports the composting and better ways towards the management of organic waste. It is important to do the air emission standards (40 CFR Part 60, Subparts Cc and WWW) of the landfills. In the landfills there are many gas emissions like that of methane, carbon dioxide along with more than 100 various nonmethane organic compounds which include vinyl chloride, toluene and benzene. According to the air emission standards the emitted gases should be collected and there must be systems to treat them. If any person in the public opposes the inception of incinerators and landfills for waste disposal then it becomes an issue of economic, society and politics. Public involvement effectively is a very significant component of a comprehensive strategy.

A widely used method to address increased MSW disposal is combustion especially waste to energy combustion. If in case the modular combustors/incinerators are not available than there is a lot of initial capital requirement and for the construction it may take 3 to 5 years. However, the air emissions must be controlled. It has been found that it is usually carbon dioxide which is a greenhouse gas. It is released when products derived from fossil fuels (plastics) are burned. Pollution concerns include the emission of particulate matter, acidic gases (particularly sulfur dioxide and nitrogen oxides), heavy metals, halogens, dioxins and products of incomplete combustion. From incineration of chlorinated polymers it has been found out that dioxins and halogens are released. These include PVC. Incomplete combustion of the organic components of MSW is also possible with suboptimal operation of an incinerator.

Plastics and colorants contribute to the heavy metal content in MWC ash which includes lead and cadmium. Under the subtitle C of the Resource Conservation and Recovery Act the disposal of ash comes under potentially hazardous material. The Clean Air Act regulates MWCs. Several regulations are currently in place for new and existing MWCs (40 CFR Part 60, Subparts Ea, Eb, Cb, AAAA, and BBBB). In 2004 nearly all MWCs were equipped with particulates and acid gas controls in compliance with state and federal standards (Kiser 2004).

Proper disposal of packaging material is a must. Legislation towards this, needs bottle bills and recycling programs including requirements for recycling levels (Raymond Communications, 2005). It has been designed for recycling and to reduce litter. The bottle bills are making a positive impact. According to the surveys a lot of reduction in total road side and beverage container litter in states has been found with the use of bottle bills (Container Recycling Inst., 2006 a,b). As per the survey there are 11 states that have bottle bills in USA. These include California, Connecticut, Delaware, Hawaii, Iowa, Maine, Massachusetts, Michigan, New York, Oregon and Vermont (Container Recycling Inst. 2006a). Some of these states are attempting to expand bottle bill programs while others are reviewing their existing programs. Other states like California, Oregon and Wisconsin have passed rigid plastics packaging container requirements. This specifies recycling rates for rigid containers. New Jersey is also proposing for similar legislation. In case of China where the economy has expanded rapidly and they are willing to buy post consumer materials. This would reduce the availability of such materials in the United States. As a result, recycling can occur on a global level, however, making it difficult for states to meet their recycling targets.

Solid waste management and packaging norms are different in USA than in other countries (Raymond Communications, 2006). Many countries have under taken container deposit legislation. These countries are Australia, Canada, Denmark, Germany, Norway and Sweden. The approach is different. In this program the companies must collect and recycle a portion of their secondary packaging which includes shipping containers and outer wrapping. Such programs are in effect in many European countries. Some companies perform the take-back themselves or may opt to join collection organizations.

The waste management programs increase the cost which is passed on to the consumer. However the beneficial impact on solid waste reduction needs to be seen. In case of Duales System Deutsch land program in Germany had to change itself to a for-profit organization because of Germany's lucrative packaging laws. There are many fee-based programs such as Green Dot which charge the companies for the right to sell packaged goods in certain locales. The fees may or may not be tied to the recycling programs. A regulatory program that imposes fees on landfill disposal affects the cost and choice of materials for packaging. The requirements which are designed to reduce the environmental impact of packaging are a veiled trade barrier. Testing must be done in a certified laboratory with national standards and international standards, take-back programs that impose greater expense on imports than domestic products and unnecessary bans on substances irrelevant in the importing country such as tropical pesticides (Marsh, 1993). Definitions can also result in trade barriers. For example, a mandate for recycled content that requires domestic sources for materials may be a trade barrier for imports.

In order to have a successful packaging system it must be selected and designed to satisfy the competing needs like product characteristics, marketing considerations (including distribution needs and consumer needs), environmental, waste management issues and the cost. This needs a different analysis for each product, considering factors such as the properties of the packaging material, the type of food to be packaged, possible food/package interactions, the intended market for the product, desired product shelf-life, environmental conditions during storage and distribution, product end use, eventual package disposal and costs related to the package throughout the production and distribution process. Some of the factors are dependant on each other. The type of food and the properties of the packaging material determines the nature of food package interactions during the storage period. For a proper package design and development there should be a thorough knowledge of product characteristics which include deterioration mechanisms, distribution needs and potential interactions with the package. These constitute the physical, chemical, biochemical and microbiological nature of the product. Packaging which provides optimum protection to the product quality and safety are most preferred. Similarly, the distribution system of the product helps to find out the type of packaging material used. Importantly, it is the interaction between the food and package that plays an important role in the proper selection of packaging materials for various food applications. Each packaging material has different inherent properties (rigidity and permeability to gases). Depending on these properties the packaging material is selected. FDA (2017) reported on the use of antimicrobials in food producing animals.

According to a survey (IFT, 1988) during the storage period there can be interaction between food and the package. This may involve the transportation of low molecular weight compounds like gases or vapors and water from the food through the package, from the environment to the package, of the food into the package. It may also include chemical changes in the food, package or both. These interactions result in food contamination posing health hazard, loss of package integrity and decrease in quality. Mostly it has been found to have food package interactions. Here the migration of low molecular weight substances like stabilizers, plasticizers, antioxidants, monomers and oligomers takes place from the plastic packaging materials into the food (Arvanitoyannis and Bosnea, 2004). Other, low molecular weight compounds which include volatile and nonvolatiles can migrate from the food into the packaging materials through the sorption mechanism (Hotchkiss ,1997). The volatile substances such as flavors and aromas directly affect food quality while the nonvolatile compounds such as fat and pigments affect the package (Texas Food Establishment Rules, 2015). The most important aspect is the marketing. It is good marketing that promotes any product in a competitive market and increases the choice of the consumer (Coles, 2003). Consumers look for packages with convenient qualities which includes resealability, container portability (light weight materials are preferred), ease of opening, convenient preparation features and product visibility. A total analysis is vital to determine the impact of environment on any package. This includes the quantitative evaluation of environmental costs. This consists of material use, energy consumption and waste generation (Smith and White, 2000). According to McDonough and Braungart (2002) the sustainability goal inherent within the concept builds on the life cycle analysis.

It also addresses material and energy recovery. Many new packaging materials are being developed which will facilitate the goal of sustainability. A food package to be purposeful must consist of a material that maintains the quality and safety of the food without any degradation with time. It should look attractive, convenient and easy to use while conveying all pertinent information. They are useful in case they are made from renewable resources, do not generate any waste for disposal and are economical. Making of a food package is not only science but also an art with acceptable standards. Being inert and barrier to gases, it is glass which is the best choice for most packaging applications. However, as it is expensive the alternative is plastic. Plastics have good properties and and have various uses in food applications. The problem is their permeability to light, moisture and air. Presently scientists are trying to solve this problem by combining different plastics using co-extrusion or lamination or by laminating plastics with foil or paper. A significant role in package design is made by the consumers. They drive the sales of the product through the package. It is important that food packaging must continue for maintaining safety, wholesomeness and the quality of food. All efforts by industry, government, and consumers need to work together for the continued improvement for which an understanding of the functional characteristics of packaging is required.

Over long term the synthetic chemicals used in the packaging, storage and also in processing of food stuffs can be harmful to human health. It may happen as these substances are not inert and can leach into the foods we eat. People who eat packaged or processed foods regularly are exposed to low levels of these substances throughout their lives and might have side effects.

Toxicants, like formaldehyde which is a cancer causing substance, are legally used in these materials. It is widely present at low levels in plastic bottles used for fizzy drinks and melamine tableware. Chemicals that disrupt hormone production are FCMs. These include bisphenol A, tributyltin, triclosan, and phthalates. The total number of known chemical substances which are intentionally used in FCMs exceeds 4000. As these are not considered in toxicology analysis it gives doubts on the adequacy of chemical regulatory procedures. Looking into this a population-based assessment and biomonitoring are urgently required. This will be informative to know if there are any potential links between food contact chemicals and chronic conditions like cancer, obesity, diabetes, neurological and inflammatory disorders or any other environmental pollutants (Muncke et al., 2014).

In a review article by Valentina, (2012) inspite of so many health issues the use of polymer materials in food packaging field is one of the largest growing market area. The optimization behaviour of packaging permeability is vital. The knowledge will help in the extension of shelf-life of the food and to reach the best engineering solution. Studying the permeability characterization of the different polymer material (homogeneous and heterogeneous polymer system) to the different packaging gases, in different environmental conditions will help in the understanding of the selected material to the chosen food contact field. Parameters are important for food quality preservation as that of temperature and humidity. The authors have reviewed the state of art on the permeability characteristics of the polymer packages used in the food field. Knowing which product group spoils easily, at what point along the chain they spoil the most, what brings about the food loss and can the losses be avoided or not with respect to packaging system. It has been observed that the fresh fruit and vegetables (FFV) are the most perishable food items. Fruits and vegetables account for the maximum food losses, followed by other perishables such as bakery and dairy products, then meat and fish (Thonissen, 2009 as cited in Parfitt, et al., 2010). If fruits and vegetables can be packed in a proper way, then it is possible to reduce this wastage. Knowing when and where the losses occur in the commodity chain, it is helpful to pinpoint the causes of food loss, which can determine the packaging solutions.

In a value chain there are some barriers to waste reduction which could be either external (not within control) or internal (within control) (World Economic Forum, 2009). External barriers, include infrastructure, regulations, competitive pressures, consumer behaviour, stake holder relationships and technology. Internal barriers include the expertise, infrastructure and technology with the addition of management support, business models and financial resource concerns. These have a direct or indirect bearing on packaging which could provide solutions or potential support areas. The identification of these areas will help the organizations to have effective programmes on food loss reduction. Development in technology, management support and business models support the packaging development. This is expressed in terms of the specific packaging solutions.

Various compositions of plastic polymers are found to be good in many of the functions. A major drawback is that they require additives to guarantee certain properties required for their function. These are plastifiers to provide sufficient plasticity, antioxidants to protect from degradation by atmospheric oxygen, light protectants to decrease UV-light dependent degradation, colors to generate the intended appearance of the package, printing inks to apply written and graphic information and also ornaments. They contain residues of mono and oligomers of the starting material and additives required for the polymerization and maturation process of the polymer. These include catalysts, cross-linkers, polymerization modifiers and stopping components. An alternative is paper and cardboard which originate from wood treated in various ways. The manufacturing process of these materials may require the use of chemicals. These include preservatives that would help against pathogenic microorganisms, plastic polymer coating to improve the barrier function of paper, printing inks and colors, UV protectants and chemicals used in the pulp and paper production. Since non-coated paper and cardboard are of porous material, aqueous or oily liquids (from the food) may have access to the body of the material eventually extracting those chemicals from the packaging material. The use of recycled paper and cardboard for the production of packaging material is a problematic issue due to the chemicals present in the recycled materials.

Incorporation of Bacteriocins in Antimicrobial Films and Coatings

Recently a new and innovative strategy for preservation of foods that are eaten raw or without further cooking is the application of edible films or coatings which contain s antimicrobial substances. The incorporation of antimicrobial compounds such as bacteriocins in edible coatings and films presents an interesting alternative to control the pathogenic microorganisms in food products (Valdés et al., 2017).

Edible coatings and films are composed of thin layers of biopolymers that modifies the surrounding atmosphere of foods, forming a barrier between the food and the environment, improves the safety, quality and functionality of food products without changing organoleptic and nutritional properties (Han, 2003; Valdés et al., 2017). The use of purified bacteriocins or bacteriocin-producing bacteria in the packaging system may be more effective in the inhibition of the growth of pathogenic and/or deterioration microorganisms throughout the extent of the latency phase (Balciunas et al., 2013). Hydrocolloids (proteins and polysaccharides) are the most extensively investigated biopolymers in edible coatings and films applied to cheese. They facilitate the incorporation of functional compounds such as bacteriocins and bacteriocin-producing bacteria and allow an increase in the stability, safety and the shelf life of dairy foods (Scannell et al., 2000a). The effectiveness of incorporating bacteriocins and/or bacteriocin-producing LAB has been studied in coatings and films applied to dairy products. The work by many scientists have shown an inhibition in the growth of pathogenic microorganisms in foods which were packed with coatings and films containing antimicrobial metabolites synthesized by LAB (Cao-Hoang et al., 2010; da Silva Malheiros et al., 2010; Ercolini et al., 2010; Aguayo et al., 2016; Malheiros et al., 2016) or containing viable LAB in the film/coating matrix (Concha-Meyer et al., 2011; Barbosa et al., 2015). Studies on the effectiveness of incorporating purified bacteriocins in edible coatings shows a limited reduction of pathogens such as *L. monocytogenes*. Cheeses, particularly fresh cheeses, are highly perishable due to their high content in caseins, lipids and water. The complexity of cheese composition and the manufacturing conditions support the development of pathogenic and deteriorating microorganisms that increase the risk of foodborne illnesses and reduce the quality and acceptability of cheese (Ramos et al., 2012). The application of edible coatings and films containing bacteriocins may be helpful in the problems associated with post-process contamination. This will enhance the safety and extend the shelf-life of the cheese. Cao-Hoang et al., (2010) worked on the incorporation of nisin into the films of sodium caseinate applied to semi-soft cheese and observed a small reduction in *L. innocua* counts (1. 1 log cfu g^{-1}) after a week of storage at 4°C. In another study, done by Scannell et al., (2000b) the incorporation of nisin and lacticin in cellulose coatings when applied to Cheddar cheese reduced the cell numbers of *L. innocua* by 2 log cycles, and *S. aureus* by 1. 5 log cycles. However, other studies have shown an effective reduction on pathogen growth. In Ricotta cheese coated with galactomannan and nisin, the growth of *L. monocytogenes* was prevented for 7 days at 4°C (Martins et al., 2010). The application of a coating in Port Salut cheese, consisting of tapioca starch and combined with nisin and natamycin, reduced the counts of *L. innocua* above 10 cfu ml^{-1} during storage. In the study the coating became a barrier to post-process contamination (Resa et al., 2014). Recently, Marques et al., (2017) used a biodegradable film incorporated with cell-free supernatant (CFS) containing bacteriocin-like substances of *Lactobacillus curvatus* P99, to control the growth of *L. monocytogenes* in sliced "Prato" cheese. These films containing the bactericidal concentration of CFS were able to control *L. monocytogenes* for 10 days of storage at 4°C.

Cheeses get contaminated with *L. monocytogenes* as they provide the appropriate growth conditions for it. Listeriosis out-breaks linked to the consumption of contaminated cheeses have been reported worldwide. More research needs to be done on the application of bacteriocins to the preservation of milk, cream, yogurts, and other dairy products.

6

Food Adulteration and the Harmful Side Effects

Food adulteration is an unethical and often criminal malpractice. It is very common in countries of the South-East Asia region. It frequently occurs where informal food production and marketing services are predominant and enforcement of food regulation is weak. The prohibited substances are either added or used to partly or wholly substitute healthy ingredients. This creates the impression of freshness in old food artificially. Adulteration in food is usually done for financial gain. The high target is spices due to the demand and also their high price. Sometimes it can happen due to lack of proper hygienic conditions of processing, storage, transportation and marketing. The consumer is either cheated financially or becomes a victim of illness or disease. In case, the consumers take a lot of precautions at the time of purchase of such products it would help them in avoiding the purchase of such adulterated food. The whole spices: Dirt, dust, other seeds; chili powder: brick powder, salt powder or talc, powder. In ghee/butter: vegetable ghee, animal fat, mashed potato, sweet potato, etc. ; ice cream and beverages:saccharin; honey: jaggery, sugar syrup. Food contaminants are not added to the foods. They are present in foods as a result of the production (crop farming, animal husbandry and aquaculture), manufacture, processing, preparation, treatment, packing, packaging, transport or holding of such food or as a result of environmental contamination. Usually it includes the presence of harmful chemicals or microorganisms in food, which if consumed can cause several debilitating illnesses. Many times, unknowingly, the contamination may occur during the production, processing, storage and marketing. Edible oil may be contaminated during processing due to the leakage of mineral oil in production line. Food contamination is an even bigger threat in countries of the South-East Asia region due to the lack of produce harvesting norms, food handling standards and environmental regulations. For ripening the usage of calcium carbide and copper sulphate in mangoes, bananas is very harmful. Oxytocin is a harmone which is used for the rapid growth of Pumpkin, watermelon, brinjal, gourds and cucumber. Protection is done by using wax which adds shine on apples and pears. Cheap green colours along with chemicals like metallic lead used in bitter gourd and leafy vegetables are harmful. Pesticides and herbicides are used for growing fruits and vegetables. These creates many health Problems especially for the digestive system, eyes and liver. It can also result in vomiting and diarrhea in children and kidney failure. Oxytoxin can lead to brain damage. However, accidental food adulteration occurs accidentally in nature, without our knowledge. These include green vegetables grown in marshy areas. These areas have high levels of industrial pollutants, including heavy metals, which are absorbed by the plants causing health problems. It can be harmful to various organs in the body. This is found to be common in almost all developing countries. The harmful effects include stomach disorder, giddiness, joint pain, diarrhea, liver disorder, dropsy, gastrointestinal problems, respiratory diseases, cardiac arrest, glaucoma carcinogenic effects and paralysis. Organic foods are the products grown and manufactured which needs to adhere to standard set by the country they are sold in. Mostly, it is presumed that organic farms do not consume or release any synthetic pesticides into the environment. However, some organic pesticides, accepted for restricted use by most organic standards, include Bt, pyrethrum and rotenone can be harmful. Food irradiation is the process of exposing food to ionizing radiation to destroy and check the multiplication of microorganisms, bacteria, viruses or insects that might be present in the food. These rays can be harmful. These are used in sprout inhibition, delay of ripening and improvement of rehydration. Cobalt-60 is the most commonly used radionuclide for food irradiation. The quantity of dosage is very much important. With over dosage many health problems can arise. These can be like disruption of internal metabolism of cells, DNA cleavage, formation of free radicals, disruption of chemical bonds, increase in capital costs, development of resistant MO, inadequate analytical procedures to detect irradiation in food and causing public resistance. Ionizing radiation also breaksdown chemical bonds within the food itself. This causes chemical changes in the food. Some are desirable, others are not. It can change the structure of certain foods to withstand the irradiation like lettuce and other leafy vegetables which turn mushy. Slow ripening and maturation in certain fruits and vegetables lengthens the shelf-life. Food is one of the basic need for every living being and is composed of carbohydrates, water, fats and proteins, which can be eaten or drunk by animals, including humans for nutrition or pleasure (FAO/WHO, 2007). There are around 2,000 plant species which are cultivated to be eaten as food and many have several distinct cultivars. Almost all foods are of plant or animal origin. Staple food that provides more food energy around the world is cereal grain. According to Awasthi et al., (2014) it is maize, wheat and rice which together account for 87% of all grain production across the globe. Animals are also used as food with their products like meat which is a direct product and is derived from either muscle systems or from organs. Other food products produced by animals are milk which in many countries is drunk or processed into many different dairy products as that of cheese and butter. In addition birds and other animals lay eggs and the bees produce honey (Awasthi et al., 2014). According to Ayalew et al., (2013) public has now become very conscious to eat safe food and therefore, food safety has become very important. It is one of the major emerging areas in the food supply chain and has attracted a lot of attention from various research, government and regulatory bodies. Food and drink products are often a target of adulteration. The supply chains deals with perishable products which are harmful to the consumers if they are not managed properly (SGS, 2013). Adulteration involves either the addition of extraneous substances (adulterants) into food items or products or reducing essential nutrients for financial gains. This is done during processing, storing, transportation or marketing. This results in many diseases to the consumer (Anita and Neetu, 2013). Recently, in the People's Republic of China there was a milk scandal (Chinese milk scandal case with melamine) where many children were killed and harmed (Lakshmi et al., 2012). According to Alauddin, (2012) adulterated foods have increased to more than several hundred billion dollars which constitutes more than 10 percent of total trade. According to Anita and Neetu (2013) the adulteration increases visible quantities and reduces the manufacturing costs. On a daily basis we are consuming adulterated food. Adulteration is usually addition of impure matter to the food or drink which is intended to be sold in order to increase the quantity of the product. Adulterated food is impure and unsafe. This is done to increase profit but adulterated food causes serious ill effects to the human life. It is very common in chili powder in which it is added with wood powder or brick powder. Most of the times it resembles the original product. In the process the seller gets more money by selling less quantity of the actual product. It is a punishable offense in India and there are laws governing it. Government had made an act which is called as 'The Prevention of Food Adulteration Act'. It was passed in 1954 and was repealed by the Food Safety and Standard Act (FSSA) and passed by the parliament in 2006. Based on the seriousness of the offense, punishments and penalties are given starting from three months to six years imprisonment. Inspite of the laws, food adulteration still persists. The system can be changed and laws can be implemented only after the corruption at all levels is put off. Presently, it is done in almost all the possible products to increase the profits. The most common products are milk, ghee, vegetable oil, masala powders, spices, ice creams, honey, coffee, tea leaves, flour, food grains and sweets etc. Apart from these products, grape wines, juices of fruits are adulterated in most of the countries. In olden days each family had a kitchen garden which was used to grow their own food stuff and mill them at home. Therefore, the chances of food adulteration were not there. Today, with both spouses working outside home has made them very busy. This has increased the dependency on instant food items. These food items are adulterated with added preservatives which are harmful to the human beings. If adulterated food is consumed for a long time it has both short term and long term impact on the human health. Therefore, it is better to produce stuff in the back yard as far as possible or to purchase from the organic centers or directly from the farmers. Adulterated food is also found to be of low quality and has low nutritional contents. Many a times the taste may also vary. Various chronic diseases like liver disorder, diarrhoea, stomach disorder, lahyrism cancer, vomiting, dysentery, cancer, joint pain, heart diseases and food poisoning may occur due to the consumption of adulterated food. It is the minerals, chemicals and/or poor quality substances added to the foods that are responsible for the bad health conditions. Sometimes the adulterated foods can even lead to abortion or a brain damage. Young girls can have issues in conceiving later in life. Simple analysis for food adulteration includes: testing of milk by pouring a small drop of milk in a slanting surface. If the drop leaves a white trail, it is pure milk and if there is no white trail, then it is adulterated. In ice cream the main adulterant used is washing soda. To find this, few drops of lemon juice can be added to it. If it starts to froth, then it indicates the presence of washing soda in the ice cream. In sugar the presence of chalk powder can be checked by taking a clear glass of water and adding a spoon of sugar to it. Unadulterated sugar will sink to the bottom while the chalk powder will remain on the surface of the water itself. For black pepper seeds usually papaya seeds are used as adulterants. To check this few pepper seeds can be added into alcohol. The pure papaya seeds will sink. To check adulteration in mustard seeds; few mustard seeds can be crushed. Actual mustard seeds will have a yellow color inner area whereas its adulterant argemone seeds will not be yellow in its inner surface. It is advisable to purchase branded products with ISI mark. Inspite of this if adulteration is seen it can be reported to consumer rights. Vegetables and fruits can be purchased from organic markets or directly from the farmer. Instead of ready made masala powders they can be prepared at home in a mixie. Olive oil, milk, honey, saffron, orange juice, coffee and apple juice are the foods which can easily get adulterated. The adulterated food must have a label of "poisonous or deleterious substance" which may render it injurious to health. Food can be adulterated intentionally and accidentally. Sometimes due to ignorance or due to the lack of facilities adulteration is done in order to maintain food quality. Food colours, chemicals and additives are also present in packaged products. Government is taking legal action to prevent food adulteration in India. Adulteration and contamination in India is found in the food consumed at the household level, in the food service establishments, in business firms and also for street foods. Contamination of various mytotoxins, metals and pesticides in daily foods and milk has been found highly toxic and 70% of deaths are due to food borne contamination. Recently, there are many genetically modified foods, nano-tech foods, functional foods which are helpful for the food processing and food production. However, the ethical committees need to check the various health issues. The food supplied in restaurants is many times cooked with poor quality ingredients rather than providing a wholesome nutritional meal. This causes a major impairment to the health. There are different ways in which the food can be adulterated. It can be intentional, accidental or as natural. Intentional food adulteration is usually done to get more money. It includes adding some thing on purpose to earn more profit. Commonly, it is the addition of water or urea to liquid milk, extraneous matter to ground spices, addition of argemone seeds to mustard seeds which can cause loss of vision and heart diseases, aluminium foil to silver foil, water to honey, chicory to coffee, coloured leaves and iron fillings into tea leaves. There are many health problems happening due to these adulterations. These include appendicitis and small intestine problems, rodamine culture (a biological stain), brick powder to red chilli powder may cause loss of vision and respiratory diseases, metanil yellow to turmeric powder, dal moong and washed channa may cause anaemia, epilepsy, neurotoxity. When malachite green is added to green vegetables like chilli, vanaspathi to pure ghee or butter, papaya seeds to black pepper, white powered stone to common salt all these may cause appendicitis. Argemone oil to groundnut oil can cause loss of vision and heart diseases. Addition of melami to infant formula and pet food can falsify the level of protein content. Removal or substation of milk solids from the natural products is essential. Many chemicals and colours which are used in fruits and vegetables are very poisonous to the health. Adulteration of food is normally observed in its most crude form, where prohibited substances are either added or used to partly or wholly substitute healthy ingredients or to artificially create the impression of freshness in old food.

Edible oil may be contaminated during processing due to leakage of mineral oil in production line. Food contamination is an even bigger threat in countries of the South-East Asia region due to the lack of produce harvesting norms, food handling standards and environmental regulations. It is widely known that pesticides / insecticides and herbicides / fungicides are used for pest and weed / spoilage control. These agrochemicals are available over the counter. However, the farmers may use them without proper understanding or supervision. These chemicals can be hazardous, and there is always a chance of contamination of agriculture produce with residues of these chemicals. Considering the possible health hazard of dietary intake of residues of these chemicals, the Joint FAO/ WHO Meetings on Pesticide Residues has provided independent scientific expert advice on pesticide residues, residual limits and the acceptable maximum dietary intake (ADI) and maximum residual limit (MRL) on the basis of evidence-based information and risk assessment.

Acceptable daily intake (ADI) of a chemical is the daily intake which, during an entire lifetime, appears to be without appreciable risk to the health of the consumer on the basis of all the known facts at the time of the evaluation of the chemical by the Joint FAO/WHO Meeting on Pesticide Residues (JMPR). It is expressed in milligrams of the chemical per kilogram of the body weight. Maximum residual limit (MRL) is the maximum concentration of a pesticide residue (expressed as mg/kg), recommended by the Codex Alimentarius Commission to be legally permitted in food commodities and animal feeds. Respective MRLs are intended to be toxicologically acceptable. MRLs which are primarily intended to apply in international trade are derived from estimations made by the JMPR following both: toxicological assessment of the pesticide and its residue. Review of residue data from supervised trials and supervised uses include the ones which reflect the national food agricultural practices. In order to accommodate variations in national pest control requirements, ADI and MRLs they are determined for veterinary drug residues and antimicrobial substances as these chemicals are used for growth promotion and animal disease prevention and control. Some raw foods contain toxins which become less toxic once cooked. Red kidney beans can contain high dose of lectin which is toxic when consumed in raw form. As few as four or five raw beans can cause severe stomach ache, vomiting and diarrhoea. Therefore, red beans must be well cooked or boiled briskly in fresh water for at least 10 minutes to destroy toxins. Similarly, the castor bean lectin ricin can cause deaths of children when consumed in raw state. Solanine, is a glycoalkaloid found in high concentration in budding or green areas of the potato, is highly toxic to humans. Glycoalkaloids are not destroyed on cooking, therefore, it is important to avoid eating the sprouts and to remove any green or damaged parts before cooking. Cucumber may occasionally contain a group of natural toxins known as cucurbitacins. These toxins give cucumber a bitter taste. Cabbage and related vegetables contain thioglucosides which may be absorbed in people with low dietary iodine, and may contribute to thyroid enlargement. Naturally occurring cyanogenic glycosides are found in raw or unprocessed cassava (*Manihot esculenta*), which can cause nerve damage or death if consumed in large quantity. To avoid exposure to these toxins, it should be cooked thoroughly before eating. Cassava should be peeled and sliced and then only it should be cooked, either by baking, boiling or roasting. Most of natural toxins found in fish are produced by species of naturally occurring marine algae. They accumulate in fish when they feed on the algae or on other fish that have fed on the algae. Ciguatera fish poisoning is characterized by numbness and tingling of the lips and tongue, vomiting, diarrhea and is associated with consumption of toxin-contaminated subtropical and tropical reef fish. Unfortunately, these toxins are not destroyed by normal cooking or processing. Food poisoning from the consumption of poisonous wild mushrooms has been reported during monsoon season in countries of South and South-East Asia. Many cases of loss of human lives due to consumption of poisonous wild mushrooms have been reported. Fatal mushroom poisoning occurs due to ingestion of *Amanita Phalloides* also called as the 'death cap'. This has high content of amatoxin which is a potent cytotoxin. One single mushroom contains enough poison to kill an adult. Cooking or peeling does not inactivate the toxin, and all parts are equally poisonous. Onset of symptoms is seen within 6-24 hours or more after ingestion of mushrooms. Fatal poisoning is usually associated with delayed onset of symptoms which are very severe, with toxic effect on liver, kidney and nervous system. Mycotoxins are a group of naturally occurring chemicals produced by certain moulds or fungi. They can grow on a variety of different crops and foodstuffs including cereals, nuts, spices and dried fruits. Mycotoxins are produced by several fungi in foodstuffs and feed during production, storage, transportation, often under warm and humid conditions. The mycotoxins which have huge concern to human beings from a food safety perspective, include, the aflatoxins, ochratoxin A, fumonisins, trichothecenes and zearalenone. The aflatoxins are most commonly found in maize, peanuts and feed as contaminants and it can also be found in the milk of animals which are fed on contaminated feed, in the form of aflatoxin M1. Some mycotoxins such as trichothecene remain toxic even after being cooked. Mycotoxins can cause a variety of adverse health effects in humans. Aflatoxins B1 are genotoxic and carcinogenic, and can cause liver cancer in humans. Other mycotoxins have a range of other health effects including kidney damage, gastrointestinal disturbances, reproductive disorders or suppression of the immune system. Dioxins are a group of chemically-related compounds that are mainly the by-products of industrial processes but can also result from natural processes, such as volcanic eruptions and forest fires. Dioxins are found throughout the world in the environment and they accumulate in the food chain, mainly in the fatty tissues of animals and pass to human body through food, mainly meat and dairy products, fish and shellfish. More than 90% of human exposure is through food, mainly meat and dairy products, fish and shellfish. Short-term exposure of humans to high levels of dioxins may result in skin lesions. Dioxins are highly toxic and can cause reproductive and developmental problems, damage the immune system, interfere with hormones and also cause cancer. Prevention of human exposure is done by a strict control of industrial processes to reduce formation of dioxins. Many national authorities have programmes which monitor the food supply. This monitoring has led to early detection of contamination and has often prevented impact on a larger scale. Another chemical called as acrylamide is found in certain foods that have been cooked and processed at high temperatures, the levels of acrylamide increase with the time of heating. Of the limited range and number of foods analysed till date, acrylamide levels are highest in potato and cereal-based products subjected to heat processing like frying, grilling or baking. However, the mechanisms of formation of acrylamide in food are not well understood. Studies with laboratory animals suggest that acrylamide has a carcinogenic potency in rats which is similar to the other carcinogens in food. A joint FAO/WHO consultation on health implications of acrylamide in foods which was held in 2002, recognized, the presence of acrylamide in food which could induce cancers and heritable mutations in laboratory animals. There are currently no regulatory maximum limits for acrylamide in food. It should be seen that the food is not cooked excessively. However, all food, particularly meat and meat products, should be cooked thoroughly to destroy the foodborne pathogens, bread should be toasted to the lightest colour acceptable. People should eat a balanced and varied diet, which includes plenty of fruits and vegetables. It is important that consumption of fried and fatty foods is moderate. FDA (2017) reported on the use of antimicrobials in food producing animals.

7

What are Food Pathogens and What is Food Contamination

When certain disease-causing bacteria or pathogens contaminate food, they can cause foodborne illness, often called "food poisoning". Foods that are contaminated may not look, taste or smell any different from foods that are safe to eat. *Salmonella, Campylobacter, Listeria* and *Escherichia coli* are common bacteria causing foodborne illness. Some foodborne illnesses are mild where most people get better within a few days, but illness can sometimes be more severe and fatal. *Bacillus cereus* can form heat resistant spores. On cooking the food with these spores the heat resistant toxin is left at room temperature. Reheating or lightly cooking the food does not destroy these toxins. Most types of foodborne illnesses cause symptoms like nausea, vomiting, watery diarrhea, abdominal pain and cramps. Fever and/or the presence of blood or mucus in the stool are signs of bacterial infection that may require a medical consultation. Proper food handling is vital to prevent most foodborne illnesses and diseases. WHO has given keys towards safer food which should be followed. Unsafe food containing harmful bacteria, viruses, parasites or chemical substances, causes more than 200 diseases which range from diarrhoea to cancers. Foodborne and waterborne diarrhoeal diseases kill around 2 million people worldwide annually, including many children. A single foodborne bacterium can grow into more than two million bacteria in just seven hours under the right conditions. Foodborne pathogen called *Listeria* sp. can survive and sometimes grow on foods being stored in the refrigerator. Hepatitis A virus can cause long-lasting liver disease and spreads typically through raw or undercooked seafood or contaminated raw produce. It is estimated that more than 250 million people worldwide suffer food allergies. The boiling temperature (smoke point) of cooking oil is above 200 °C. The boiling point of water is 100 °C.

According to the World Health Day (WHO,2015) food contamination happens when food are corrupted with another substance. It can happen during the process of production, transportation, packaging, storage, sales and cooking process. The contamination can be physical, chemical and biological. Faulde et al., (2013) studied the infection caused to patients by mothfly.

Physical Contamination

Physical contaminants (or 'foreign bodies') are objects such as hair, plant stalks or pieces of plastic and metal (Aladjadjiyan, 2005). When the foreign object comes into the food, it is a physical contaminant. According to Aladjadjiyan, (2005) in case of foreign objects being bacteria, the case will be physical and biological contamination. Chemical contamination happens when food is contaminated with a natural or artificial chemical substance (Cascales et al., 2007). These include pesticides, herbicides, veterinary drugs, contamination from environmental sources (water, air or soil pollution), cross-contamination during food processing, migration from food packaging materials, presence of natural toxins or use of unapproved food additives and adulterants. Contamination can happen at one or more stages in food production. According to Rather et al., (2017) illness is likely to result if consumers ingest them in large quantities. Biological contamination refers to food that has been contaminated by substances produced by living creatures, such as humans, rodents, pests or microorganisms. These include bacterial contamination, viral contamination or parasite contamination which is transferred via saliva, pest droppings, blood or faecal matter (Cast,1994). Bacterial contamination is the most common cause of food poisoning worldwide (Food and Drug Administration; FDA,1998). If an environment is high in starch or protein, water, oxygen and has neutral PH level along with temperature between 5°C and 60°C (danger zone), bacteria are able to survive (Hernandez, 2000). Tainted Romaine Lettuce in 26 states of the United States confirmed the outbreak of bacteria strain *E. coli* O157:H7 (FDA,2018). The U.S. Food and Drug Administration, along with the Centers for Disease Control and Prevention (CDC) and state and local partners, investigated a multistate outbreak of *E. coli* O157:H7 illnesses linked to romaine lettuce from the Yuma growing region. (FDA, 2018). The most common symptoms of *E. coli* include diarrhea, bloody diarrhea, abdominal pain, nausea and vomiting (Melton-Celsa, 2014). Raw foods of animal origin, that is, raw meat and poultry, raw eggs, unpasteurized milk and raw shellfish usually get contaminated. Fruits and vegetables can also be contaminated with animal waste when manure is used to fertilize produce in the field or unclean water is used for washing the produce. Caution needs to be taken with raw sprouts because the conditions under which they are sprouted are also good for growing microbes. Unpasteurized fruit juices or cider can also be contaminated if there are pathogens on the fruit that is used to make it. Any food item that is touched by a person who is ill with vomiting or diarrhea, or recently had such an illness, can become contaminated. When these food items are not subsequently cooked (salads, cut fruit) they can pass the illness to other people.

During the past century, the microbiological safety of the US food supply has improved. Still, there are many foodborne illnesses and outbreaks which occur annually. Hence, opportunities for the food industry to improve the safety of both domestic and imported food exist through the adoption of risk-based preventive measures. Challenging food safety issues that are on the horizon include demographic changes to a population whose immune system is more susceptible to foodborne and opportunistic pathogens, climate changes that will shift where food is produced and consumers' preferences for raw and minimally processed foods. Increased environmental and product testing and anonymous data sharing by the food industry with the public health community would aid in identifying system weaknesses and will enable more targeted corrective and preventive actions. Clinicians will continue to play a major role in reducing the foodborne illnesses by diagnosing and reporting cases and in helping to educate the consumer about food safety practices.

It has been found that both organic and conventional animal products are contaminated with *Salmonella* and *Campylobacter*. In a study done by the scientists it was seen that almost all chicken samples (organic /inorganic), were contaminated with *Campylobacter.* One third of the samples had *Salmonella*, but the risk of exposure to multidrug-resistant bacteria was lower in organic meat. The difference is that the food poisoning from organic meat is little easier for the doctors to treat it. Exposure to pesticides has shown to increase the rate of chronic diseases like many types of cancers, diabetes, neuro degenerative disorders which include Parkinson's, Alzheimer's and ALS, birth defects and reproductive disorders. In Salinas Valley of California it has been found that they spray heavy pesticides. This has affected the childhood with a defective brain development and lowered IQ. Twenty-six out of 27 studies showed negative effects of pesticides on brain development in children. In urban areas a survey was made with mothers having higher levels of a common insecticide in their umbilical cord blood. It was found that those with higher levels were born with brain defects. Using insecticides may have a risk factor for childhood leukemia. Pregnant women farm workers may have high risk of their child getting leukemia and/ or brain tumor. It is also harmful for the pregnant farm women who picks them.

Researchers have measured the levels of two pesticides running through children's bodies by measuring specific pesticide breakdown products in their urine. The kids were later fed only organic diet for five days and then back to the conventional diet. It was found that eating organic food provides a protective effect against exposure to pesticides used in agricultural production. It was found that consumption of organic foods provides protection against pesticides. However, scientists are working on the role of organic food against diseases.

Organic eggs can lower the risk of *Salmonella*, which is an egg-borne epidemic every year in the US (In health, MichaelGreger, M. D., 2017) Nutrition facts. org . In another study (Hafstro et al., 2001) a vegan diet free of gluten improves the signs and symptoms of rheumatoid arthritis: the effects on arthritis correlate with a reduction in antibodies to food antigens.

Mosupye et al., (1999) have studied the safety of ready to eat street vended foods. Looking into the safety of street foods the first step is to check the safety of foods which are usually wrapped in newspapers. It has been found out that wrapping fried food in newspapers or any other printed papers by street vendors is very unhealthy. The consumption of such foods are injurious to health, even if the food has been cooked hygienically. Street food vendors are very common in South-East Asia Region as these foods are very economical. It is difficult to say how safe street food is as there are many contributing factors associated with the safety of street food. A survey was done by WHO in 1993; where almost 100 street foods were taken from 100 countries. It showed a major health threat facing the public from raw and undercooked food, infected food handlers and inadequate hygiene during processing and storage of these food. The chances of food poisoning may be less as the majority of street foods are cooked well and served hot. The water quality, hygienic conditions and the level of cleanliness especially in the summer season contributes to the chances of food poisoning. However improvements in physical infrastructure and hygienic conditions are growing as consumers are also demanding quality food. Usually the focus is on a relatively short list of well-studied pathogens which include the strains of *Listeria, E. coli* and *Salmonella*, along with chemical contaminants and physical hazards like wood or metal fragments. Some of these contaminants are inherently found in the environment or production systems. For these there are good inspection and detection methods available in the literature. Many food processing systems include treatments like oxidizing, heat treatments or compositions (low water activity, pH) that can inactivate the pathogens. The food pathogens cause diseases especially when the food is not cooked properly or stored. Sometimes the microorganisms in fermented foods such as cheese, yogurt, bread, beer are also responsible to cause food diseases. Foodborne illness (also foodborne disease and colloquially referred to as food poisoning) is any illness resulting from the food spoilage of contaminated food, pathogenic bacteria, viruses or parasites that contaminate food, as well as toxins such as poisonous mushrooms and various species of beans. The most common types of foodborne illness include bacteria. The most common bacterial pathogens are *Salmonella, Campylobacter, E. coli and Listeria.* Mostly the foodborne illnesses are caused by pathogenic bacteria or viruses in food. Other foodborne illnesses occur from accidental chemical poisoning and natural contaminants.

The common viruses causing food diseases are Norovirus, Rotavirus and Hepatitis A. Intoxication can be caused by toxins which are produced by these pathogens. These are *Staphylococcus aureus, Bacillus cereus* and *Clostridium perfringens*. Symptoms and the most frequent food vehicle varies depending on the type of pathogen. Infected foods have been found with different pathogen source. However, it is not that all these foods will always have the infection. It is important that the food should be prepared in a safe and hygienic way following all GMP practices. Florida Press Educational Services Inc. (FPES, 2016) have written an excellent article on the partnership for Food Safety Education (USA) and found that the food supply in the United States is among the safest in the world. However, when certain disease-causing bacteria or pathogens contaminate food, they can cause foodborne illness, which is known as food poisoning. Many diseases are named after the causal organism. (*Campylobacter*), Cryptosporidiosis (*Cryptosporidium*),*Escherichia coli* O157:H7 infection, Giardiasis (*Giardia*), Listeriosis (*Listeria monocytogenes*), Norovirus infection (aka Norwalk virus, calicivirus, viral gastroenteritis), Salmonellosis (*Salmonella*), Scombroid fish poisoning, Shigellosis (*Shigella*), Toxoplasmosis (*Toxoplasma gondii*),Vibrio infection (*Vibrio parahaemolyticus*) and Yersiniosis (*Yersinia* species). Foodborne illnesses are caused by consuming contaminated foods or beverages. Many different disease-causing microbes or pathogens can contaminate foods, so there are many different types of food borne illnesses. Many a times the food borne pathogens are present in drinking water or swimming pool water which can infect the humans. Other ways of getting infection are by the contact of animals, the environment or through person-to-person. The food infections have major symptoms like diarrhea and/or vomiting which usually takes 1 to 7 days to appear. Symptoms also include abdominal cramps, nausea, fever, joint/back aches and fatigue. Commonly known as stomach flumay is a foodborne illness caused by a pathogen (virus, bacteria or parasite) in contaminated food or drink. The incubation period is the time between exposure to the pathogen and onset of symptoms which usually ranges from several hours to a week.

The microbes growing inside the refrigerator needs to be prevented. They usually grow inside the refrigerator if dryness and cleanliness are not maintained. They deteriorate the foods and develop unpleasant odors, tastes, textures and therefore they have become a major cause of concern. This brings down not only the quality but also the wholesomeness of the food. With these pathogenic bacteria there is a chance of cross-contamination especially if the raw and cooked foods are kept together. Some pathogenic bacteria such as *Listeria monocytogenes* thrive at cold temperatures and can also multiply in the refrigerator over time and could cause illness. There are many power cuts for a long time especially in India. In such cases the door of the refrigerator should be kept closed as much as possible to maintain internal temperature safe. It has been found that uptil 4 hours the food is safe. The perishable foods which include meat, poultry, fish, eggs, and other left over foods should be discarded which are kept above 5 °C for more than 2 hours. Items in the refrigerator that have come into contact with raw meat juices should also be discarded. As a result of food spoilage different food odors can develop. This can happen due to power failure, overloading of refrigerator with food items or poor hygiene. Strong food odors are frequent after power failures. In order to get rid of the unpleasant, undesirable food odors it is important to empty and unplug the refrigerator while cleaning, all removable parts must be taken out and should be washed with mild soap and water and then dried. It is advisable to use vinegar solutions (One cup vinegar per gallon of water) to wash the interior walls of the refrigerator or freezer. This helps in removing the bad odor. In case it persists it is important to change vinegar every eight hours. In case of canned foods there is no time limit on how long unopened cans will be safe. Usually the canned foods with low acidity such as vegetables and meats can be kept and used from 2 to 4 years. Acidic canned foods such as fruits, fruit juices and tomatoes stay in good quality from 1 to 2 years. It is observed that the long storage will change the quality of the food before the safety is affected. Temperatures over 38°C for an extended time will speed loss of quality. Cans that are rusted or badly dented should be thrown away, no matter of their age. Cans and jars with bulging sides or tops should also be thrown away. This usually has an infection with botulinum toxin which can be fatal if consumed. Unopened jars or bottles with visible bubbles or discoloration should also be thrown away. Canned or packaged foods should have a proper label of dates on packages.

8

Genetically Modified Foods and Safety

Even after public concern the scientists feel that the GMO foods are quite safe. According to them the genetically-engineered crops are very safe to eat as much as the non-GE of the same crop. According to them the environment has improved by reducing the use of pesticides. GMO stands for genetically modified organism. Soy, corn, or other crops are very common as GM crops which are made by genetically engineered DNA. Many have come to the conclusion that GM foods are safe, healthy and sustainable. However, some people are against them. There are many benefits of GM crops like less water or pesticides to be used. Genetically engineered crops fight diseases to save the foods. At the same time the industries/ farmers that grow them are also benefitted. These foods have higher nutritional profiles, both in the US and abroad like corn which is one of the most prominent GMO foods. Soy is found in tofu, vegetarian products, soybean oil, soy flour and many other products. The other major crops taken for genetic modification are Papayas, Canola and Cotton. For dairy products a lot of work is being done on the genetic engineering of animals in a big way. This is going to be the most intractable environmental challenge of the 21st Century. Presently, around 92% of the U. S. corn (GE), 94% of soybeans and 94% of cotton are genetically engineered.

There have been advances in molecular methods, surveillance systems and quality assurance programs which will revolutionize the foodborne outbreak investigations. In order to ensure the successful implementation of FSMA, the evaluation of the effectiveness of its individual components by the ongoing monitoring of human illness is vital. This is done by the CDC's foodborne disease surveillance systems that gets their data from local, state and the health departments. There is a list of diseases which must be reported by clinicians or clinical laboratories. There are many important parts of the CDC foodborne disease surveillance system. The most fundamental is PulseNet in which molecular subtyping of the foodborne pathogens can be applied to detect clusters of human illnesses caused by similar strains. PulseNet is a new tool in which the complete genome sequencing (WGS) can be done. This method of analysis has revolutionized the detection of foodborne outbreak. This has also reduced the cost to sequence an entire bacterial genome which had come down to $50–$70 in 2014 as compared to $3500 which was in 2007 (Land et al., 2015). Earlier it was diagnostic subtyping of foodborne pathogens based on pulsed-field gel electrophoresis (PFGE) and multilocus sequence typing. However, these molecular profiling techniques were very laborious, time consuming and very difficult to standardize between different laboratories. As these were based on pure cultures it had been difficult to use culture independent diagnostics (Cronquist et al., 2012). Though WGS technique requires highly trained personnel and expensive equipment. However, it provides a lot of genetic information as compared to the current techniques and also has features for culture independent subtyping through metagenomic approaches. Recently, Canadian and CDC investigators used WGS to identify *Listeria* isolates from contaminated lettuce (CDC; centre for disease control and prevention, 2016). The use of WGS can be done from human, food and environmental sources, when paired with epidemiologic investigation. This gives rapid identification of pathogens in food or other exposure that is the source of contamination. Recently, the GenomeTrakr network (http://www. fda. gov/Food/Food Science Research/Whole Genome Sequencing ProgramWGS/ucm363134. htm) has assembled a database of WGS which results on the food isolates that can be compared with similar data on clinical isolates obtained by PulseNet. With this, the effective control measures can be implemented quickly. It reduces the number of illnesses in relation to the outbreak. Recently, the public health and hospital laboratories have coordinated to conduct WGS which has expanded nationally and internationally. With this method there has been an increase in the number of outbreaks recorded and the causative agent is identified. Now even the new outbreaks caused by known agents that would have otherwise got escaped can be detected. The new and enhanced surveillance system must be communicated to the public health community and communicated to the general public. This will also convey to the public that the food safety system has not deteriorated but has improved in sensitivity and in the ability to detect causative agents. There is a Surveillance System Pulse Net network for the public health and food regulatory agency laboratories (50 states and 82 countries). This performs standardized molecular subtyping (fingerprinting) of foodborne disease caused due to bacteria (*Escherichia coli* O157) and other Shiga toxin–producing *E. coli*, *Campylobacter jejuni*, *Clostridium botulinum*, *Listeria monocytogenes*, *Salmonella*, *Shigella*, *Vibrio cholerae*, and *Vibrio parahaemolyticus*). FoodNet which is an active surveillance system with >650 clinical laboratories provides an input of laboratory-confirmed cases of *Campylobacter, Listeria, Salmonella*, Shiga toxin–producing O157 and non-O157 *E. coli*, *Shigella*, *Vibrio, Yersinia, Cryptosporidium* and *Cyclospora*. It assesses the impact of food safety initiatives on the burden of foodborne illnesses with the surveillance area encompassing 15% of the US population. Another database known as CaliciNet is used by the public health laboratories where the genetic sequences of norovirus strains can be submitted. It studies the epidemiology data from norovirus gastroenteritis outbreaks. The norovirus strains can be compared with other strains in the database. This helps the Centers for Disease Control and Prevention to link outbreaks with a common source. It monitors norovirus strains that are circulating and identifies newly emerging norovirus strains. Another network known as Food CORE Coordinated Outbreak Response and Evaluation Network streamlines the decision making and responds quickly to the outbreaks. It can also standardize post response activities like environmental assessments and to know the root cause of the diseases. With this, the information, can be utilized to make new strategies in order to prevent outbreaks in the future. The National Outbreak Reporting System (NORS) collects reports of the foodborne pathogen outbreaks from state, local and territorial public health agencies. The data improves the understanding of the human health along with the impact of these out breaks. The system that preceded NORS was called the electronic Foodborne Outbreak Reporting System (eFORS) or Outbreak Net. NVEAIS (National Voluntary Environmental Assessment Information System) is used to establish a detailed characterization of food vehicles and their trends identify and monitor contributing factors and their environmental antecedents. It gives a hypothesis generation regarding factors that might contribute to foodborne outbreaks. Selected programs are designed to build capacity for whole-genome sequencing in future projects. In the University of California, a scientist, Davis, has been sequencing pathogen isolates which are commonly transmitted by food, that are stored in culture collections of FDA, USDA and other worldwide partners which are working to make a genetic catalog. This is helping in identifying the biomarker genes which are associated with persistence, serotype diversity, antibiotic resistance, pathogenesis and host association. This will be an innovative method to manage foodborne diseases (http://100kgenome. vetmed. ucdavis. edu/). Molecular sequencing and bioinformatics are being upgraded at national and state levels so that these techniques could be used consistently as an aid to detect disease outbreaks. http://www. cdc. gov/amd/. There is another identification method known as Global Microbial Identifier. This is a global database network which can be used for the identification of all types of microorganisms (bacteria, parasites and fungi) that includes both genomic information and metadata (epidemiologic or environmental details). These methods can be accessed by academicians, industrialists, government officials and clinical laboratories (clinicians, veterinarians, epidemiologists, microbiologists). It can be used for clinical tasks and for national and international public health surveillance along with the outbreak investigation and its response. The following website: http://www.

globalmicrobialidentifier. org gives the details. Scientists have also invented a Genome Trakr which is a new pilot network program for the state and federal public health laboratories. It collects and shares genomic data from pathogen isolates which gets transmitted by the food. Work has been done in 10 FDA field labs and 10 state public health laboratories(http://www. fda. gov/Food/FoodScienceResearch/WholeGenomeSequencingProgramWGS/ucm363134. htm.). The aim of Center for Genomic Epidemiology is to provide the tools for analysis and to extract information from sequence data and internet/Web interfaces for using the tools in the global scientific and medical community (http://www. genomicepidemiology. org.). According to Mandal et al.,(2011) the food industries have quality assurance programs that can test foodborne pathogens and microbes in their products. When the samples are taken for WGS analysis from food production sites and this data is shared it assists in the identification, and, traces back during the outbreaks. It also helps the food processors to understand the ecology of the pathogens involved. VoluntaryNet (VolNet) is a new joint initiative between the CDC, the University of Georgia's Center for Food Safety and the food industry of meat, poultry and egg product processing establishments. Scientists found many *Salmonella* and *L. monocytogenes* isolates from food companies in 2014 (Pedro et al., 2018). To minimize liability issues, foodborne pathogens and information on their source (environmental or food type) are given to the Center for Food Safety to check for PFGE analysis. The molecular patterns are submitted to CDC which is then included in the PulseNet databases. This information will help the food industry in the identification of emerging pathogen strains of concern and understand the public health significance of these pathogens in specific foods. It will enable the companies to perform better by conducting more thorough food safety risk assessments with respect to food products, pathogens of concern and country of origin. It will also enable CDC and state health departments to prepare targeted questionnaires for case-control studies used in the investigation of illness clusters and outbreaks. It is envisioned that WGS profiles will slowly replace the present PFGE fingerprints. With all these advantages GM foods have some disadvantages like allergic reactions, toxin production and less nutritional value. Release of toxins into the soil and toxins on the surface of the soil have also been found. The pests develop resistance to toxins. Biodiversity is disrupted. Sometimes genetically modified ingredients are also found to be carcinogenic. Another problem is the dependency of farmers on the corporations which is not healthy. The bioengineered foods are not good for the Environment (Amanda, 2019). Pesticide-Resistant GMO's can also lead to "Super" Pests. GM crops are made through a process known as genetic engineering. During the process the genes of commercial interest are transferred from one organism to another. There are two primary methods to introduce transgenes into plant genomes. There is a device called a 'gene gun'. However, it has been noticed that some GM crops need fewer pesticides improving the environment. Bt Cotton that's engineered for being pest-resistant can help farmers to use fewer chemical pesticides. Likewise, the growth of Bt corn in the United States since 1996 has allowed farmers to use fewer insecticides in cornfields. The National Research Council has warned on the improper use of GM technology. Farmers who plant herbicide-resistant GM crops often use a limited range of herbicides on their fields, which can give rise to herbicide-resistant "superweeds. " Similarly, there is an evidence that overplanting of Bt corn has fostered a new breed of resistant insects in some fields. There are many factors concerning the environment. In North America the increased use of herbicide spray on herbicide-tolerant crops has shown a decline in the monarch butterfly. The genetically engineered traits that are still in the testing phase could escape into the nature. GM is found in infant food, which is sold by US pharma firm, Abbott Laboratories for toddlers with lactose intolerance which caused allergies. However on the label no warning label of GM ingredients has been mentioned. One of the health concerns of GM food is that it could lead to allergic reactions. In 2008 (updated in 2012), the Indian Council of Medical Research had issued guidelines for determining safety of such food, as it cautioned that there is a possibility of introducing unintended changes, along with intended changes which may in turn have an impact on the nutritional status or health of the consumer. Due to this Australia, Brazil, the European Union and others have regulation on GM in food. People are concerned about the possible toxicity of eating this food. India is still more or less GM-free. The one food that did positive test is cottonseed edible oil. This is because Bt-cotton is the only GM crop that has been allowed for cultivation in India. There has been no permission given for the use of GM cottonseed oil for human consumption. Second, cottonseed oil is also mixed in other edible oils, particularly in vanaspati. The Environment Protection Act in India has prohibition on the import, export, transport, manufacture, process, use or sale of any genetically engineered organisms unless the approval has been taken by the Genetic Engineering Approval Committee (GEAC) under the Ministry of Environment, Forests and Climate Change. In 2006 Food Safety and Standards Act (FSSA) and the Food Safety and Standards Authority of India (FSSAI) have been made incharge of these regulations. According to the Legal Metrology (Packaged Commodities) Rules 2011 mandate that GM must be declared on the food package and the Foreign Trade (Development and Regulation) Act 1992 says that GM food cannot be imported without the permission of the concerned committee or otherwise the importer will be liable to be prosecuted under the Act for violation.

The data improves the understanding of the human health along with the impact of these. The system that preceded NORS was called the electronic Foodborne Outbreak Reporting System (eFORS) or OutbreakNet. NVEAIS (National Voluntary Environmental Assessment Information System) is used to establish a detailed characterization of food vehicles and their trends identify and monitor contributing factors and their environmental antecedents. It gives a hypothesis generation regarding factors that may contribute to foodborne outbreaks. Selected programs are designed to build capacity for whole-genome sequencing in Future Projects. The University of California, Davis is sequencing pathogen isolates commonly transmitted by food that are stored in culture collections of FDA, USDA and other worldwide partners which are working to make a genetic catalog. This helps in biomarker genes which are associated with persistence, serotype diversity, antibiotic resistance, pathogenesis and host association. This will be an innovative method to manage foodborne diseases. http://100kgenome. vetmed. ucdavis. edu/.

Molecular sequencing and bioinformatics are being upgraded at national and state levels so that these techniques could be used consistently as an aid to detect disease outbreaks. http://www. cdc. gov/amd/. There is another identification method known as Global Microbial Identifier. This is a global database network which can be used for the identification of all types of microorganisms (bacteria, parasites, and fungi) that includes both genomic information and metadata (epidemiologic or environmental details). These methods can be accessed by academicians, industrialists, government officials and clinical laboratories (eg, clinicians, veterinarians, epidemiologists, microbiologists). It can be used for clinical tasks and for national and international public health surveillance along with outbreak investigation and its response. The following website: http://www. globalmicrobialidentifier. org gives the details.

Scientists have also invented a GenomeTrakr which is a new pilot network program for the state and federal public health laboratories. It collects and shares genomic data from pathogen isolates which gets transmitted by the food. Work has been done in 10 FDA field labs and 10 state public health laboratories (http://www. fda. gov/Food/Food Science Research / Whole Genome Sequencing Program WGS/ucm363134. htm). The aim of Center for Genomic Epidemiology is to provide the tools for analysis and to extract information from sequence data and internet/Web interfaces for using the tools in the global scientific and medical community(http://www. genomicepidemiology. org).

Food industries have quality assurance programs that can test foodborne pathogens and microbes in their products(Mandal et al., 2011). When the samples are taken for WGS analysis from food production sites and this data is shared it would assist in the identification and trace back during the outbreaks. It will also help the food processors to understand the ecology of the pathogens involved. VoluntaryNet (VolNet) is a new joint initiative between the CDC, the University of Georgia's Center for Food Safety and the food industry of meat, poultry and egg product processing establishments. Scientists found many (429) *Salmonella* and *L. monocytogenes*(8) isolates from food companies in 2014. To minimize liability issues, foodborne pathogens and information on their source (environmental or food type) are given to the Center for Food Safety to check for PFGE analysis. The molecular patterns are submitted to CDC which is then included in the PulseNet databases. This information will help the food industry in the identification of emerging pathogen strains of concern and understand the public health significance of these pathogens in specific foods. It will enable the companies to perform better by conducting more thorough food safety risk assessments with respect to food products, pathogens of concern and country of origin. It will also enable CDC and state health departments to prepare targeted questionnaires for case-control studies used in the investigation of illness clusters and outbreaks. It is envisioned that WGS profiles will slowly replace the present PFGE fingerprints (Tauxe et al., 2010).

With all these advantages GM foods have some disadvantages like allergic reactions, toxin production and less nutritional value. Release of toxins to soil and toxins on soil have also been found. The pests develop resistance to toxins. Biodiversity is disrupted. Sometimes genetically modified ingredients are also found to be carcinogenic. Another problem is the dependency of farmers on the corporations which is not healthy. The bioengineered foods are not good for the Environment. Pesticide-Resistant GMOs can also lead to "Super" Pests. GM crops are made through a process known as genetic engineering. During the process the genes of commercial interest are transferred from one organism to another. There are two primary methods to introduce transgenes into plant genomes. There is a device called a 'gene gun'. However, it has been noticed that some GM crops need fewer pesticides improving the environment. Cotton that's engineered to be pest-resistant can allow farmers to use fewer chemical pesticides. Likewise, the growth of Bt corn in the United States since 1996 has allowed farmers to use fewer insecticides in cornfields. The National Research Council has warned on the improper use of GM technology. Farmers who plant herbicide-resistant GM crops often use a limited range of herbicides on their fields, which can give rise to herbicide-resistant "superweeds. " Similarly, there is an evidence that overplanting of Bt corn has fostered a new breed of resistant insects in some fields. There are many factors concerning the environment. In North America the increased use of herbicide spray on herbicide-tolerant crops has shown a decline in the monarch butterfly. The genetically engineered traits that are still in the testing phase could escape into the nature. GM is found in infant food, which is sold by US pharma firm, Abbott Laboratories for toddlers with lactose intolerance which caused allergies. However on the label no warning label of GM ingredients is mentioned. One of the health concerns of GM food is that it could lead to allergic reactions. In 2008 (updated in 2012), the Indian Council of Medical Research had issued guidelines for determining safety of such foods, as it cautioned that there is a possibility of introducing unintended changes, along with intended changes which may in turn have an impact on the nutritional status or health of the consumer. Due to this Australia, Brazil, the European Union and others have regulation on GM in food. People are concerned about the possible toxicity of eating this food. India is still more or less GM-free. The one food that did positive test is cottonseed edible oil. This is because Bt-cotton is the only GM crop that has been allowed for cultivation in India. There has been no permission given for the use of GM cottonseed oil for human consumption. Second, cotton seed oil is also mixed in other edible oils, particularly in vanaspati. The Environment Protection Act in India has prohibition on the import, export, transport, manufacture, process, use or sale of any genetically engineered organisms unless the approval has been taken by the Genetic Engineering Approval Committee (GEAC) under the Ministry of Environment, Forests and Climate Change. In 2006 Food Safety and Standards Act (FSSA) and the Food Safety and Standards Authority of India (FSSAI) have been made incharge of these regulations. The Legal Metrology (Packaged Commodities) Rules 2011 mandate that GM must be declared on the food package and the Foreign Trade (Development and Regulation) Act 1992 says that GM food cannot be imported without the permission of the concerned committee or otherwise the importer will be liable to be prosecuted under the Act for violation. It was till 2016 that GEAC was in charge however the court has made FSSAI to be responsible. The law is clear that everything is illegal with respect to GM ingredients. In 2018, FSSAI has issued a draft notification on labelling including GM food. There is no way that government can quantify the percentage of GM ingredients in the food. The same FSSAI has issued another notification that it will have to be mandatory to write that it is "certified" and does not contain residues of insecticides.

9

Role of WHO in Food Safety and Legislation

Due to the globalization of the food supply there has been a need to strengthen food safety systems in and between the countries. In this regard the World Health Organization (WHO) has promoted efforts in improving the food safety from farm to plate which started on World Health Day, 7 April 2015. The World Health Day 2015 slogan was "From farm to plate, make food safe". The informal food production and marketing system is still strong in most countries of the South-East Asia Region, for the enforcement of food safety regulations. Traditionally, the preferences have been for hot and well-cooked food. These habits are partially responsible for preventing foodborne infections. In many countries street food has become very popular as it is cheap. Hygienic conditions are improving, provided that potable water supply and clean facilities are ensured by municipal authorities. Many waterborne and diarrhoeal diseases could be prevented with the introduction of bottled drinking water. The awareness and education on food safety by the consumers has strengthened in the enforcement of food standards. This has improved the hygienic practices and has helped in the prevention of foodborne illnesses. The "WHO Five keys to safer food" serves as the basis for educational programmes to train food handlers and educate the consumers. These help in preventing the foodborne illnesses. The important points are to keep food surfaces clean, washing of all utensils, plates and cutlery, keeping the raw food separate from the cooked food. Cooking food thoroughly to the appropriate temperature, keeping food at safe temperatures, both for serving and storage and using safe water and raw materials.

Food safety has to be ensured from the production stage at the farm level. During farming misuse of agro-chemicals which includes pesticides, growth hormones and veterinary drugs may have harmful effects on human health. The microbial and chemical risks could be introduced at the farm-level. Good agricultural practices should be applied in order to reduce microbial and chemical hazards. Organic farming (without the use of pesticides) has been promoted in many countries of the South-East Asia region, as people are becoming more health-conscious. Even with high cost of organic products as compared to commonly available food items the health conscious consumers buy organic foods. For proper food safety, attention is needed during the harvest, transport, processing, storage and finally during food preparation and storage by the consumers. In recent years processed, frozen and ready-to-eat foods have gained popularity as the food habits, product diversification, lifestyle and mass production practices have changed. In most of the cities with both spouses working outside home purchase meat, milk and vegetables on the weekend and store these items in the freezer or refrigerator. Microwave ovens are often used for reheating the food. Therefore, it is important to safely store and reheat food. A proper survey analysis has been done in USA in which a questionnaire with questions including topics of food adulteration and falsification (food fraud), unhealthy and unsafe food-handling practices in food markets and at home and food allergies were included to know the experiences from the people. However, more people friendly survey's are required to be done in the developing countries also.

For the food quality positive and negative attributes are included that influence a product's value to the consumer. Positive attributes include the origin, colour, flavour, texture and processing method of the food. The negative attributes include visible spoilage, contamination with filth, discolouration, bad odour and taste. Sometimes, unsafe food may appear to be of good quality, like tainted meat prepared using bleach or strong spices. This distinction between safety and quality has implications for public policy and influences the nature and content of the food control system most suited to meet predetermined national objectives (FAO/WHO, 2008, 2016). Importantly, it is creating awareness among individuals to ensure that the food for consumption is safe. In this context, World Health Organization (WHO) has been promoting 'five keys' to food safety (WHO, 2019; Five keys to safer food Manual). WHO (2017) have looked into the super bugs in foods.WHO (2017 a) explains the strategies for effective communication. WHO (2017b) explains guide for emergency risks.

WHO is supporting establishment of a network for food safety authorities in partnership with countries to promote the exchange of food safety information and improve collaboration among food safety authorities at national and international levels. WHO is also helping countries to prevent, detect and respond to foodborne disease outbreaks using the Codex Alimentarius, a collection of international food standards, guidelines and codes of practice covering all the main foods and processes. WHO is also advocating that food safety should be properly incorporated into national disaster-management programmes and emergencies. During floods, earthquakes and other natural disasters it is the facility of safe water and quality of food which are the major problems. Many a times the food in the affected area gets contaminated and it becomes the cause of the outbreak of food-borne disease. As part of its Regional Food Safety Strategy, WHO is assisting countries in different regions to initiate, develop and sustain multisectoral approaches. It will also measure for promotion of food safety among all population groups. This has initiated the use of mobile food courts in Bangladesh, the establishment of a Food Standard and Safety Authority in India, and certification of street food vendors with a "Clean Food, Good Taste" logo in Thailand.

One of the World Health Day themes has focused on the food safety. Food safety includes national health, food, agriculture, animal husbandry, fisheries, water, sanitation and environment-related programmes. It is a shared responsibility of all the avenues. In order to make our food safe, working together, is a must that will contribute towards better health of the people. The addition of one or more essential nutrients to a food is known as food fortification. Food fortification has the advantage of being able to deliver nutrients to large segments of the population without many changes in the pattern of food consumption. The rapidly changing lifestyles and usage of highly processed foods has brought the need for the addition of nutrients to the foods in order to have better nutrition in the diet. Fortification of food with micronutrients is a valid technology for reducing micronutrient malnutrition as part of a food-based approach in the diet. Powell, (2011) has given the ways to enhance food safety culture to reduce rates of foodborne illness. This will contribute towards the better health of the people by food fortification. During this decade food technologists have developed many cereal based products, milk and milk products, fats and oils for the fortification of vitamins and micronutrients in the cereals. The World Health report (2000) identified iodine, iron, vitamin A and zinc deficiencies as being among the world's most serious health risk factors. A major accomplishment in public health nutrition over the last 30 years has been in the area of controlling iodine deficiency through salt iodization. The focus of the international community for promotion of food fortification has been on the three most prevalent deficiencies which include vitamin A, vitamin B complex, iodine and iron (West et al., (2012). The workers have studied the global health diseases, their programs, systems and policies. Use of pesticides / insecticides and herbicides / fungicides for pest and weed spoilage control is widely known. These agrochemicals are available in the market and the farmers use them without proper understanding. These chemicals can be hazardous, and there is a chance of contamination of agriculture produce with residues of these chemicals. The dietary intake of residues of these chemicals causes many health hazards. In the Joint FAO/WHO (2019) meetings on pesticide residues, the scientific experts, have adviced on the pesticide residues, residual limits, the acceptable maximum dietary intake (ADI) and maximum residual limit (MRL) based on the information of risk assessment. WHO (2013) gives the detailed Geneva report : WHO (2018) which explains the whole genome sequencing for food borne disease surveillance.

In a Joint FAO/WHO Meeting on Pesticide Residues (JMPR) it was decided on the evaluation of the chemical by the acceptable daily intake (ADI) of a chemical which is the daily intake. This should not have risk on the health of the consumer during an entire lifetime. It is expressed in milligrams of the chemical per kilogram of body weight. Maximum residual limit (MRL) is the maximum concentration of a pesticide residue (expressed as mg/kg), recommended by the Codex Alimentarius Commission to be legally permitted in food commodities and animal feeds. Respective MRL's should be toxicologically acceptable. MRL's which are intended to go for international trade are derived from estimations made by the JMPR following both the toxicological assessment of the pesticide and its residue. To study the variations in national pest control, the ADI and MRLs are determined for veterinary drug residues and antimicrobial substances. These chemicals are used for the promotion of growth, animal disease prevention and their control. *Shigella* bacteria, hepatitis A virus and Norwalk virus have been found in foods prepared with unwashed hands. However, the pathogens like *Campylobacter* and *Escherichia coli* are mainly found in raw and undercooked meat, poultry or raw milk and untreated water born bacteria. The pathogens like *Listeria monocytogenes* and *Yersinia enterocolitica* grow at refrigerated temperatures. Essential virulence determinants of *Listeria monocytogenes* pathogenicity are well known as bacterial toxins which are very dangerous as they produce food poisoning substances. Therefore, elimination of pathogenic microorganisms from foods is a solution. In some food processes it is not practical and sometimes it is not desirable in many processes. Risk assessment and microbial monitoring plays an important role in food safety. Technological advancement have the ability of delivering detection systems that cannot only monitor pathogenic microorganisms, but the total microbial populations in the food matrix.

Food safety regulation is an administrative activity conducted by the government to ensure food safety. Both the law-making activities and law enforcement activities of the relevant competent authorities are responsible for the food safety regulation.

International Organization for Standardization dealing with food safety has developed standard HACCP and GMP practices under ISO 22000(2005) which is a derivative of ISO 9000. According to the international standard it specifies the requirements for a food safety management system that involves interactive communication, system management, prerequisite programs and HACCP principles. According to this it provides a food product that is safe from pathogens and other contaminants. The standards are reviewed every five years so that the standards remain relevant and are also useful to business people (WHO ;World Health Organization,1999).

As per a report of World Health Organization (WHO) 30% of food poisoning outbreaks which were reported in the WHO European Region was detected from the private homes (WHO,2004). According to the WHO and CDC it is in the USA alone in which every year 76 million people suffered from foodborne illness. It leads to 3,25,000 people getting hospitalized and 5,000 deaths (WHO. (2015). It was in 1963, that the WHO and FAO published the Codex Alimentarius which serves the guidelines for food safety (Codex Alimentarius,2012). According to a communication of Health & Consumers Directorate General of the European Commission (SANCO): 'The Codex, while being recommendations for voluntary application by members, in many cases the Codex standards serve as a basis for national legislation'. The Codex food safety standards in the World Trade Organizations Agreement on Sanitary and Phytosanitary measures (SPS Agreement) include far reaching implications for resolving trade disputes in Codex system. Sometimes, WTO members want to have stricter food safety measures than those set by Codex. However, they can do that only after justifying these measures scientifically. An agreement was made in 2003 which was signed by all the member states. This included EU (codex Stan Codex 240 – 2003) for coconut milk, sulphite containing additives like E 223 and E 224. They are allowed till 30 mg/kg as per the guidelines given in Rapid Alert System for Food and Feed (RASFF) entries from Denmark: 2012. 0834; 2011. 1848; en 2011. 168. According to the federal level regulation, the Food and Drug Administration (FDA) needs to publish the Food Code, a model set for the guidelines and procedures that assists food control jurisdiction by providing a scientifically sound technical and legal basis for the regulation of the retail including the food service industries, restaurants, grocery stores and institutional food service providers like nursing homes. In the United States the regulatory agencies use the FDA Food Code at all levels of government to develop or update food safety rules in their jurisdictions that are consistent with national food regulatory policy. According to the FDA, 48 of 56 states and territories, representing 79% of the U. S. population, have adopted food codes patterned after one of the five versions of the Food Code, beginning with the 1993 edition FDA, Food Code (2001). In the United States, federal regulations governing food safety are fragmented and complicated, as per a report from the Government Accountability Office FDA, USF and DA (2017). Altogether 15 agencies are sharing the responsibilities in the food safety system. The primary agencies being the U. S. Department of Agriculture (USDA) Food Safety and Inspection Service (FSIS). These are responsible for the safety of meat, poultry and processed egg products and the Food and Drug Administration (FDA).

FSIS enforces the Federal Meat Inspection Act, the Poultry Products Inspection Act and the Egg Products Inspection Act, portions of the Agricultural Marketing Act, the Humane Slaughter Act, and the regulations that implement these laws. Under the inspection program every animal is inspected before the slaughter and also each carcass after slaughter. This ensures public health requirements. At U. S. borders imported meat and poultry products are also inspected (Hogan and Hartson, (2007). Even with all these practices, there have been concerns over the efficacy of the safety practices which puts pressure on the U. S. regulators. As per a report it was found that the food industries are jeopardizing the U. S. public health by withholding information from food safety investigators or pressurizing the regulators to withdraw or alter the policy designed to protect consumers. It has been found out by the scientists that corporate management removes the field inspectors who stand up for food safety against industry pressure (Reuters, 2010). Recently, it has been observed that many food and beverage manufacturers are improving food safety standards. The manufacturers are incorporating a food safety management system which automates all the steps in the food quality management process (International Organization Standards;(ISO) 2017). It is a subject of great concern all over the world. It has lead the healthcare institutes and the governments of several countries in finding ways to monitor production chains. It is important to include all quality management tools which must emphasize the standardization of products and processes, product traceability and food-safety assurance. The basis of the food-safety system to be adopted in the food industry consists of a combination of good manufacturing practices (GMP), sanitation standard operating procedures (SSOP) and hazard analysis and critical control point (HACCP) system. The need of the day is that the Food Safety Programme (FSP) must address the needs of the customers for food safety covering both organisational and technical issues where improvements can be made. The concept covers continuous improvement along with the participation of all the members of the organisation which needs a consistent and disciplined approach. Usually, a food safety policy includes the core values and guides principles of food safety, their effect with other policies and guidance on best management practices and technology. The company must have commitment to compliance with legislation and regulations. It should provide a framework which could reduce the risk. It must evaluate the performance and involve education and training of internal and external interested parties. Management should ensure that the food safety policy is fully understood by all employees to carry them in their activities. Under the policy if issues are prioritised it will help in taking action in order to achieve full implementation. The company should have resources to improve the food safety performance. Formation of cross-functional teams with team's on specific activities (HACCP team or teams providing assistance or support to suppliers and customers) could be built that would develop the opportunity for everyone to contribute. Awareness of the FSP and the linkage of food safety to the company activities should be developed. Correct presentation should ensure support from everyone in the organisation. There needs to be a collective proactive attitude towards food safety issues. For this, the Food Safety Programme should have good evidence and regulatory compliance contributing to the development of a food safety conscious culture within the company. This will provide consumer satisfaction and enhance competitiveness. Grande et al., (2018) have discussed on the potential risks from bacterial communities in fish farming.

From outside the contaminated food does not look, taste or smell different from foods that are safe to eat. Common bacteria causing foodborne illnesses are *Salmonella, Campylobacter, Listeria* and *Escherichia coli (E. coli)*. These illnesses sometimes can be very severe and many may be fatal. *Bacillus cereus* forms heat resistant spores which releases a heat resistant toxin in cooked rice if left at room temperature. Reheating the food does not destroy this toxin. They cause symptoms like nausea, vomiting, watery diarrhea, abdominal pain and cramps. Fever and/or the presence of blood or mucus in the stool are other signs of bacterial infection. This needs a medical consultation. It is therefore, important to follow the five keys to safer foods given by the WHO.

Public policy sets standards for food safety which reflect policy decisions about the acceptable risks and costs of risk avoidance. Many a times the consumers are not able to detect the hazard at the time of purchase. They are also unable to measure or guarantee a particular level of safety. These pathogens are a threat to human health and drug resistant pathogens appear many a times which affects the food safety. Many changes have been done in the US food industry in order to change the production and processing practices and also the microbiological safety of the products. The surveillance system for foodborne illnesses are improving continuously. Under the Improved surveillance tools (WGS and bioinformatics) for better identification and tracking of earlier contamination of outbreaks there will be various tools available at the disposal of public health and regulatory authorities. Other factors like that of climate change, pathogen evolution, an increase in the elderly and immunocompromised populations and lack of knowledge on factors contributing to unsafe food in homes by the consumers could lead to increased foodborne illnesses. Improvements towards the microbiological safety of foods requires the cooperation of the food industry, clinicians, educators, government and consumers to reduce contamination and minimize exposure to foodborne pathogens.

10

What is the Hazard Analysis of Critical Control Point System (HACCP) and How Does HACCP Work in Safe Food Production? What are GMP Practices

A process control system that identifies the hazardous points in the food production process is known as HACCP (Hazard Analysis Critical Control Point). It puts into place many stringent actions to prevent any hazards from occurring. In any food process if there is strict monitoring and controlling steps, the chances for hazards are minimum. Today, HACCP is an integral part of modern food industry that identifies and controls the major food risks. These are like microbiological, chemical and physical contaminants. These can also be practiced at home scale by checking proper storage, handling, cooking and cleaning procedures. These steps have resulted in industries taking greater responsibility for and controlling the food safety risk factors. This helps in better consumer protection, agriculture, the food processing industries and promotes domestic and international food trade.

The temperature of food storage is important. Below 4°C the food pathogens do not grow. Above 60°C some colonies begin to die. However, almost all die at 71°C. How hot is the food served is important for food safety. The size of the piece is important to how quickly it will cook all the way through. The smaller the piece the faster it will cook. Large or thicker pieces take a longer time to be cooked. Otherwise cooking for a longer time is better so that pathogenic microorganisms do not grow. In case of a thicker piece the color must be checked inside to outside after cooking which should be consistent. If the meat has the bone, check the color/temperature closest to the bone in the thickest part. In case of meat the contamination usually gets introduced while butchering and the practices used. The street food on a cart is perhaps the most contaminated as it has some level of bacteria in it. The meat in the decorated malls also has pathogenic bacteria. The environment plays a very pivotal role which can recontaminate the food even after proper cooking. The amount of harmful pathogens keeps growing in number with the passage of time. Fresh from the grill is the best, an hour or two is not ideal, half a day or overnight could make a person sick. Due to the reason that consumers do not know the time and temperature of cooking it is important to eat food which is freshly prepared.

Cooking time is very important to have safe food. If it is cooked for a long time (above $60^{0}C$) it is usually safe. The food items are like soup and noodles. The difficult foods are ice and clean equipment (knives, blenders, etc.). Many times ice is made with bad water, or otherwise it is stored in an old contaminated cooler. The cut fruits are also unsafe if not washed properly. Fruits which can be peeled and eaten are good. These include Bananas and oranges and many others. With fruits of which the skin is eaten like berries, apples, etc. a lot of caution needs to be taken. These can be disinfected with a concentrated disinfectant that mixes with water. Killing bacteria by the use of chemicals is not advisable. Many green leafy vegetables are also dangerous as they are grown in soil rich with pathogenic bacteria and are eaten raw. Another important parameter is the equipment. Other important things to check are broken or very old coolers as they can contaminate the ingredients, blenders not being broken down and cleaned often, cutting boards used for both cooked and uncooked items. Dried veggie/plantain chips are safer as pathogens need water in the food to grow. Dried foods do not allow the pathogen population to grow. However, if they are soggy it may contain pathogens. According to the Hazard Analysis of Critical Control Point (HACCP) is a control system of a process that identifies in which place of the process line any hazards might occur in the food production process and puts into place stringent actions to prevent the hazards from occurring. By strictly monitoring and controlling each step of the process, there is less chance for hazards to occur. HACCP is an integral part of a modern food industry used to identify and control major food risks which include microbiological, chemical and physical contaminants. These can be implemented even at home by following proper storage, handling, cooking and cleaning procedures. The introduction of this preventive approach of HACCP has brought in more responsibility in controlling the food safety risks. This integrated approach facilitates improved consumer protection, effectively stimulates agriculture and the food processing industry. It also promotes domestic and international food trade. In industrialized countries it is mandatory to have a good food labelling as it helps the consumer and protects them. The globalization creates new challenges in conducting the key activities associated with food recalls which includes the trace-forward and trace-back activities that are required for a food suspected to be unsafe. Recently in a referenece by (Essays, UK. (November 2018) the stages of the development of food safety concept throughout the food chain have been provided. Likewise, shifting to the concept of ISO 22000, much has been given like; the date of issue, the urgent need that derived its issue and what it includes from the requirements of food safety, management systems and the fundamentals that the system is based upon. It details on the HACCP and its effectiveness, the ISO 9000 system and why many institutions applied both for HACCP and ISO 9000, and why the HACCP system did not hold on by its own keeping the quality of the food safety management system. Demonstrating the urgent need that lead institutions concerned with applying food safety systems to the necessity of running a system such as the ISO 22000 system instead of the HACCP system alone has also been covered. The important role of applying this system in the development and advancement of the institutions, the benefits of applying the ISO 22000 on the institutions running on this system as well as the fields of its applications starting from the farm to the table are important to be discussed. It also includes all the processes such as transportation, circulation and supplying the institutions with food that is to be processed. When food safety is discussed it deals with the prevention, elimination and control of food borne hazards at the point of consumption. Globally, the consumers need and deserve assurance that the food sold to them is safe to consume. The fact is that the food safety hazards can enter at any stage of the food supply chain. Therefore, every company in the supply chain must exercise adequate hazard controls. In fact, food safety can be only ensured through the combined efforts of all parties in the food chain. The food supply chain ranges from farmers through food processor, storage and transportation operators, subcontracts to the retail outlets (groceries and restaurants) and industries in between. Their products could also be makers of processing equipment, packaging materials, cleaning agents, additives/ ingredients and even service provider (equipments testers). Food safety is therefore, a complex system engineering, which involves raw materials, activities of production and product test. If food safety is done properly it gives an oppotunity for improvement in preventing illness from known food pathogens and in responding to new and emerging food borne illnesses and threats (The ASQ the Quarterly Quality Report June 2007). A similar description of food safety as protection of food against chemical, biological and physical factors which can endanger human health has been used by Codex Alimentarius commission,(2003). Food safety as a concept means that foodstuffs should not be harmful to the consumer and recognizes that food safety hazards can be introduced at any stage of the food chain (GFSI Technical Committee – September 2007). In this their was a major breakthrough called as "GFSI Breakthrough": 7 major retailers (Tesco, Wal-Mart, Metro, Carrefour, Migros, Royal Ahold and Delhaize) announced acceptance of the GFSI benchmarked food safety certification programmes. The World Health Organization (WHO,

2004) and Food and Agriculture Organization (FAO) define food safety as food that is free from all hazards, whether chronic or acute, that may make food injurious to the health of the consumer. Food safety relates to the conditions and practices that preserve the quality of food and prevent contamination and food borne illnesses. IT is related to how safe the food we eat is. It covers from transportation, manufacturing and processing, consumer safety to the production of equipment for food safety, storage, delivery exportation and importation. (WHO, 2003) defines food safety as all conditions and measures that are necessary during the production, processing, storage, distribution and preparation of food to ensure that it is safe, sound, wholesome and fit for human consumption.

The process of laws of food safety started in November 2001 with voting on the final draft in August 2005. All 34 national standard bodies that voted were positive and there were no rejections. The standard was published in September 2005 and subsequently translated for publication by national standard bodies, which are replacing national standards by ISO 22000. It has also been published as an European (CEN) standard: EN-ISO 22000 and is currently the standard in over 40 countries. ISO 22000:2005 provides a framework of internationally harmonized requirements for the global approach that is needed. The standard has been developed within ISO by experts from the food industry, along with representatives of specialized international organizations and in close cooperation with the Codex Alimentarius Commission. The body has been jointly established by the United Nations, Food and Agriculture Organization (FAO) and World Health Organization (WHO) to develop food standards (ISO 22000) for safe food supply chains, ISO 22000:2005 are for Food safety management systems. This will emphasize on the requirements for any organization in the food chain to ensure that there are no weak links in the food supply chain. Since its publication in September 2005, the standard has been well received by the food industries and is becoming a global standard. ISO 22000 has been designed with flexibility to enable a tailor-made approach to food safety for all segments of the food chain. It is possible that the standards and procedures required for high risk areas in one food sector may not be appropriate in another. Therefore, this standard has not prepared a checklist methodology. In 2005, ISO 22000, was published to be the first international food safety management standard applicable to the whole food supply chain. ISO 22000 requires an organization to demonstrate its ability to manage food safety hazards and provide consistently safe products that meet both customer requirements and food safety regulations. It was hailed as the ultimate opportunity to harmonies global food safety approaches. ISO 22000 standard is considered to be the first international quality standard designed to work with all cultural prescription, statutes and regulation. ISO 22000 is dedicated to improve consumer confidence in the food product and the process. It applies to every link in the food supply chain from the farm to the table (Essays, UK.,November 2018). ISO 22000 is an international, auditable standard that specifies the requirement of food safety management by incorporating all the elements of Good Manufacturing Practices (GMP) and Hazard Analysis Critical Control Points (HACCP) together with a comprehensive management system (Almeida et al., 2014). Food safety management system (FSMS) combines between Good Management Practices, Hazard Analysis and Critical Control Point (HACCP) principles and effective supplier verification. It also includes validation, ensuring that all possible actions are taken, recorded and verified to ensure safe food, which is based on the HACCP principles. This requires a company policy definition and quality manual, with definition of responsibilities for management and employees, prerequisite programs and HACCP plan implementation, and preparing pre-request programs and measures for implementing the food safety program. Preparing the HACCP team and effective recording systems, and a combination of self assessment with application of internal auditing, management review, application of all legal requirements and supplier evaluation, are other concerns in this system. (Essays, UK. November 2018). The food safety management systems principally controls the specific food safety hazards associated with the product and ensure compliance with food safety legislation (Manning and Baines,2004). A systemic approach has been adopted by the ISO 22000 standard, which is based on the application of process management principles. Effective management of these processes ensures effective management of the whole organization (Armistead et al., 1999). In ISO 22000, as mentioned in ISO/TS 22004 (giving guidelines for applying the standard), processes are considered in terms of food safety (ISO 22000:2005). Plan-Do-Check- Act (PDCA) deals with the design of processes, in a way that fully specifies which activities are to be done (when, by whom and how) ensuring that it is repeatable and consistent. Do cover the implementation of these activities, in accordance with the plan. Measurements of end-to-end process performance and assessment of these measurements in order to facilitate target setting are actually part of the Check. Finally, it refers to process improvement and ensures that the critical activities are executed in the most efficient and effective manner. Processes standardization is very important (Davenport, 2005). For the application of PDCA cycle, ISO 22000 has adapted a requirement presentation scheme directly analogous to the ISO 9001:2000 quality systems standard. Specifically, after three initial clauses (scope, references and definitions) the ISO 22000 requirements are grouped into five clauses as given below:

(1)   Food safety management system

(2)   Management responsibility

(3)   Resource management

(4)   Planning and realization of safe products

(5)   Validation, verification and improvement.

Under the first clause, the organization needs to establish and document a food safety management system and must define its scope which includes products, processes and sites. In the management responsibility clause the specific requirements are covered which include safety policy definition, safety planning (through objectives and targets), communication issues and management review. Provision of all resources necessary for the implementation of the system is the scope of the resource management clause. In the planning and realization of safe products clause, all production processes affecting the safety of the products need be designed and the respective safety plans to be developed accordingly. It includes technical requirements of classical HACCP (this is the only clause drastically different from its ISO 9001 counterpart). Finally, the last clause specifies requirements which ensures system verification (the systems ability to reliably deliver expected safety outcomes) and continuous improvement. This International Standard specifies the requirements for a food safety management system that combines the following generally recognized key elements to ensure food safety along the food chain, up to the point of final consumption: interactive communication, system management, prerequisite programmes and HACCP principles. Communication along the food chain is very vital and it can ensure that all relevant food safety hazards are identified and adequately controlled at each step within the food chain. Communication with customers and suppliers about identified hazards and control measures helps in clarifying customer and supplier requirements (with regard to the feasibility and need for these requirements and their impact on the end product). Recognition of the organization's role and position within the food chain is essential to ensure effective interactive communication throughout the chain in order to deliver safe food products to the final consumer. The most effective food safety systems are established, operated and updated within the framework of a structured management system and incorporated into the overall management activities of the organization. This provides maximum benefit for the organization and also for the interested parties. This International Standard has been aligned with ISO 9001 in order to enhance the compatibility of the two standards. This International Standard can be applied independently of other management system standards. Its implementation can be aligned or integrated with existing related management system requirements, while organizations may utilize existing management system(s) to establish a food safety management system that complies with the requirements of this International Standard. This International Standard integrates the principles of the Hazard Analysis and Critical Control

Point (HACCP) system and application steps developed by the Codex Alimentarius Commission. This has both the HACCP plan along with prerequisite programs (PRPs). Hazard analysis is the key to an effective food safety management system, since conducting a hazard analysis assists in organizing the knowledge required to establish an effective combination of control measures. This International standard requires that all hazards that may be reasonably expected to occur in the food chain, including hazards that may be associated with the type of process and facilities used, are identified and assessed. In the hazard analysis, the organization determines the various points to ensure hazard control by combining the PRP(s), operational PRP(s) and the HACCP plan. It provides any organization (small and/or less developed organization) to implement an externally developed combination of control measures for the organizations to apply towards a more focused, coherent and integrated food safety management system. The ISO 9000 quality management system standards have become a major element of supplier management strategy for many multinational corporations. Manufacturers implement the ISO 9000 standards with the intention of reaping the benefits, while customers perceive ISO 9000-registered plants as being more capable of delivering products of consistent quality (Nguyen et al., 2004). Many manufacturers and customers have indicated that companies using quality systems like ISO 9000 standards have several advantages over competitors that have not implemented such systems. In various studies done (Nguyen et al., 2004) found out that it improved product quality and reliability, increased customer satisfaction, reduced scrap and rework, increased manufacturing efficiency, superior delivery times, rapid systematic response to change and increased interdepartmental communication leading to increased teamwork. However, in ISO 9000, there are standards which contain clauses that are directed at the quality management process of an organization. It defines a quality framework in which a registered company must operate as a minimum criterion for a quality management system. These are reviewed regularly, with the most recent revision having been published in December 2000. Attaining ISO 9000 registration does not equate with achieving a world class quality system since the ISO standards describe only the minimum criteria for a quality management system. SABS (ISO 9000: 2000) have studied the quality management systems and also the fundamentals about the system. These are generic and can be applied to any industry. Their purpose is to establish the existence of a documented quality system. ISO 9000 standards do not describe how a company should manage its quality system, rather it focuses on whether a company is complying with its own written policies and procedures or not. Both HACCP and ISO 9000 systems are management philosophies that rely on disciplined operator control and teamwork. Both focus on prevention rather than retrospective inspection. HACCP is process and product oriented and is totally focused on food safety therefore quality factors should not be a part of HACCP program (Newslow, 2002). In contrast, ISO 9000 is more systems-oriented and designed to manage quality (National Food Processors Association, 1992). It provides confidence for a quality system in place (Bennet and Steed, 1999). HACCP puts control mechanisms in place to ensure that the product is safe and manufactured to standards that are formulated internationally, whereas ISO 9000 requires that an organization define its own system and demonstrate that it can comply with it. HACCP and ISO focus on prevention. HACCP assures food safety by controlling the process. ISO 9000 ensures system conformance to the standards. These two programs have complimentary systems that reinforce and strengthen an organization in the overall quality system (Newslow, 2002). According to National Food Processors Association, (NFPA;1992) for maximum effectiveness of these plans they should be tailored to the manufacturing facilities. This may need management leadership and commitment, expert knowledge in program development, employee training and operator control. At the international level the food quality and food safety are immersing critical issues since outbreaks of food borne illnesses and can damage trade and tourism, and lead to loss of earnings, unemployment and litigation. Food spoilage is wasteful, costly and can adversely affect trade and consumer confidence. For this to happen, HACCP (Hazard analysis critical control points) was established in the USA 3 decades ago as the preventive mechanism for safety control of foods and has been worldwide adopted into the production and service food industries (Prasert, 2007). There has been a lot of emphasis on preventative food safety since 1930s, but it is only since the 1970's that this approach has been adopted leading to the use of HACCP (Bauman,1974). Many food companies have been developing their own HACCP plans for about a decade, following the seven HACCP principles and applying them to their circumstances in order to produce safe foods. However, HACCP plans have to be so specific to the type of business and the physical layout of each site that it is not possible to have one set of HACCP standards for all companies to follow in all situations. And while HACCP plan requirements have been codified in many localities, HACCP regulations cannot be-made uniform. Management system needs to be applied beside HACCP which makes the company combined between policy definition and quality manual, with definition of responsibilities for management and employees, prerequisite programs and HACCP plan implementation, along with measures for implementing the food safety program. Preparing the HACCP team and effective recording systems, with a combination of self assessment and application of internal auditing, management review, application of all legal requirements and supplier evaluation, are other concerns in this system (Mehrdad, 2007). Bryan et al., (1992) have studied the HACCP of street foods in Pakistan. For the implementation of any standardized management system, a company needs to identify and redesign its processes so as to incorporate the specifications of the respective standard. Processe interactions also need to be determined. In most cases, additional processes related with various internal operations (such as targets setting, internal audits etc.), often not previously identified and standardized, will need to be designed. As a final step, written standard operating procedures (SOPs) need to be developed, effectively describing all activities for implementing the processes (as designed) together with the respective managerial responsibilities. For the ISO 22000 standard, a safety plan according to given specifications is also required, as described in the next section. When developed, the safety plan needs to be integrated within respective SOPs for actual use. It can be stated that ISO 22000 implementation provides a food safety system designed, operated and continuously updated (improved) as an integral part of overall organization n management. Note that classical HACCP, practically designed to operate as an effective stand-alone system, may lead to inefficient implementations, with food safety not integrated within but operating in parallel with other management systems such as ISO 9000. Due to this, in order to ensure appropriate action all measures should be included that may directly (or indirectly) impact food safety in the HACCP plan. For quantitative objectives ISO 22000 requirement should be kept necessary as it provides a solid basis for improvement and the determination of acceptable hazards levels. According to Gorris , (2005) in the Food Safety Objectives (FSO) the concept on the public health goals have been given which can be systematically translated into quantitative operational targets for food safety management. FSIS gives details on food safety.

In the classical HACCP two safety control levels (Prerequisite Programs (PRPs) and the HACCP plan) have been given. The differentiation between these two levels has not been understood. It is the ISO 22000 standard that has imposed an additional control level, and created a three-level safety control. These are PRPs, Operational Prerequisite Programs (O-PRPs) and HACCP plan. Each of these control levels, provide specific interpretations when necessary. The PRPs define all basic conditions and activities that are necessary to maintain a hygienic environment throughout the food chain, by enforcing the implementation of the appropriate Good Manufacturing Practices (GMP) and Good Hygienic Practices (GHP) specifications throughout the organization. PRPs are the control measures covering the design and the basic operations of all infrastructures deployed and which impose specifications for the development of the system SOPs. Typical PRPs examples include cleaning and sanitation of production equipment, maintenance, personnel selection and training. PRPs may control serious hazards and fully complement safety control at operational level. Operational safety control, which directly relates to the product and production process used, is accomplished by the O-PRPs and the HACCP plan control levels.

Both the O-PRPs and the HACCP plan are specified as the outcome of the hazard analysis that defines the measures to control the hazards essential to food safety other that those covered by the PRPs. In case of a production flow design where some hazard cannot possibly be timely controlled should be controlled by an O-PRP. In case the hazard impact on public health is very high, it needs to be a part of the HACCP plan. In that case a redesigning of the production process is needed. ISO 22000 has many advantages. There is a benefit in implementing a systematic and effective food safety management system (FSMS) to have one common system throughout the supply chain, better communication, integration of quality management and food safety management, controlling /reducing food safety hazards, legal compliane, provides recognition throughout the food supply chain as a single standard approach to food safety and can be applied independently. It integrates the principles of HACCP and application steps of CODEX, allowing small and /or less developed organizations to implement an externally developed system. It is also possible to get certification with one audit that can cover both the food safety management system and quality management system (IFS, 2014). Lancaster (1971) has looked into the consumer demand in modern times.

As per the Food Standards Agency Annual Report (2007/08) the FSMS gives the ability towards increase in business effectiveness and profitability. It gives the ability to streamline processes, reduces the inefficiencies due to paper-based forms and provides greater visibility into the system. Using quantitative risk assessment tools, it is possible to identify hazards more effectively, make the process more efficient and mitigate any unforeseen risks in the food chain. It provides the recent steps in the evolution of food safety systems beyond HACCP involving all the five preliminary steps like documentation, management responsibilities, resources management, planning and realization of safe product and validation, verification and improvement of FSMS and seven principles of HACCP perform hazards assessment, identify critical control points, establish critical limits, establish monitoring procedures, detail corrective actions and effectiveness.

11

Role of Bacteriocins in Food Safety, Biopreservation and Food Allergies

Jay et al., (1995) and Joerger et al., (2003) have written the role of bacteriocins as an alternate to antibiotics. Bacteria produce bacteriocins which are proteinaceous in nature or peptidic toxins. These inhibit the growth of similar or closely related bacterial strain(s). There are different types of bacteriocins which are classified based on the producing strain, common resistance mechanisms and mechanism of killing. Ahmad et al., (2013) have isolated antifungal compound for food preservation. Martin et al., (2018) and Batish et al., (1997) found antibacterial compounds from fungus. Mahboubi et al., (2019) studied the potential of Zataria in IBS. There are bacteriocins from gram-positive bacteria known as the colicins, the microcins and the bacteriocins from Archaea (Cascales et al., 2007). The bacteriocins from *E. coli* are called colicins (formerly called 'colicines,' meaning 'coli killers'). These have been studied widely. Earlier called colicin V and is now known as microcin V. It is very small and is produced and secreted in a different manner than the classic colicins. Adrienne et al., (2016) have studied the colicin resistance.

The name of the bacteriocin is usually based on the organism that produces the bacteriocin. However, this has made the naming system very problematic and has given rise to alternative classification systems. Bacteriocins that contain the modified amino acid lanthionine in their structure are called lantibiotics. In order to reorganize the nomenclature of the family of ribosomally synthesized and post-translationally modified peptide (RiPP), the natural products, have led to the differentiation of lantipeptides from bacteriocins based on biosynthetic genes(Arnisonet al., 2013). Therefore, alternative methods of classification have come into the system which includes: method of killing (pore-forming, nuclease activity, peptidoglycan production inhibition, genetics (large plasmids, small plasmids, chromosomal), molecular weight and chemistry (large protein, peptide, with/without sugar moiety, containing a typical amino acid such as lanthionine), and method of production (ribosomal, post-ribosomal modifications, non-ribosomal). Bacteriocins from Gram negative bacteria are classified on the basis of size. Microcins are less than 20 kDa in size; colicin-like bacteriocins are 20 to 90 kDa in size and tailocins or so called high molecular weight bacteriocins which are multi subunit bacteriocins that resemble the tail of bacteriophages. This size classification also coincides with genetic, structural and functional similarities. According to Pag et al., (2002) these multiple lantibiotics are novel antibiotics. Incidence of *E.coli* on beef caracasses was studied by Siragusa et al., 1998).

Colicins are bacteriocins (CLBs) which are found in the Gram-negative *E. coli* and in other Gram-negative bacteria. These CLBs are distinct from Gram-positive bacteriocins. They are modular proteins between 20 and 90 kDa in size. They often consist of a receptor binding domain, a translocation domain and a cytotoxic domain. Combinations of these domains between different CLBs occur frequently in nature and can be created in the laboratory. Further subclassification are based on either import mechanism (group A and B) or on cytotoxic mechanism (nucleases, pore forming, M-type, L-type) based on these combinations (Maartenet al., 2014). Most well studied are the tailocins of *Pseudomonas aeruginosa* which can be subdivided into R-type and F-type pyocins (Cotter et al., 2006). Bacteriocins from Gram positive bacteria are typically classified into Class I, Class IIa/b/c, and Class III (Heng et al., 2007).

The class I bacteriocins are small peptide inhibitors and include nisin and other lantibiotics,the class II bacteriocins are small (<10 kDa) heat-stable proteins. It is subdivided into five subclasses. The class IIa bacteriocins (pediocin-like bacteriocins) are the largest subgroup with an N-terminal consensus sequence -Tyr-Gly-Asn-Gly-Val-Xaa-Cys across this group. The C-terminal is responsible for species-specific activity, causing cell-leakage by permeabilizing the target cell wall. Class IIa bacteriocins have a large potential for use in food preservation as well as in medical applications due to their strong anti-Listeria activity and broad range of activity. Eijsink et al., (2002) have worked in lactic acid bacteria. One example of Class IIa bacteriocin is pediocin PA-1(Nissen-Meyer et al., 2013). The class IIb bacteriocins (two-peptide bacteriocins) require two different peptides for their activity. As in lactococcin G, this permeabilizes cell membranes for monovalent sodium and potassium cations, but not for divalent cations. Almost all of these bacteriocins have GxxxG motifs. This motif is also found in transmembrane proteins, which is involved in helix-helix interactions. The bacteriocin GxxxG motifs can interact with the motifs in the membranes of the bacterial cells by killing the cells (Meyer et al., 2010). Class IIc consists of cyclic peptides, where the N-terminal and C-terminal regions are covalentely linked. The prototype of this group is Enterocin AS-48. Class IId cover single-peptide bacteriocins, which are not post-translationally modified and is not like the pediocin, like the highly stable aureocin A53. This bacteriocin is stable under highly acidic conditions, high temperatures, and is not affected by proteases (Netz et al., 2002). A recent subclass is Class IIe. This includes the bacteriocins which consists of three or four non-pediocin like peptides as aureocin A70 which is a four-peptide bacteriocin, highly active against *Listeria monocytogenes*, with potential biotechnological applications(Netz et al., 2001, 2002). Class III bacteriocins are large, heat-labile (>10 kDa) protein bacteriocins. This class is subdivided into two subclasses: subclass IIIa or bacteriolysins and subclass IIIb. Subclass IIIa consists of peptides that kill the bacterial cells by cell wall degradation causing cell lysis. Scientists have studied bacteriolysin lysostaphin, a 27 kDa peptide that hydrolyses the cell walls of several *Staphylococcus* species, principally *S. aureus* (Bastos et al., 2010). However, Subclass IIIb consists of peptides that do not cause cell lysis, killing the target cells by disrupting the membrane potential, which causes ATP efflux. Class IV bacteriocins are defined as complex bacteriocins which contain lipid or carbohydrate moieties. Confirmation by experimental data was established with the characterisation of sublancin and glycocin F (GccF) by two independent groups (Bastos et al., 2011). Two databases of bacteriocins are available: Anne et al.,(2006) and de Jong et al., (2006) BAGEL [BAGEL, 10. 1093/nar/gkl237] and BACTIBASE (Hammami et al., 2007). As of 2016, nisin was the only bacteriocin generally recognized as safe by the FDA and was used as a food preservative in several countries (Fahim et al., 2016). Generally bacteriocins are not useful as food preservatives as they are expensive. These get broken down in food products and are harmful to some proteins in the food. They target too narrow a range of microbes (Fahim et al., 2016). Bacteriocins prepared from the non-pathogenic *Lactobacilli* helps to maintain the stability of the vaginal microbiome (Nardis et al., 2013). These are a good replacement to the antibiotics for which the pathogenic bacteria have become resistant. Deshmukh et al., (2013) isolated a bacteriocin from *Lactobacillus rennanquilfy*. The bacteriocins could be produced by bacteria mainly to be introduced into the patient in order to prevent infection (Cotter et al., 2013). There are several strategies by which new bacteriocins can be discovered. In the past, bacteriocins had to be identified by intensive culture-based screening for antimicrobial activity against suitable targets and then purified using fastidious methods before testing. However, since the advent of the genomic era, the availability of the bacterial genome sequences has revolutionized the approach in identifying the bacteriocins. Dicagno et al., (2010) studied the quorum sensing in *Lactobacillus plantarum*. Newer methods like silico-based can be applied for the rapid screening of thousands of bacterial genomes in order to identify the novel antimicrobial peptides (Rezaei Javan Reza et al., 2018). Some bacteriocins have been studied in *in vitro* in order stop viruses from replicating as staphylococcin 188 against Newcastle disease virus, influenza virus and coliphage HSA virus; each of enterocin AAR-71 class IIa, enterocin AAR-74 class IIa, and erwiniocin NA4 against coliphage HSA virus; each of enterocin ST5Ha, enterocin NKR-5-3C, and subtilosine against HSV-1; each of enterocin ST4V and enterocin CRL35 class IIa against HSV-1 and HSV-2; labyrinthopeptin A1 against HIV-1 and HSV-1; and bacteriocin from *Lactobacillus delbrueckii* against influenza virus (Al Kassaa et al., 2014). Some of the bacteriocins namely, cytolisin, pyocyin S2, colicins A and E1, and the microcin MccE492, have been tested on cancer cell lines and in a mouse model of cancer and found to be preventive (Lagos et al.,2009). Some of the common bacteriocins are: acidocin, actagardine, agrocin, alveicin, aureocin, aureocin A53, aureocin A70, bisin, carnocin, carnocyclin, caseicin, cerein (Naclerio, et al., 1993) circularin A (Kawai et al., 2004), colicin, curvaticin, divercin, duramycin, enterocin, enterolysin, epidermin/gallidermin, erwiniocin, gardimycin, gassericin A (Pandey, 2013), glycinecin, halocin, haloduracin, klebicin, lactocin S (Mørtvedt, etal., 1991), l actococcin, lacticin, leucoccin, lysostaphin, macedocin, mersacidin, mesentericin, microbisporicin, microcin, mutacin, nisin, paenibacillin, planosporicin, pediocin, pentocin, plantaricin, pneumocyclicin (Bogaardt et al., 2015), pyocin (Michel-Briand et al., 2002), reutericin 6 (Kabuki et al., 1997), sakacin, salivaricin (Wescombe et al., 2006), sublancin, subtilin, sulfolobicin, tasmancin (Müller et al., 2012), thuricin 17, trifolitoxin, variacin, vibriocin, warnericin and warnerin. Meyer et al., (2010) worked on the class II bacteriocins. The bacteriocins are mostly antimicrobial peptides which are synthesized by the ribosomes produced by bacteria. They can kill /inhibit bacterial strains closely-related or non-related to the bacteria producing it. It does not harm the bacteria themselves due to the presence of specific immunity proteins. Bacteriocins become one of the weapons against microorganisms due to the specific characteristics of large diversity of structure and function, natural resource and being stable to heat. A lot of research has been done towards the purification and identification of bacteriocins for the application in food technology towards extending food preservation time, in preventing pathogen disease/ cancer therapy and in maintaining human health. Bacteriocins have the potential to replace antibiotics in future.Castellano et al., (2008) have written a review on lactic acid bacteria as being bacteriocinogenic. Celia et al., (2019) have studied the application of bacteriocins in dairy foods. Many antibacterial substances are produced by animals, plants, insects and bacteria which include hydrogen peroxide, fatty acids, organic acids, ethanol, antibiotics and bacteriocins. Antimicrobial peptides (AMPs) or proteins produced by bacteria are categorized as bacteriocins. These have a large diversity, the genes encode antimicrobial peptides or proteins which are ribosomally synthesized, which kill other related (narrow spectrum) or non-related (broad spectrum) microbiotas which are the inherent defense methods of the bacteria (Cotter et al., 2005). Almost 99% of bacteria have the ability to produce at least one bacteriocin, most of which are not identified (Riley and Wertz, 2002a,b). Antibiotics are secondary metabolites whereas the bacteriocins are sensitive to proteases and are generally harmless to the human body and surrounding environment. In order to extend the shelf-life of any food either the antibiotics or food preservatives are added (nitrite and sulfur dioxide) to delay the microbial growth. Most of the commercial preservatives are developed by chemical synthesis. Long-term consumption of the synthetic preservatives can have an adverse impact on the human body. Legally antibiotics cannot be used in the food products. The bacteriocins produced by Gram-positive or Gram-negative bacteria are gene encoded peptides or proteins, which are suitable as natural preservatives in food products. As the bacteriocins are very sensitive to some proteases the harmless bacteriocins get digested. Unfunctional small peptides and amino acids are bacteriocin loaded foods which are digested in the gastrointestinal tract (Cleveland et al., 2001). Bacteriocins are basically safe food additives after ingestion by the gastrointestinal system. Bacteriocins are natural food additives present in many foods like cheeses, yogurts and fermented meat (Yang et al., 2012; Todorov et al., 2013). Nisin is produced by *Lactococcus lactis* and was the first antibacterial peptide found in LAB (Rogers, 1928). It is marketed as Nisaplin®. It is the only bacteriocin which is approved for the utilization as a preservative in many foods by the U. S. Food and Drug Administration (USFDA). It is licensed as a food additive in over 45 countries (Settanni and Corsetti, 2008). Pediocin PA-1, is also commercially available as Alta® 2341. It inhibits the growth of *Listeria monocytogenes* in meat products (Settanni and Corsetti, 2008). In traditional European cheeses, the contamination in milk can be prevented by using the bacteriocinogenic Enterococci as starter cultures or co-cultures to reduce microbiota contamination (Foulquié Moreno et al., 2006). Settanni and Corsetti (2008) have written a good review article on the bacteriocinogenic LAB strains as a co-culture/ protective or starter cultures in fermented and non-fermented vegetables. These include olives, sourdough, miso, sauerkrauts, refrigerated pickles and mungbean sprouts. They have shown the incorporation of bacteriocins in foods such as nisin, which is used in kimchi, mashed potatoes and fresh-cut products. Enterocin AS-48 has been tried in cider, fruit and vegetable juices and canned vegetables to inhibit contamination. Enterocin CCM4231 and EJ97 are used in soy milk and zucchini purée for suppression of contamination, respectively (Settanni and Corsetti, 2008).

It was Alexander Fleming who had discovered the use of antibiotics in the year 1928, in order to treat pathogens. Antibiotics were first approved by USFDA in 1951, and used in animal feed. This significantly reduced the number of deaths due to bacterial infection cases. In recent days, the problem with multiple drug resistance pathogens has become very serious. This has brought the awareness about the harmful effects of antibiotics (Christine et al., 2003). Bacteriocins inhibit animal and plant pathogens, like Shiga toxin-producing *E. coli* (STEC), enterotoxigenic *E. coli* (ETEC), methicillin-resistant *Staphylococcus aureus* (MRSA), VRE, *Agrobacterium*, and *Brenneria* spp. (Grinter et al., 2012; Cotter et al., 2013). The bactericidal mechanism of bacteriocins is mainly located in the receptor-binding of bacterial surfaces and then through the membrane, it causes bacterial cytotoxicity. In addition, bacteriocins are low-toxic peptides or proteins sensitive to proteases, such as trypsin and pepsin (Cleveland et al., 2001). Jordi et al. (2001) found that 20 strains of *E. coli* expressed the formation of colicin, which inhibited five kinds of Shiga toxin-producing *E. coli* (O26, O111, O128, O145, and O157: H7). These *E. coli* can cause diarrhea and hemolytic uremic syndrome in humans. In a simulated cattle rumen environment, colicin E1, E4, E8-J, K, and S4 producing *E. coli* can significantly inhibit the growth of STEC. Stahl et al. (2004) utilized purified colicin E1 and colicin N for effective activity against enterotoxigenic *E. coli* pathogens F4 (K88) and F18 *in vitro*, which caused post-weaning diarrhea in piglets. It was found that the incidence of post-weaning diarrhea reduced by F-18 positive *E. Coli* (Cutler et al., 2007). The growth of the piglets was also better. Józefiak et al. (2013) used the nisin supplemented bird diet to feed broiler chickens, and found a reduced number of Bacteroides and Enterobacteriacae in ileal digesta of nisin-supplemented chickens. The action of nisin was similar to that of salinomycin. The average body weight gain of nisin supplemented chickens was higher than the non nisin supplemented chickens after 35 days of growth period. Stern et al. (2006) reported that class II low molecule mass bacteriocin OR-7 was purified from the *Lactobacillus salivarius* strain NRRL B-30514. This bacteriocin exhibited an ability against human gastroenteritis pathogen *Campylobacter jejuni*. The purified bacteriocin OR-7 was encapsulated in polyvinylpyrrolidone for chicken feed and the results were found to be good which suggests that nisin, OR-7, and other bacteriocins, have the potential to replace the antibiotics in poultry and other animal feeds. Benjakul et al., (2013) have shown the antibacterial activities of bacteriocins. Camargo et al., (2017) have studied the presence of *Listeria monocytogenes* in food processing and human listeriosis.

In the recent times cancer has become a serious problem among people. According to the WHO, around 8. 2 million people died from cancer and 14. 1 million new cancer cases were diagnosed in 2012. Most of the cases (60% of world's total new annual cases) occur in Africa, Asia and Central America and South America. Maximum cases include lung cancer (2. 7% including trachea and bronchus cancer). In 2013, the cancer cases increased in the United States which increased to 1,660,290 new cancer cases, with 580,350 cancer deaths. According to the scientist's, the deaths caused due to cancer are higher than the deaths due to the heart diseases (Center, 2013; Siegel et al., 2014). Some studies have shown that bacteriocins can prevent tumors by killing the tumor cells. The bacteriocins are therefore suitable to be added into food as anti-tumorous. Some bacteriocins, like pore-forming colicin A and E1 can reduce the growth of human standard fibroblast line MRC5 and 11 human tumor-cell lines (Chumchalová and Smarda, 2003). Bacteriocins like colicin U and RNAase activity colicin E3 which are pore-forming did not have the growth inhibition ability. Colicin D, E2, E3, and pore-forming colicin A could inhibit the viability of murine leukaemia cells P388, while pore-forming colicin E1 and colicin E3 suppressed v-myb-transformed chicken monoblasts (Fuska et al., 1979). Bures et al., (1986) isolated *E. coli* from the faeces of 77 colorectal carcinoma patients, where 32 patients (41. 6%) had barteriocins-producing *E. coli*. In the faeces of 160 healthy people, 102 people (63. 8%) had barteriocins-producing *E. coli*, which also showed that colicins from bacteria in the intestine may be one of the factors in reducing human colorectal carcinoma. They act as an anti-cancer drug. Supplementation of probiotics which produce such bacteriocins has been found to prevent the occurrence of cancer. Joo et al., (2012) have shown the ability of nisin to prevent cancer cell growth. It reduced cell proliferation of head and neck squamous cell carcinoma. The floor of mouth oral cancer xenograft mouse model was taken up, to check, the anti-HNSCC function of nisin. The results showed that nisin provided a safe and novel therapy for treating HNSCC. Di et al., (2017) studied the inhibition of *E.coli* and shiga toxin with pectic oligosaccharides.

Usually, it is the genes of bacteriocins which are encoded on the same plasmid / adjacent regions of a chromosome. It is also possible for the bacteriocin gene to enter into other bacterial cells via the method of conjugation (Ito et al., 2009; Burton et al., 2013). During the process, the bacterial pathogen can get some immune genes of bacteriocins. Therefore, for prevention of pathogens the use of purified bacterial peptide or protein is better as compared to the use of bacteria that produces bacteriocins. There are many different kinds of pathogens and mutants in nature. If the use of a particular bacteriocin is very specific than it cannot eliminate all bacterial pathogens. In such cases, there could be a mix of bacteriocin proteins for the prevention of certain human or animal pathogens. Bacteriocins can solve the most challenging problems of multi-drug resistant pathogens. Identification of new antimicrobial substances is a difficult task. With the use of biotechnology a consortium of bacteriocins could be beneficial. Acuña et al. (2012) integrated enterocin CRL35 and microcin V genes and obtained a new bacteriocin; Ent35-MccV by using recombinant PCR techniques. This has bactericidal capacity against clinically isolated enterohemorrhagic *E. coli* along with *Listeria monocytogenes*. Torres et al. (2005) have studied the differential binding of *E.coli* to alfa alfa. It can also be used against some Gram-positive pathogens and Gram-negative bacteria. Such a technique can crease new or multi-functional bacteriocins, which are more powerful in functionality and are germicidal also. They are widely used in food, animal husbandry and medicines. Bacteriocins are bactericidal peptides that regulate the competitive interactions in natural microbial systems. From the perspective of human health, they have narrow target range, high activity, stability and low toxicity. Twomey et al., (2002) have studied the lantibiotics produced by lactic acid bacteria. Suryachandran et al., (2007) studied the role of oxygen scavengers in increasing shelf life of viable lactic acid bacteria.

Nisin

It is a class-Ia bacteriocin or lantibiotic and is the most characterized and commercially important bacteriocin (Ross et al., 2002). Nisin is licensed as a food preservative (E234) and is recognized to be safe according to the Joint Food and Agriculture Organization/World Health Organization (FAO/WHO) Expert Committee on Food Additives (FAO and WHO, 2006; Favaro et al., 2015). To date, eight types of nisin variants were discovered and characterized: nisins A, Z, F, and Q produced by *Lactococcus lactis* and nisins U, U2, P, and H produced by some *Streptococcus* strains (O'Connor et al., 2015). The commercially available form of nisin for use as a food preservative is NisaplinTM, with the active ingredient nisin A (2. 5%) and other ingredients such as NaCl and non-fat dry milk (Chen and Hoover, 2003). It has antimicrobial activity against many Gram-positive bacteria, LAB, pathogens like *Listeria* and *Staphylococcus* and the spore forming bacteria, *Bacillus* and *Clostridium* (Chen and Hoover, 2003). One of the earliest applications of nisin was to prevent the bloating in cheese caused by gas-producing *Clostridium* spp. (Galvez et al., 2008). According to Abee et al., (1995) it can replace nitrate to prevent the outgrowth of *Clostridia* spores. In a study of Arqués et al. (2011), nisin was shown to reduce *L. monocytogenes* and *S. aureus* in milk stored at refrigeration temperatures. Studies have shown that the addition of nisin to diverse cheese types (cottage cheese, cheddar and ricotta-type cheeses) is very effective in the reduction of *L. monocytogenes* growth to almost 1–3 log cycles (Chen and Hoover, 2003). Several reports indicate that nisin A is not very active against *L. monocytogenes*, but its anti-listerial effect is enhanced by the reduction of pH and addition of NaCl (Khan and Oh, 2016). In Minas Frescal cheese, *S. aureus* counts were reduced by approximately 1. 5 log cycles after addition of nisin (Felicio et al., 2015). *Bacillus cereus* and *Bacillus subtilis* could be inhibited by nisin (Sobrino-López and Martín-Belloso, 2008). The use of nisin as an antimicrobial treatment extended the shelf-life of a Greek soft acid–curd cheese (Galotyri) by >21 days (Kykkidou et al., 2007). Incorporation of nisin (at a level of 2. 5 mg l^{-1}) has successfully increased the shelf-life of Ricotta-type cheese by inhibiting the growth of *L. monocytogenes* for >8 weeks (Davies et al.,

1997). In products like pasteurized dairy products which include chilled desserts, flavored milk, clotted cream and canned evaporated milks, reduced the levels of contaminating bacteria such as *L. monocytogenes* (Galvez et al., 2008). Pimentel-Filho et al., (2014) have studied the efficacy of a combination of nisin and bovicin HC5 against *L. monocytogenes* and *S. aureus* in fresh cheese. They observed a reduction of *L. monocytogenes* to undetected levels after 9 days of storage at 4°C. When tried with a combination of bacteriocins it failed to prevent the growth of *S. aureus*. Nisin activity was not affected by proteases in a study on Emmental cheese (Favaro et al., 2015). Tasara and Stephan (2006) have studied the cold stress tolerance of *Listeria monocytogenes*.

The application of nisin in dairy products is also limited as it loses the activity at higher pH (de Arauz et al., 2009). Other studies also mention several limitations on the use of nisin in dairy products, due to its interaction with fat and other components in the food matrix (Favaro et al., 2015). However, the role of fats in the activity of nisin is not entirely clear, as studies on heat-treated cream show the inhibition of *B. cereus* growth by low concentrations of nisin (Nissen et al., 2001). Interestingly, it was found to prevent spoilage bacteria and was able to extend the shelf-life in high-fat milk-based pudding (Oshima et al., 2014). Homogenization of milk has shown to reduce the anti-listerial effects of nisin, demonstrating that the treatment of foods may play an important role in the efficacy of bacteriocins such as nisin (Bhatti et al., 2004). According to Abee et al., (1995) the wide spectrum of inhibition associated with nisin includes LAB themselves. Therefore, in dairy foods which require LAB for fermentation processes, the application of nisin presents a great limitation. An alternative consists in employing other bacteriocins with a highly specific activity range (Abee et al., 1995). Scientists are in the search of new bacteriocins with a widespread range of antibacterial activity, stability in different food environments, tolerance to heat and resistance to proteolytic enzymes. The bacteriocin nisin is classified as a class-Ia bacteriocin or lantibiotic and is the most characterized and commercially important bacteriocin (Ross et al., 2002). Nisin is licensed as a food preservative (E234) and is recognized to be safe by the Joint Food and Agriculture Organization/World Health Organization (FAO/WHO) Expert Committee on Food Additives (FAO and WHO, 2006; Favaro et al., 2015). Till today, many nisin variants have been discovered and characterized which are known as nisins A, Z, F, and Q produced by *Lactococcus lactis* and nisins U, U2, P, and H produced by some *Streptococcus* strains (O'Connor et al., 2015). The commercially available form of nisin for use as a food preservative is Nisaplin™, with the active ingredient nisin A (2. 5%) and other ingredients such as NaCl and non-fat dry milk (Chen and Hoover, 2003). Nisin has antimicrobial activity against numerous Gram-positive bacteria, including LAB, pathogens such as *Listeria* and *Staphylococcus*. Also with the spore forming bacteria, *Bacillus* and *Clostridium* (Chen and Hoover, 2003). Initially, it was known to prevent late blowing in cheese caused by gas-producing *Clostridium* spp. (Galvez et al., 2008). Nitrate has been replaced by nisin for preventing the outgrowth of *Clostridia* spores (Abee et al., 1995). In dairy products it inhibits both *L. monocytogenes* and *Staphylococcus aureus* (Sobrino-López and Martín-Belloso, 2008). Arqués et al.,(2011) has shown the reduction of *L. monocytogenes* and *S. aureus* in milk stored at refrigeration temperatures. However, nisin A has not been found to be very active against *L. monocytogenes*, but its anti-listerial effect is enhanced by the reduction of pH and addition of NaCl (Khan and Oh, 2016). Gomez et al., (2014) studied the prevention of late blowing defect by reuterin produced in cheese by *Lactobacillus reuteri*.

In Minas Frescal cheese, *S. aureus* counts were reduced by approximately 1. 5 log cycles after addition of nisin (Felicio et al., 2015) and in processed cheese, *Bacillus cereus* and *Bacillus subtilis* were inhibited by nisin (Sobrino-López and Martín-Belloso, 2008). The use of nisin as an antimicrobial treatment extended the shelf-life of a Greek soft acid–curd cheese (Galotyri) by >21 days (Kykkidou et al., 2007). Incorporation of nisin (at a level of 2. 5 mg l^{-1}) has also shown to increase the shelf-life of Ricotta-type cheese by inhibiting the growth of *L. monocytogenes* for >8 weeks (Davies et al., 1997). Maximum retention of nisin was observed in 10 week storage period (10–32% of nisin loss). Ferreira and Lund (1996) also studied the effect of nisin on survival of the most resistant strains of *L. monocytogenes* in cottage cheese and showed a reduction to approximately 3 log cycles in 3 days. When tried in other pasteurized dairy products like chilled desserts, flavored milk, clotted cream and canned evaporated milks it was found to reduce post process contaminating bacteria such as *L. monocytogenes* (Galvez et al., 2008). The efficacy of a combination of nisin and bovicin HC5 against *L. monocytogenes* and *S. aureus* in fresh cheese was studied by Pimentel-Filho et al., (2014). They observed a reduction of *L. monocytogenes* to undetected levels after 9 days of storage at 4°C. But the combination of bacteriocins did not prevent the growth of *S. aureus*. In case of proteolysis during the cheese-making process can affect the activity of nisin and could limit the antimicrobial efficacy. Nisin loses the activity at higher pH (de Arauz et al., 2009) and in dairy products, due to its interaction with fat and other components in the food matrix (Favaro et al., 2015). The role of fat in the activity of nisin is not clear. Studies on heat-treated cream show the inhibition of *B. cereus* growth by low concentrations of nisin (Nissen et al., 2001). It was found to prevent spoilage bacteria and extend shelf-life in high-fat milk-based pudding (Oshima et al., 2014). LAB themselves (Abee et al., 1995) are associated with inhibition of pathogens. The application of nisin presents a great limitation in dairy foods which require LAB for fermentation processes.

Other bacteriocins with a highly specific activity range can be a good alternative (Abee et al.,1995). Scientists are working towards new bacteriocins which will have a widespread range of antibacterial activity, stability in different food environments, tolerance to heat and resistance to proteolytic enzymes. Heron et al., (2016) have associated diseases with food and heart disease.

Lacticins

Certain strains of *Lc. lactis* which has lacticin 3147 and lacticin 481 produce lacticins (Piard et al., 1992; McAuliffe et al., 1998). Lacticin 3147 was isolated from an Irish kefir grain used for making buttermilk. It is a two-component lantibiotic and requires both structural proteins for biological activity. This bacteriocin exhibit antimicrobial activity against a wide range of food pathogenic and food spoilage bacteria in addition to other LAB (Sobrino-López and Martín-Belloso, 2008; Martínez-Cuesta et al., 2010). In a study done by Morgan et al., (2001) lacticin 3147 powder preparation is found to be effective in controlling *Listeria* and *Bacillus* in infant milk formulation, natural yogurt, and cottage cheese. In case high concentrations of lacticin powder was used (10% of product weight) then it was non-economical. Lacticin 481 is a single-peptide lantibiotic that exhibits a medium spectrum of inhibition which is active against other LAB, *Clostridium tyrobutyricum* (O'Sullivan et al., 2003) and *L. monocytogenes* (Ribeiro et al., 2016). Arqués et al., (2011) found that even the non-purified lacticin 481 has shown a mild bacteriostatic activity in milk when stored at refrigeration temperatures. The application of semi-purified lacticin 481 to fresh cheeses at refrigeration temperatures was shown to reduce the counts of *L. monocytogenes* by 3 log cycles in 3–7 days (Ribeiro et al., 2016). The scientists found to it to be active against other LAB, *Clostridium tyrobutyricum* and *L. monocytogenes*. The application of semi-purified lacticin 481 to fresh cheeses stored at refrigeration temperatures reduced *L. monocytogenes* by 3 log cycles in 3–7 days (Ribeiro et al., 2016). It has antimicrobial activity against many food pathogens and food spoilage bacteria in addition to other LAB (Martínez-Cuesta et al., 2010). A lacticin 3147 powder preparation was shown to be effective for the control of *Listeria* and *Bacillus* in infant milk formulation, natural yogurt and cottage cheese (Morgan et al., 2001). According to them use of high concentrations of lacticin powder was impracticable and uneconomical. In a study done by O'Sullivan et al., (2003) Lacticin 481 is a single-peptide lantibiotic that exhibits a medium spectrum of inhibition. Nevertheless, the application of lacticins in food systems is not likely to ensure complete elimination of pathogens such as *L. monocytogenes*. Hillman et al., (1998) have analysed mutacin 1140 from *Streptococcus mutans*.Shobha rani and Renu Agrawal (2007) studied the therapeutic importance of volatile compounds produced by *Leuconostoc mesenteroides*. Shobha rani and Renu

Agrawal (2008) studied the effect on cellular membrane fatty acids in the stressed cells of *Leuconostoc mesenteroides*. Shobha rani and Renu Agrawal (2010 a) studied the enhancement of cell stability and viability of *Leuconostoc mesenteroides* on freeze drying to be used in food formulations. Shobha rani and Renu Agrawal (2010b) studied the interception of quorum sensing signal molecule by furanone to enhance shelf life of fermented milk. Shobha rani and Renu Agrawal (2011) isolated and characterized a lactic acid bacteria from cheddar cheese as a probiotic potent strain. Shobha rani et al., (2006) improved the *Leuconostoc mesenteroides* by mutagenesis.

Enterocins

According to Egan et al., (2016) enterocins are produced by *Enterococcus* species that are a diverse group of bacteriocins in terms of their classification and inhibitory spectrum. The use of purified enterocins may be considered more suitable for food consumption. Enterocin AS-48 is produced by *Enterococcus faecalis* and is a class IIc cyclic bacteriocin that is active against a number of *Bacillus* and *Clostridium* sp. (Egan et al., 2016). As it has high stability to pH and heat, it has wider applications in food. LAB come under the GRAS status and can be safely used in food applications,however, bacteriocinogenic enterococci raise some safety concerns (EFSA, 2007). Enterococci are associated with a number of infections in humans (Moreno et al., 2006). Some may have virulence factors and antibiotic resistance genes (EFSA, 2007). It has been observed that the food enterococci have less virulence than clinical strains. They are known for their capacity to exchange genetic information (Eaton and Gasson, 2001). The efficacy of the enterocin AS-48 for controlling *Staphylococci* in skimmed milk has been worked out by Muñoz et al., (2007). In a study conducted by Yildirim et al., (2016) on the antimicrobial effects of enterocin KP toward *L. monocytogenes* in skim, half fat and full fat milks it was found that the bactericidal effect decreased as both the fat content of milk and inoculation amount of *L. monocytogenes* increased. In another study done by Lauková and Czikková (1999) it was found that an inhibitory effect of purified enterocin CCM 4231 (3200 AU ml^{-1}) was observed on the growth of *S. aureus* and *L. monocytogenes* in skimmed milk and yogurt. The activity of enterocins in food systems does not prevent the re-growth of pathogens throughout the storage time (Zacharof and Lovitt, 2012; de Souza Barbosa et al., 2015). Arqués et al., (2011) observed that the counts of *L. monocytogenes* in milk reduced below the detection limit after 4 and 24 h when bacteriocins, reuterin and enterocin AS-48 were used. Looking into the safe concerns of using bacteriocinogenic enterococci, it is important to use the purified form which is considered more suitable for food consumption. Muñoz et al. (2007) tested the efficacy of the enterocin AS-48 for controlling

Staphylococci in skimmed milk. They found the bactericidal effect which was proportional to the bacteriocin concentration (10–50 ìg ml⁻¹), but complete elimination of staphylococci was not achieved for any of the concentrations tested. This lowers the effectiveness of AS-48 in milk when compared to culture medium. Recently, Ribeiro et al.,(2017) reported the efficacy of a semi-purified enterocin produced by an *E. faecalis* strain, in reducing the contamination of *L. monocytogenes* in fresh cheese in a dose-dependent manner. Moreover, the highest dose applied to cheeses (approximately 2000 AU g⁻¹ of cheese) resulted in reduction of this pathogen very low and this effect remained for all the storage time (72 h). Maria et al., (2014) have isolated enterocin AS-48 for food applications.

Pediocins

Pediocins are a class IIa bacteriocins produced by *Pediococcus* spp. and are commercially available under the name Alta 2341TM or Microgard TM (Garsa et al., 2014). Against some food-borne pathogens such as *L. monocytogenes* and *S. aureus* (Cintas et al., 1998; Eijsink et al., 1998) and Gram-negative organisms such as *Pseudomonas* and *Escherichia coli* (Jamuna and Jeevaratnam, 2004) it has been more effective than nisin. The application of pediocins to dairy products is higher due to its stability in aqueous solutions, its wide pH range and high resistance to heating or freezing (Sobrino-López and Martín-Belloso, 2008). Many studies have investigated the addition of pediocins to milk or dairy foods. Pediocin (PA-1) was found to reduce *L. monocytogenes* counts in cottage cheese, cream, and cheese sauce (Pucci et al., 1988). They have shown that pediocin (PA-1) can reduce *L. monocytogenes* counts in cottage cheese, cream and cheese sauce. The anti-listerial effect was noticeable over a wide temperature and pH range and was effective at low initial *L. monocytogenes* contamination (10² cfu ml⁻¹). Recently, Verma et al., (2017) described the production of food-grade pediocin from supplemented cheese in whey medium. This semi-purified pediocin containing fermented cheese whey was shown to be effective in reducing *S. aureus* counts and enhancing the shelf-life of raw buffalo milk. They have worked on the production of food-grade pediocin using supplemented cheese whey medium. Vanita and Renu Agrawal (2012) purified a bacteriocin from *Leuconostoc* against *V. cholerae*. Vidyalaxme et al., (2012) studied the synergistic effect of *Leuconostoc* and *Bacillus* against *V. cholarae*.

Other Bacteriocins

Some strains of *Lc. lactis* spp produce Lactococcin BZ (lactis BZ) which has antibacterial activity against Gram-positive and Gram-negative bacteria (ªahingil et al., 2011). Yildirim et al., (2016) have studied the effect of the partially purified lactococcin BZ (400–2500 AU ml^{-1}) which could reduce *L. monocytogenes* counts to an undetectable level both in skim and full-fat milk during storage at 4 and 20°C. In addition, this anti-listerial activity was stable until the end of the storage period (25 days) and was not adversely affected by milk fat content. Dinesh and Renu Agrawal (2008) studied the preparation of fermented milk with *Pediococcus pentosaceous*.

From *S. aureus* strain Aureocin A70 was isolated which is a class II bacteriocin (Fagundes et al., 2016). It has bactericidal effect on a wide range of Gram-positive bacteria, including *L. monocytogenes* and was tested in UHT-treated skimmed milk by Fagundes et al. (2016). Aureocin A70 caused a time-dependent reduction in *L. monocytogenes* counts (5. 51 log units) up to 7 days of incubation, but was insufficient to achieve the total elimination of the pathogen, resulting in <0. 001% survivors (Fagundes et al., 2016). El- Agamy et al., (2009) studied the bioactive compounds in camel milk. Gayathri and Renu Agrawal (2012) studied the functional properties in camel milk with *Pediococcus*.

In recent times with molecular biology being on the top, genetically engineered bacteriocins have been proposed to overcome the narrow range of activity of most bacteriocins. Acuña et al., (2012) reported the construction of a chimerical bacteriocin named Ent35-MccV. This hybrid bacteriocin combines in a single molecule the anti-listerial activity of enterocin CRL35 and the anti-*E. coli* activity of microcin V. This hybrid wide-spectrum bacteriocin was active against pathogenic strains of *L. monocytogenes* and *E. coli* O157:H7 and was effective in controlling the growth of both pathogens in skim milk (Acuña et al., 2015). Gibson et al., (1997) studied the effect of probiotics in improving intestinal infection.

Application of Purified Bacteriocin-Producing Bacteria to Dairy Products

Awasthi et al., (2012) studied the contamination in milk. The application of a bacteriocin alone does not provide sufficient protection against microbial con-tamination of dairy products. A reduction of *L. monocytogenes* counts was observed in raw milk cheese by the use of two nisin-producing *Lactobacillus lactis* subsp. *lactis* as starter cultures (Rodrígez et al., 2001). In contrast, the use of nisin-producing strains to control *E. coli* and *L. monocytogenes* in Feta and Camembert cheeses has shown limited results (Ramsaran et al., 1998). *E. coli* O157:H7 survived the manufacturing process of both cheeses and was present in cheese made with nisin-producing strains at the end of 75 days of storage in greater numbers than the initial inoculum. The Feta cheese that con-tained nisin was the only cheese in which *L. monocytogenes* remained at the level of the initial inoculum after 75 days of storage. The use of bacteriocinogenic *Pediococci* in dairy products also exhibited limited results due to their inability to ferment lactose (Renye et al., 2011). Bouttefroy et al., (2000) studied nisin for avoiding the regrowth of bacteriocin resistant cells of *Listeria monocytogenes*.

Mills et al., (2011) tested a plantaricin-producing *Lactobacillus plantarum* strain as an anti-listerial adjunct in the presence and absence of nisin-producing starters for the manufacture of cheese. The combination of *Lb. plantarum* strain (at 10^8 cfu ml^{-1}) with a nisin producer reduced *Listeria*. Carnio et al., (2000) studied the anti-listerial potential of a food-grade strain *Staphylococcus equorum*, producing a bacteriocin named as micrococcin P1. A remarkable reduction of *L. monocytogenes* growth was achieved, but this effect was dependent on the contamination level. In addition, a regrowth of the viable *Listeria* could be observed after 10–16 days of maturation. Dal Bello et al., (2012) detected a reduction of *L. monocytogenes* by 3 log units in cottage cheese, after incubation with a lacticin-481-producing *Lc. lactis* strain. However, after an initial period, *L. monocytogenes* counts increased to values comparable to the control.

The heterologous production of bacteriocin by genetically engineered LAB was tested in dairy foods (Leroy and De Vuyst, 2004). A lacticin 3147 transconjugant generated via conjugation of bacteriocin-encoding plasmids was used successfully against *L. monocytogenes* in cottage cheese (McAuliffe et al., 1999). In another study, a lacticin 3147 transconjugant presented a protective effect when applied to the cheese surface, reducing *L. monocytogenes* by 3 log units (Ross et al., 2000). The use of *Enterococcus* spp. in foods may represent a risk for consumers and requires a safety assessment by the European food authorities (EFSA, 2007). However, enterococci are found naturally in some dairy foods, as they are used as starter cultures and are often part of the microbiota of artisanal cheeses (Domingos-Lopes et al., 2017). De Vuyst et al., (2003) suggested that *Enterococcus* species can be used safely in food in case the virulence genes were absent. Consequently, several studies have also employed enterocin-producing *Enterococcus* in food systems. Celia et al., (2019) studied the effect of bacteriocins in dairy foods. daSilva Sabo et al., (2017) studied the inhibitory substances by *Lactobacillus* in cheese.

Several authors (Giraffa, 1995; Giraffa and Carminati, 1997; Nunez et al., 1997; Lauková et al., 1999 a,b: Lauková et al., 2001) have worked to show the inhibitory effect of enterocin-producing *Enterococci* against *L. monocytogenes* and *S. aureus* in dairy foods like milk and cheeses. Bacteriocinogenic *E. faecalis* strain has the ability to reduce *L. monocytogenes* counts in Manchego cheese by 1 and 2 log units after 7 and 60 days, respectively (Nunez et al., 1997). Although inoculation of milk with *E. faecalis* strain reduced the rate of acid production in the curd, the flavor and bitterness of the final product were not influenced. Other *Enterococci* able to reduce *L. monocytogenes* in dairy foods have been investigated by Pingitore et al., (2012). Vandera et al., (2017) has studied the use of a multiple enterocin-producing bacterial strains possessing the structural entA, entB and entP enterocin genes. Achemchem et al., (2006) studied the effectiveness of an *E. faecium* strain in controlling *L. monocytogenes* in goat's milk and goat's traditional cheese (jben). Coelho et al., (2014) examined the inhibitory effect of enterocin-producing *E. faecalis* strains against *L. monocytogenes* in fresh cheese. Martínez-Cuesta et al., (2010) have shown that the *Lc. lactis* strain which produces lacticin 3147 is helpful in accelerating cheese ripening and also in preventing the late blowing in cheese by inhibiting *Clostridia* growth.

It has been observed that when the cheese is prepared with the bacteriocinogenic adjunct an increased cell lysis is seen with higher concentrations of free amino acids, which is associated with higher sensory evaluation scores (Morgan et al., 1997). Another bacteriocin-producing starter *Lc. lactis* (a lacticin 3147 producer) was tested for controlling the proliferation of undesirable microorganisms during cheese manufacture. Cheeses made with lacticin 3147-producing starters exhibited lower levels of NSLAB that remained constant over 6 months of ripening (Ryan et al., 1996). A three-strain starter system was tested on cheese by Morgan et al., (2002).

Although there have been many advances in bacteriocin research for food applications, the use of purified bacteriocins in the dairy industry has been very limited. The application of a bacteriocin by itself does not provide sufficient protection against microbial contamination of dairy products. The purification cost limits the commercial exploration of new bacteriocins. Along with this the restrictive food legislation of the health regulatory authorities (FDA and EFSA) limits the approval of new bacteriocins as food preservatives. It is therefore that only two bacteriocins (nisin and pediocin) are available commercially till date. Many LAB genera and species come under GRAS and QPS status. In this regard, incorporation of such bacteria into foods offers a viable solution for controlling contamination of microorganisms as also being the starter cultures in food fermentations. For the *in situ* production of bacteriocins, the protective cultures are added that grow and produce bacteriocins during the manufacture and storage of dairy foods. The selection and development of bacteriocinogenic cultures which are used as cell lysis-inducing agents can improve the cheese maturation and flavor (Beshkova and Frengova, 2012). Late blowing defect is a major cause of spoilage in ripened cheeses, resulting in the appearance of texture and flavor defects, due to the ubiquitous presence of *Clostridium* spores (Gómez-Torres et al., 2015). Garde et al., (2011) has suggested the use of bacteriocinogenic LAB. www. frontiersin. org. Havlicek et al., (2011) have worked on glycopeptide bacteriocins.

In traditional fermented milk (Iben), Benkerroum et al., (2002) studied the effect of *in situ* bacteriocin production of *L. lactis* ssp. lactis against *L. monocytogenes*. It was observed that *L. monocytogenes* decreased to below the detectable level within 24 h of storage at 7°C in Iben when it was fermented with the bacteriogenic starter culture. Moreover, the pathogen was efficiently inactivated from contaminated samples despite the high level of contamination (10^7 cfu ml^{-1}) for up to 6 days of storage at 7°C. The application of nisin-producing *Lactococcus* sp. in dairy foods needs LAB starters. It has wide spectrum of inhibition (Abee et al., 1995). Yamauchi et al.,(1996) produced a yogurt by incorporating a nisin-producing strain, *L. lactis* subsp. lactis, in raw milk. Kondrotiene et al., (2018) reported a reduction in *L. monocytogenes* when three nisin A producing *Lc. lactis* strains were added to fresh cheese. However, the reduction on *Listeria* contamination was limited to 2 log units within 7 days of cheese storage. Nisin A-producing *Lactococcus diacetylactis* in a mixed starter culture was not found to be effective in inhibiting *L. innocua* growth in Cheddar cheese (Benech et al., 2002). Maisnier-Patin et al., (1992) examined the inhibitory effect of another nisin-producing *Lc. lactis* on *L. monocytogenes* in Camembert cheese. In the presence of nisin-producing starter, the numbers of the pathogen decreased until the end of the second week, leading to a reduction of 3 log cfu g^{-1}, but a regrowth of *Listeria* was observed in cheeses throughout ripening (6 weeks). Mills et al., (2011) worked on a plantaricin-producing *Lactobacillus plantarum* strain to work as an anti-listerial adjunct in the presence and absence of nisin-producing starters for the manufacture of cheeses. The combination of *Lb. plantarum* strain (10^8 cfu ml^{-1}) with a nisin producer reduced *Listeria* to very low levels by 28 days which could not be even detected. Moreover, they found that *Lb. plantarum* was much more effective at inhibiting *Listeria* than the nisin producer alone.

Working on the anti-listerial potential of a food-grade strain of *Staphylococcus equorum* the scientists, Carnio et al., (2000) found a bacteriocin named as micrococcin P1, in soft cheese. They found a remarkable reduction in the count of *L. monocytogenes* growth. However, it was dependent on the contamination level. They also observed that *Listeria* grew again after 10–16 days of maturation. Similarly, Dal Bello et al., (2012) found a reduction of *L. monocytogenes* by 3 log units in cottage cheese which was incubated with a lacticin-481-producing *Lc. lactis* strain. However, it was found that *L. monocytogenes* counts increased to values comparable to the control after an initial period. Rekhif et al., (1994) selected mutants of *Listeria* which were resistant to bacteriocins by lactic acid bacteria.

Leroy and De Vuyst, (2004) studied on the heterologous production of bacteriocin by genetically engineered LAB in dairy foods. McAuliffe et al., (1999) worked on lacticin 3147 which was a transconjugant and was generated via conjugation of bacteriocin-encoding plasmids. They were successful to show the activity against *L. monocytogenes* in cottage cheese. Reductions of 99. 9% were seen in cottage cheese which was held at 4°C after 5 days. Ross et al., (2000) found that lacticin 3147 transconjugant had a protective effect when applied to the cheese surface. It was able to reduce *L. monocytogenes* by 3 log units. In contrast, this protective effect was not evident when a nisin-producing culture was used, possibly due to pH instability (Ross et al., 1999).

However, the use of *Enterococcus* spp. in foods has a risk for the consumers and requires a safety assessment by the European food authorities (EFSA, 2007). *Enterococci* are found naturally in some dairy foods, as they are used as starter cultures and are also a part of the microbiota of artisanal cheeses (Domingos-Lopes et al., 2017). Many strains of bacteriocinogenic enterococci lack virulence determinants (Jaouani et al., 2015). De Vuyst et al., (2003) suggested that *Enterococcus* species could be safely used in food in case the virulence genes were absent. Renu Agrawal and Shylaja (2011) isolated an anti shigella bacteriocin from *Pediococcus*. Renu Agrawal et al., (2010) studied the functional properties of *Lactobacillus plantarum*. Schneider et al., (2015) studied the prevention of *Clostridium*. Servin et al., (2004) studied the antagonistic properties of lactic acid bacteria and *Bifidobacterium* against pathogens.

The inhibitory effect of enterocin-producing enterococci against *L. monocytogenes* and *S. aureus* in dairy foods has been worked out by many scientists (Giraffa, 1995; Giraffa and Carminati, 1997; Nunez et al., 1997; Lauková et al., 1999a, b and Lauková et al., 2001). Khan et al., (2010) worked on the keeping quality in cheese with the use of enterocin. Bacteriocinogenic, *E. faecalis* strain was found to be helpful in reducing the counts of *L. monocytogenes* in Manchego cheese by 1 and 2 logs after 7 and 60 days (Nunez et al., 1997). The rate of acid production in the curd was found to reduce but the flavor and bitterness of the final product were not changed.

Pingitore et al., (2012) investigated two bacteriocinogenic enterococci strains (*Enterococcus mundtii* CRL35 and *Enterococcus faecium* ST88Ch) which were isolated from cheeses. Growth of *L. monocytogenes* was inhibited in Minas cheeses containing *E. mundtii* up to 12 days of storage at 8°C. This bacterium displayed a bacteriostatic effect since *Listeria* counts remained similar to the initial inoculum. *E. faecium* strain was found to be less effective, as the bacteriostatic affect occurred only after 6 days at 8°C. Vandera et al., (2017) studied the use of a multiple enterocin-producing bacterial strains which had the structural entA, entB, and entP enterocin genes. Some strains exhibited a bacteriostatic effect on *L. monocytogenes* in raw milk incubated at 37°C for 6 h. With the raw milk cultures when incubated at 18°C the viable counts of the pathogen came down after 24 h and up to 72 h. Pulido et al., (2015) studied the application of bacteriophages. Rosch et al., (1998) studied the tolerance of *Listeria* to nisin. Rayman et al., (1983) studied the failure of nisin to *Clostridium* in meat.

Achemchem et al., (2006) has looked into the effectiveness of an *E. faecium* strain to control the strain of *L. monocytogenes* in goat's milk and goat's traditional cheese (jben). In an experiment *E. faecium* and *L. monocytogenes* were cocultured in milk and it was observed that *Listeria* was not detected after 130h. Coelho et al. (2014) examined the inhibitory effect of enterocin-producing *E. faecalis* strains against *L. monocytogenes* in fresh cheese. Bacteriocin-producing LAB strains have also been assessed to improve cheese maturation and flavor. *Lc. lactis* producing lacticin 3147 was shown to accelerate cheese ripening and also prevent late blowing in cheese by the inhibition of *Clostridium* growth (Martínez-Cuesta et al., 2010). Many bacteriocin-producing strains possess a lytic effect on starter cultures. It was found that the use of bacteriocin-producing *Lc. lactis* ssp. *cremoris* as a starter adjunct in Cheddar cheese manufacture increased the rate of starter lysis. Cheese manufactured with the bacteriocinogenic adjunct increased the lysis of cells. The concentration of free amino acids increased along with the sensory quality (Morgan et al., 1997). Cheeses made with lacticin 3147-producing starters significantly reduced the levels of pathogenic microorganisms for more than 6 months of ripening (Ryan et al., 1996). Reale et al., (2015) studied the tolerance to stress in lactic acid bacteria while food processing and GI tract.

Morgan et al. (2002) tested a three-strain starter system on cheese in which a bacteriocin producer causes lysis of a second strain which is sensitive to bacteriocin and there is a third strain which is resistant to the bacteriocin activity. This produces the amount of acid required during the manufacture of cheese.

Other Bacteriocins

Lactococcin BZ is produced by some strains of *Lc. lactis* spp. (lactis BZ) which has a wide antibacterial activity against Gram-positive and Gram-negative bacteria. It had strong anti-listerial activity in milk. The partially purified lactococcin BZ (400–2500 AU ml^{-1}) reduced *L. monocytogenes* counts to an undetectable level in both skim and full-fat milk during storage at 4 and 20°C (Yildirim et al., 2016). The activity was found to be stable till 25 days and was not adversely affected by milk fat content.

Aureocin A70 is a class II bacteriocin produced by a *S. aureus* strain isolated from pasteurized commercial milk (Fagundes et al., 2016). It has bactericidal effect on a wide range of Gram-positive bacteria like *L. monocytogenes*

(Fagundes et al., 2016). Use of Aureocin A70 gives a time-dependent reduction in *L. monocytogenes* counts (5. 51 log units) up to 7 days of incubation. Total elimination of the pathogen was found to be difficult (Fagundes et al., 2016).

Recently, the use of genetically engineered bacteriocins is being proposed to overcome the narrow range of activity of most bacteriocins. Acuña et al., (2012) reported the construction of a chimerical bacteriocin named Ent35-MccV which is a hybrid bacteriocin and combines in a single molecule the anti-listerial activity of enterocin CRL35 and the anti-*E. Coli* activity of microcin V. It is found to be active against *L. monocytogenes* and *E. coli* O157:H7 and also effective in controlling the growth of both pathogens in skim milk (Acuña et al., 2015).

Combining Bacteriocins with Other Hurdles

Along with bacteriocins if chemical additives (EDTA, sodium lactate and potassium diacetate), are added followed by treatments like heating and high-pressure (Egan et al., 2016) and it was found that the effect was enhanced. Narayanan and Ramana (2013) found out that the incorporation of pediocin along with eugenol into the polyhydroxybutyrate films synergized the effect preventing the contamination of food. Prudêncio et al., (2015) used the mixture of bacteriocins and EDTA in the sensitization of Gram-negative bacteria. Gram-negative bacteria becomes sensitive to bacteriocins if the permeability of their outer membrane is changed with chelating agents, EDTA (Chen and Hoover, 2003). Zapico et al., (1998) used nisin and the lactoperoxidase system (LPS) to control *L. monocytogenes* in skim milk and found a synergistic effect. A listericidal effect (5. 6 log units lower than the control milk) was observed in treatments containing nisin (10 or 100 IU ml^{-1}) with LPS, after 24 h at 30°C. However, when the two preservatives were added in two steps (LPS was added after 3 h and nisin after 5 h of growth) the difference in *L. monocytogenes* counts increased by 7. 4 log units.

According to a study done by Prudêncio et al., (2015) the synergistic effect of bacteriocins was observed after temperature treatments. In a study done by Kalchayanand et al., (1992) many bacteria were resistant to the bacteriocins, at low or high temperatures and also when treated with nisin and pediocin. High pressure processing is a common technique for inactivating microorganisms at room temperature, but this treatment does not ensure the complete inactivation of microorganisms (Prudêncio et al., 2015). Garriga et al., (2002) and Zhao et al., (2013) have shown the synergistic effect of bacteriocins like nisin under high pressure processing in which the food microorganisms were reduced. Rodriguez et al., (2005) demonstrated the efficacy of reduced pressures when combined with bacteriocin-producing LAB to improve cheese safety. Marques et al., (2017) found that when a biodegradable film was incorporated with cell-free supernatant (CFS) containing bacteriocin-like substances of *Lactobacillus curvatus* P99 it was able to control the growth of *L. monocytogenes* in sliced "Prato" cheese. These films containing the bactericidal concentration of CFS were able to control *L. monocytogenes* for 10 days of storage at 4°C.

In the 21st century it is important to know the killing strategies and applications of microcins, their mode of action, the role of the van der Waals zone in the design of a new family of bacteriocins, the use of pyocins in the treatment of infections, the role of oral probiotics, veterinary applications of bacteriocins (nisin) in treating diseases especially mastitis and the genetics of bacteriocin resistance.

Bacteriocins have a role in mediating inter- and intra-specific interactions among bacteria and in bacterial community diversity. They can provide viable alternatives to traditional antibiotics. Bacteriocin production and use needs an understanding to know the production of multiple bacteriocins by a single strain in species of *Esherichia coli*. Looking into the ecological role of bacteriocins these are secreted from the cell and there is experimental evidence to suggest that more understanding is required on the secretion of bacteriocins.

Sylvie (2016) studied the Microcins and other bacteriocins which were the bridging gaps between the killing stategies, ecology and their applications. These ribosomally-synthesized antimicrobial peptides or proteins which are produced by bacteria use these to thrive in the microbial wars. The bacteriocin producing strains bear a quality not to get killed by their own toxins. Most bacteriocins are active in the pico or nanomolar range and target bacterial species that are phylogenetically close to the producing strain, although some exhibit broader spectra of activity. Bacteriocins have been widely studied in Gram-positive (lantibiotics, pediocin-like bacteriocins) and Gram-negative bacteria (colicins, microcins). Bacteriocin production is found widely in nature which includes the archaea. However, they can differ significantly both in size and in chemical properties. They can be small peptides (the Gram-positive bacteriocins and microcins are peptides below 10 kDa) or large proteins (colicins are 30-80 kDa proteins). Many are post-translationally modified using dedicated enzymes. It leads to highly complex peptide-derived structures. The structural diversity is related to the killing strategies.

André Gratia discovered the first bacteriocin in 1925. In a study done by Justyna et al., (2016) the mode of action, immunity and mechanism of import into *Escherichia coli* were studied. It was found that one strain of *Escherichia coli* producing a toxic diffusible substance killed the neighbouring *E. coli*. The mechanism of killing pattern are of two types; enzymatic colicins cleave either nucleic acids or peptidoglycan precursors while pore-forming colicins depolarize the cytoplasmic membrane. Lot of work has been done on various aspects of biochemical, structural and biophysical parameters. Today, there is a good understanding of molecular aspects of colicin nucleases towards the binding to their specific immunity proteins. The process of colicin engagement to the cell surface, periplasm, inner membrane and cytoplasm were also studied. Understanding of the translocation between the cellular compartments nuclease colicins and the extent to which the toxins must be unfolded during import is being sudied by the scientists. These will help to exploit the biomedical and biotechnological potential of bacteriocins. It will also provide insight into the working of the Gram-negative cell envelope.

Xiao and Margaret (2016) studied the power of Van der Waals zone in the creation of a novel family of bacteriocin-based antibiotics. It has been found out that the clinical potential of bacteriocins is limited by three concerns namely; the narrow range of killing activity, relative to conventional antibiotics, the ability of these proteins not to reach or function at the site of most bacterial infections and the presumed antigenicity and subsequent toxicity of protein antimicrobials. Riley (2015) have shown that colicin Ia shows no toxicity to human or animal tissues, they do not induce an immune response by their presence and exhibit high *in vivo* activity in all animal tissues tested, including the circulatory system. According to them the bacteriocins are a new generation in bacterial infection control.

Suphan Bakkal (2016) has studied the use of Pyocins in treating *Pseudomonas aeruginosa* infections. *Pseudomonas aeruginosa* is a Gram-negative opportunistic pathogen which is usually present in the urinary tract, skin, eye, ear and lungs. It is associated with life-threatening hospital acquired and community acquired infections. Antibiotics are used to fight *P. Aeruginosa* infections. However, they are very ineffective in the face of increasing *Pseudomonas* antibiotic resistance. Recently studies are being focussed on the potential of pyocins which are protein bacterial toxins of *P. aeruginosa*. These serve as as novel antibiotics for the treatment of *P. aeruginosa* infections. Pyocins kill closely related species and are classed into three categories as: the S-type protease-sensitive pyocins which are similar in size and in mode of action to colicins of *Escherichia coli*. Another one is the R and F type protease resistant pyocins that resemble bacteriophage tails. These will be the next generation antimicrobials for the treatment of *P. aeruginosa* infections. John et al., (2016) have studied strains of *Streptococcus* which produce bacteriocin and can be used orally. Most diverse of the bacterial genera is the genus *Streptococcus* which comprises of 70 defined species inhabiting a wide variety of ecological habitats. Certain species are used in the production of food products but the majority are pathogens of humans and other animals. Many are producers of bacteriocins, especially of the lantibiotic class the *Streptococcus salivarius*. It is one of the more prolific bacteriocinogenic species. Some strains of *S. salivarius* harbor especially large (>100 kb) megaplasmids which have not been reported amongst other oral bacteria. Due to the low pathogenic potential, *S. salivarius* acts as a source of oral probiotics in targeting infections of humans caused by other Streptococci including pharyngitis (*Streptococcus pyogenes*) and dental caries (*Streptococcus mutans*). Hale et al., (2012) have written an article on both the lantibiotic and heat-labile properties of salivaricins produced by *S. salivarius*. They have also highlighted the potential applications of bacteriocin producing Streptococci as oral probiotics, including a profile of *S. salivarius* probiotic products which are available in the market. Sandra et al., (2016) have studied the effect of nisin on bovine mastitis. Dairy cattle suffer from the disease called as mastitis. It affects around 56% of U. S. commercial dairy herds. Presently, the treatment is broad-spectrum antibiotics. However, there is resistance to these drugs. They have reviewed the current status and novel approaches for the treatment of mastitis. They have discussed on a new technology produced by Immu Cell, a U.S. veterinary health company. The company has developed a bacteriocin-based strategy to combat mastitis, using the lantibiotic Nisin. This is the only bacteriocin-based product available in the U.S. Kicza et al., (2016) have studied the resistance in a sensitive strain of *E. coli* (BZB 1011) to bacteriocins, a class of naturally occurring protein antimicrobials. They have studied the resistance mechanisms of the plasmid-encoded bacteriocins of *E. coli*. These colicins have been selected to represent the diversity of mechanisms underlying colicin binding, translocation and killing. The patterns of cross-resistance to other colicins exhibited by these resistant strains were examined to infer the probable mechanistic basis of the selected resistance. The sequence level changes were also seen to underlie selected cases of colicin resistance. These pathways require one of mutational steps minimizing the costs and maximizing the benefits of resistance (Perez et al., 2014). Zendo, (2013) has recently developed a rapid screening method to know the production of LAB bacteriocins based on the antibacterial spectra and molecular masses. Makras et al., (2006) found that the cell number of *Salmonella* could be reduced with lactic acid bacteria. Many novel bacteriocins have been identified by this method. This includes a nisin variant, nisin Q, a two-peptide bacteriocin, lactococcin Q, lacticin Q and a circular bacteriocin; lactocyclicin Q. It was found that some LAB isolates had the ability to produce multiple bacteriocins. They have been characterized on the basis of their structures, mechanisms of action and biosynthetic mechanisms. These can be utilized as food preservatives.

Due to the increasing emergence of antibiotics resistance it was Chi and Holo (2018) who have shown a synergistic combination to combat resistant bacteria.

They have studied the activity of garvicin KS, a novel bacteriocin produced by *Lactococcus garvieae*. The bacteriocin has a broad inhibitory spectrum, inhibiting members of all the 19 species of Gram-positive bacteria tested. Unlike other bacteriocins from Gram-positive bacteria, garvicin KS inhibits *Acinetobacter* but not other Gram-negative bacteria. The synergy with polymyxin B was shown against *Acinetobacter* spp. and *Escherichia coli*, but not with *Pseudomonas aeruginosa*. They found similar effects when the mixtures of nisin and polymyxin B were taken. The synergistic mixtures of all the three components caused rapid killing of pathogens like *Acinetobacter* spp. and *E. coli*. Garvicin KS and nisin was found to be effective against *Staphylococcus aureus*. This shows the difference in modes of action between the two bacteriocins. The two bacteriocins had synergy with farnesol. It appears that garvicin KS may be a promising antimicrobial due to the broad inhibitory spectrum, rapid killing and having synergy with other antimicrobials. Ovchinnikov et al., (2016) isolated ten *Lactococcus garvieae* isolates from raw milk which could produce a new broad-spectrum bacteriocin. Even when the milk was procured from different farm areas it was found to have the identical inhibitory spectra, fermentation profiles of sugars, and repetitive sequence-based PCR (rep-PCR) DNA patterns. Therefore, it was known that raw milk produces the same bacteriocin. Purification and peptide sequencing combined with genome sequencing of *L. garvieae* KS1546 showed that the antimicrobial activity was due to a bacteriocin unit composed of three similar peptides of 32 to 34 amino acids. The three peptides are produced without leader sequences, and their genes are located next to each other in an operon-like structure, adjacent to the genes normally involved in bacteriocin transport (ABC transporter) and self-immunity. The bacteriocin, termed garvicin KS (GarKS), showed sequence homology to four multipeptide bacteriocins in databases. These are staphylococcal aureocin A70, consisting of four peptides, and three unannotated putative multipeptide bacteriocins produced by *Bacillus cereus*. Bacterial resistance to antibiotics is a very serious global problem. As Garvicin KS has very broad antibacterial spectra it has a great potential for antimicrobial applications in the food industry and medicine. Netz et al., (2001) have shown that *Staphylococcus aureus* A70 produces a heat-stable bacteriocin designated aureocin A70. It is encoded inside a mobilisable 8 kb plasmid, pRJ6. This is active against *Listeria monocytogenes*. Experiments of transposition mutagenesis and gene cloning had shown that aureocin A70 production and immunity were associated with Hind III-A and B fragments of pRJ6. By DNA sequencing and genetic studies it has been found that the three transcriptional units are involved in bacteriocin activity. The first transcriptional unit contains a single gene, aurT, which encodes to a protein that resembles an ATP-dependent transporter. This is required for aureocin A70 production and for the mobilisation of pRJ6. The second putative operon contains two open reading frames (ORFs); the first gene, orfA, may encode to a protein similar to small repressor proteins found in some Archaea. The second gene, orf B, codes for an 138 amino acid residue protein which shares a number of characteristics (high pI and hydrophobicity profile) with proteins associated with immunity. They found four other genes in the third operon, aurABCD. On analysis of purified bacteriocin by mass spectrometry it was found that all the four peptides encoded by aur ABCD operon were produced, expressed and excreted without post-translational modifications. Therefore, aureocin A70 is a multi-peptide non-lantibiotic bacteriocin which can be transported without processing. dos Santos Nascimento et al. (2002) isolated fifty strains of *Staphylococcus* spp. from bovine mastitis cases and were screened for antimicrobial substances. Out of these twelve strains had high antagonistic activity against the pathogen strain (*Corynebacterium fimi*) and were therefore further characterised. This was thought to be bacteriocins (Bac) as it was sensitive to proteolytic enzymes. The identification was done by the amplification of femA gene as *S. aureus*. Plasmid profile analysis showed the presence of a plasmid of a size similar to that of pRJ6 (8. 0kb), which encodes aureocin A70, a bacteriocin produced by the Brazilian *S. aureus* strain A70 which was isolated from commercial milk. PCR experiments, using specific primers for amplification of the bacteriocin operon found the presence of 525bp amplicon. With this it was inferred that the bacteriocin produced may be related to aureocin A70. The genomic DNA of all Bac (+) strains were then analysed by pulsed-field gel electrophoresis (PFGE) in order to investigate clonal relationships amongst the strains. Based on the results of PFGE experiments, 10 out of the 12 Bac (+) strains belonged to the same clone. dos Santos Nascimento (2005) found that 188 coagulase-negative *Staphylococcus* (CNS) strains were isolated from bovine mastitis cases from 56 different Brazilian dairy herds and were tested for antimicrobial substance production. Twelve CNS strains (6. 4%) exhibited antagonistic activity against a *Corynebacterium fimi*. Most antimicrobial substances were sensitive to proteolytic enzymes suggesting that they might be bacteriocins (Bac). Amongst the CNS producers, six were identified as *S. epidermidis*, two as *S. simulans*, two as *S. saprophyticus*, one as *S. hominis* and one as *S. arlettae*. At least one plasmid was also present in all the strains. The Bac(+) strains had very few antibiotic resistance phenotypes and some produced novel bacteriocins and antimicrobial peptides. They were able to inhibit *Listeria monocytogenes*, an important food-borne pathogen, and several strains of *Streptococcus agalactiae* associated with bovine mastitis, suggesting a potential use of these bacteriocins either in the prevention or in the treatment of streptococcal mastitis. Carson et al., (2017) studied the non-aureus staphylococci (NAS), the bacteria most commonly isolated from the bovine udder, potentially protect the udder against infection by major mastitis pathogens due to bacteriocin production. The bovine NAS isolates were studied against bovine *Staphylococcus aureus* and the positive strains were tested against a human methicillin-resistant *S. aureus* (MRSA) isolate using a cross-streaking method. The scientists found that strains like *S. capitis, S. chromogenes, S. epidermidis, S. pasteuri, S. saprophyticus, S. sciuri, S. simulans, S. warneri, and S. xylosus* had the ability to inhibit the growth of *S. aureus in-vitro*. Strains of *S. capitis, S. chromogenes, S. epidermidis, S. pasteuri, S. simulans,* and *S. xylosus,* could also inhibit MRSA. NAS from bovine mammary glands are a source of potential bacteriocins, with >21% being possible producers, represent the potential for future characterization and along with clinical applications. Mastitis disease is the major cause for the use of antibiotics on dairy farms. As the use of antibiotics is showing an increase in antibiotic resistance and regulations there is an increase in the search for natural bacteriocins (bacterially produced antimicrobial peptides). These help in the prevention of bacterial infections. dos Santos Nascimento et al., (2002) isolated fifty strains of *Staphylococcus* spp. from bovine mastitis cases in several herds from different Argentinian provinces. They screened for antimicrobial substances and found that twelve strains exhibited a high antagonistic activity against *Corynebacterium fimi*. These were found to be sensitive to proteolytic enzymes and were bacteriocins (Bac). These strains were identified as *S. aureus* by the amplification of the femA gene. Plasmid profile analysis of these strains revealed the presence of at least one plasmid. All the strains had 525bp amplicon which suggests that the bacteriocin produced may be related to aureocin A70. The genomic DNA showed the clonal relationships amongst strains. Based on the results of PFGE experiments, 10 out of the 12 Bac (+) strains were of the same clone.

Lin et al., (2018) worked on the prolonged vancomycin usage which might cause methicillin-resistant *Staphylococcus aureus* to become vancomycin-intermediate *S. aureus* (VISA) and heterogeneous VISA (hVISA). Many single-nucleotide variations were found. Out of this six were found to be novel and resulted in the amino acid changes. Some of the amino acid variations are not associated with vancomycin resistance. The other three amino acid variations (T278I in HtrA, K101E in Upps and Q692E in FmtC) were present in the majority (87.5%-93.8%) of hVISA and VISA isolates, but only in a small number (22.9%-25.7%) of VSSA isolates, which were associated with vancomycin resistance. Howden (2005) studied the vancomycin resistance in *Staphylococcus aureus* which has recently emerged as an important clinical problem with implications for laboratory detection and clinical management of patients infected with resistant strains. It has been found that low-level vancomycin resistance in the form of vancomycin-intermediate *S. aureus* (VISA) and heterogenous vancomycin-intermediate *S. aureus* (hVISA) are common. These strains have been found in patients with antibiotic exposure. Despite the difficulties in laboratory detection, there are increasing data linking VISA and hVISA for the failure of glycopeptide antimicrobial therapy. Aggressive surgical intervention and non-glycopeptide-based antimicrobial therapy appears to improve the conditions for patients which have got infected with the low-levels of vancomycin-resistant strains. It is important that the clinicians and diagnostic laboratories are aware of VISA and hVISA as a clinical problem, and consider aggressive surgical debridement and non-glycopeptide-based therapy where infections with such strains have been proved.

Ruef (2004) studied the strain of *Staphylococcus aureus* towards the resistance to glycopeptide antibiotics that have been considered rare in clinical infections. It was found that these infections were caused by *S. aureus* with high-level resistance to vancomycin (VRSA). This happens due to the insertion of the vanA gene, which gets transferred from enterococci with vancomycin resistance. Infections by *S. aureus* causing resistance to glycopeptides (VISA), or glycopeptide resistance (hVISA), have become very common. The genetic basis of this is not known. Certain genetic regulatory elements such as agr II are associated with reduced susceptibility of *S. aureus* to glycopeptides. Presently,some of the infections are treated using high doses of vancomycin. However, many times, other antibiotics like linezolid are used during surgery to treat infections. Haddad et al., (2018) studied the infection of *Staphylococcus aureus*, which causes many diseases, and is found in the skin and mucosal membranes. It is of major clinical importance as a nosocomial pathogen. Multidrug-resistant strains are very common and therefore the recent work has been oriented towards the novel strategies to combat multidrug-resistant pathogens. The enzymes from bacteriophages called as endolysins and peptidoglycan hydrolases have the ability to disrupt cell walls. These provide alternatives to the use of antibiotics with high host specificity and have the ability to either replace or be utilized in combination with antibiotics. This specifically treats infections caused by Gram-positive drug-resistant bacterial pathogens such as methicillin-resistant *S. aureus*. LysK is one of the best-characterized endolysins with activity against multiple Staphylococcal species. In order to enhance the antibacterial efficacy and applicability of endolysins scientists have constructed a recombinant endolysin derivative along with novel delivery strategies for various applications like their conjugation to nanoparticles. Briers and Lavigne (2015) studied the emergence and spread of antibiotic-resistant bacteria which has driven the search of novel classes of antibiotics active against bacterial infections particularly for Gram-negative pathogens. They found a novel class of antibacterials known as endolysins. These enzymes encoded by bacterial viruses hydrolyze the peptidoglycan layer efficiently. This results in a sudden osmotic lysis and cell death. Due to the outer membrane in

Gram-negative bacteria it gives a formidable barrier to the use of endolysins. However, the scientists have shown the development in the use of endolysins to kill Gram-negative species with a special focus on endolysin-engineered Artilysins. Lee et al., (2008) studied the genomic analysis of *Bifidobacteria*. Liliam et al., (2018) studied the application of bacteriophages. Lucy et al., (2006) looked into the biopreservation with bacteriocins. Lorena et al., (2018) could increase shelf life of pears after harvest with *Lactobacillus rhamnosus*.

Bacteriocins in packaging film

In the last decade active research has taken place to incorporate bacteriocins into packaging films in order to control food spoilage and pathogenic organisms. This helps in preventing microbial growth on food surface by direct contact of the package with the surface of foods especially the meats and cheeses. However, the antimicrobial packaging film should be in touch with the surface of the food. This helps the bacteriocin to diffuse to the food surface. This has been found to be better than dipping/spraying foods with bacteriocins. This happens as the antimicrobial activity could have been lost or reduced due to the inactivation of the bacteriocins by food components or dilution below active concentration due to migration into the foods (Appendini and Hotchkiss 2002). They have found two different methods which could be used in the preparation of packaging films with bacteriocins (Appendini and Hotchkiss 2002). In one case they have tried to incorporate bacteriocins directly into polymers like incorporation of nisin into biodegradable protein films (Padgett et al.,1998). In the other one, the heat-press and casting were used to incorporate nisin into films made from soy protein and corn zein. Both cast and heat-press films formed excellent films and inhibited the growth of *L. lantarum*. Compared to the heat-press films, the castfilms exhibited larger inhibitory zones when the same levels of nisin were incorporated. The inhibitory effect of nisin against *E. coli* was found to increase when incorporated with EDTA. Siragusa et al., (1999) incorporated nisin into a polyethylene-based plastic film that was used to vacuum-pack beef carcasses. Nisin retained the activity against *Lactobacillus helveticus* and *B. thermosphacta* inoculated in carcass surface tissue sections. *B. thermosphacta* populations from nisin-impregnated plastic-wrapped samples were significantly less than control (without nisin). Coma et al., (2001) incorporated nisin into edible cellulosic films made with hydroxypropyl methyl cellulose by adding nisin to the film-forming solution. Inhibitory effect could be demonstrated against *L. innocua* and *S. aureus*, but film additives such as stearic acid, used to improve the water vapor barrier properties of the film, significantly reduced inhibitory activity. Desorption from the film and diffusion into the food requires optimization from food to food for the effective functioning of nisin as a preservative agent in the packaged food. Bacteriocins can also be incorporated into packaging films so as to coat or adsorb bacteriocins into the polymer surfaces. Examples include nisin/methylcellulose coatings for polyethylene films and nisin coatings for poultry, adsorption of nisin on polyethylene, ethylene vinyl acetate, polypropylene, polyamide, polyester, acrylics, and polyvinyl chloride (Appendini and Hotchkiss, 2002). Bower et al., (1995) demonstrated that nisin adsorbed onto silanized silica surfaces inhibited the growth of *L. monocytogenes*. Nisin films were exposed to medium cotaining *L. monocytogenes* and the contacting surfaces were evaluated at 4-h intervals for 12 h. Cells on surfaces that had been in contact with a high concentration of nisin (40000 IU/ml) exhibited no signs of growth and many displayed evidence of cellular deterioration. Surfaces contacted with a lower concentration of nisin (4000 IU/ml) had a smaller degree of inhibition. When the surfaces were covered with films of heat-inactivated nisin then the growth of *L. monocytogenes* was observed. Scannell et al., (2000 a,b) utilized the cultures of *L. innocua* and *S. aureus* (with *L. lactis* subsp. lactis) along with cellulose-based bioactive inserts and antimicrobial polyethylene/polyamide pouches. Lacticin 3147 and nisin were tested as bacteriocins. The nisin bound film was found to be stable for 3 months with / without refrigeration. Bacterial reductions of upto $2 \times \log^{10}$ CFU/g cycles in vacuum-packed cheese were seen in combination with modified atmosphere packaging (MAP) with storage at refrigeration temperatures. Inserts with immobilized nisin reduced *L. innocua* (starting inocula of 2 to 4 x 10^5 CFU/g) by >3 log10 CFU/g in cheese after 5 d at 4 °C, and by approximately 1. 5 log10 CFU/g in sliced ham after 12 d, while *S. aureus* (starting inocula of 2 to 4 x 10^5 CFU/g) was reduced by 1.5 and 2.8 $\log^{10}$ CFU/g in cheese and ham, respectively. IFT (1998) have shown the migration of compounds into packaging systems. It has been found that the coating of pediocin on the cellulose casings and plastic bags inhibits the growth of inoculated *L. monocytogenes* in meats and poultry during a storage period of 12-wk at 4^0C (Ming et al.,1997). Coating of solutions containing nisin, citric acid, EDTA and Tween 80 onto polyvinyl chloride, linear low density polyethylene and nylon films reduced the counts of *Salmonella typhimurium* in fresh broiler drumstick skin by 0.4- to 2.1-log10 ^{6} cycles after incubation at 4 °C for 24 h (Natrajan and Sheldon ,2000). It controlled the growth of the pathogens and enhanced the shelf life of the food products. According to a communication (Anil et al., 2017) Rhodia, Inc., is developing a casing with a combination of bacteriocins, enzymes, and botanicals to be used in hot dog and other cooked meats for preservation for which it has obtained a regulatory clearance. It is found to be very cost effective. Vazquez and Kollef (2014) studied the treatment of Gram-positive infections in critically ill patients. Methicillin-resistant *Staphylococcus aureus* (MRSA), methicillin-susceptible *Staphylococcus aureus* (MSSA) and enterococci needs to be included as vancomycin-resistant enterococci (VRE). The resistance and virulence factors have a role in infections of the critically ill. Recently, infections with these pathogens have increased which are resistant to the available antimicrobial agents. A good understanding of the infections caused by Gram-positive bacteria is required. Blaskovich et al., (2018) have developed a glycopeptides antibiotic. Glycopeptide antibiotics (GPAs) are important to fight against drug resistant bacteria as they are the main therapy against Gram-positive infections. Dalbavancin and oritavancin can treat vancomycin-resistant infections. Many have been characterized having *in-vivo* efficacy as well as good PK/PD profiles. They do not have preclinical toxicity and have been found to be suitable for preclinical development. Rodriguez-Pardo et al., (2016) have studied the effectiveness of sequential intravenous-to-oral antibiotic switch therapy in patients which are hospitalized with gram-positive infection. Oral antibiotic therapy can reduce gram-positive infection. Isadore et al., (2010) have shown the role of 3rd parties in implementation of USDA.Ingrid et al., (2008) have worked on the production of bacteriocins from pathogens.

Application of edible films or coatings containing antimicrobial substances is a common strategy for preservation of foods that are eaten raw or without cooking. The incorporation of antimicrobial compounds such as bacteriocins in edible coatings and films helps in the control of pathogenic microorganisms in food products (Valdés et al., 2017). The films and coatings are composed of thin layers of biopolymers. This modifies the surrounding atmosphere of the foods and forms a barrier between the food and the environment. This improves the safety, quality and functionality of food products without changing the organoleptic and nutritional properties (Han, 2003; Valdés et al., 2017). These are also effective in the inhibition of the growth of pathogenic microorganisms throughout the extent of the latency phase (Balciunas et al., 2013). Among the coatings hydrocolloids (proteins and polysaccharides) are the most extensively investigated biopolymers in edible coatings and films applied to cheese. They facilitate the incorporation of functional compounds such as bacteriocins and bacteriocin-producing bacteria and allow an increase in the stability, safety and shelf-life of dairy foods (Scannell et al., 2000 a,b). Investigations to this effectiveness of incorporating bacteriocins and/or bacteriocin-producing LAB in coatings and films is being worked out by many scientists for dairy products (Cao-Hoang et al., 2010; da Silva Malheiros et al., 2010; Ercolini et al., 2010; Aguayo et al., 2016; Malheiros et al., 2016) or containing viable LAB in the film/coating matrix (Concha-Meyer et al., 2011; Barbosa et al., 2015).

Studies of the effectiveness of incorporating purified bacteriocins in edible coatings show a limited reduction of pathogens such as *L. monocytogenes*. Cheeses, particularly fresh cheeses, are highly perishable due to their high content in caseins, lipids, and water. The complexity of cheese composition and its manufacture support the development of pathogenic and deteriorating microorganisms that increase the risk of foodborne illness and reduce cheese quality and acceptability (Ramos et al., 2012). The application of edible coatings and films by the incorporation of bacteriocins may be helpful and may overcome problems associated with post-process contamination. This will ultimately, enhance the safety and may extend the shelf-life of the cheese.

In a study done by Cao-Hoang et al., (2010) in semi-soft cheese the counts of *L. innocua* could be reduced by the incorporation of nisin in the films of sodium caseinate. In another study by Scannell et al., (2000) the incorporation of nisin and lacticin in cellulose coatings was applied to Cheddar cheese. It showed reduced levels of *L. innocua* by 2 log cycles and *S. aureus* by 1. 5 log cycles. In Ricotta cheese coated with galactomannan and nisin, the growth of *L. monocytogenes* was prevented for 7 days at 4°C (Martin et al., 2012). The application of a coating in Port Salut cheese, consisting of tapioca starch combined with nisin and natamycin, reduced *L. innocua* counts above 10 cfu ml^{-1} during storage, acting as a barrier to post-process contamination (Resa et al., 2014). Recently, Marques et al., (2017) used a biodegradable film incorporated with cell-free supernatant (CFS) containing bacteriocin-like substances of *Lactobacillus curvatus* P99, to control the growth of *L. monocytogenes* in sliced "Prato" cheese. These films containing the bactericidal concentration of CFS were able to control *L. monocytogenes* for 10 days of storage at 4°C.

The purified and concentrated bacteriocins have been found to be more efficient as food additives. Presently, it is the preferred method than the direct application of bacteriocinogenic cultures. In food systems the effectiveness of bacteriocins is low as several factors like adsorption to food components, enzymatic degradation, poor solubility and uneven distribution is found in the food matrix. There is a difficulty in the application as it lacks the compatibility between the bacteriocin-producing strain and other cultures required in the fermentation of dairy foods.

Cheeses get contaminated with *L. monocytogenes* as they provide the appropriate growth conditions for it. Listeriosis out-breaks linked to the consumption of contaminated cheeses have been reported worldwide. More research needs to be done on the application of bacteriocins to the preservation of milk, cream, yogurts, and other dairy products.

A lot of efforts have been put, in order to discover novel bacteriocins with unique properties. However; more studies are required towards the effective application of bacteriocins in dairy foods in order to understand bacteriocin performance in the complex environment of food matrices. These can also be combined with other protection tools as part of hurdle technologies for proper bio-preservation and shelf-life extension of dairy foods. If combined with other preservation techniques, or if incorporated into biofilms and active packaging it may give better results. Further studies are necessary to ensure the adaptation of edible coatings and films to bacteriocin activity and efficacy in dairy as well as other products.

Around the globe there is a FAO/WHO International Food Safety Authorities Network (INFOSAN) which coordinates communication of the contaminated food distributed internationally. This helps for food recalls in importing countries. The alerts are sent to national INFOSAN Emergency contact points in each country for their attention and action.

Genesis of diseases

The study on patients with culture-confirmed methicillin-resistant gram-positive infection /methicillin-susceptible gram-positive infection and beta-lactam allergy. No differences were observed regarding the incidence of catheter-related bacteraemia. Cunha (1997) attempted a study in switching the hospitalized infectious-disease patients from intravenous to oral antibiotic therapy. This decreases the number of nosocomial infections along with shorter hospital stay, and also lowers the intravenous-line infections. Chang et al., (2017) have shown the use of oral therapy to be used which will be helpful to the patients. Ahlstrom et al., (2014) have studied the improved short-sequence-repeat genotyping of *Mycobacterium avium* subsp. *paratuberculosis* by using matrix-assisted laser desorption ionization-time of flight mass spectrometry. Accurate sequence analysis of mononucleotide repeat regions is difficult, complicating the use of short sequence repeats (SSRs) as a tool for bacterial strain discrimination. Although multiple SSR loci in the genome of *Mycobacterium avium* subsp. *paratuberculosis* allow genotyping of *M. avium* subsp. *paratuberculosis* isolates with high discriminatory power, further characterization of the most discriminatory loci is limited due to inherent difficulties in sequencing mononucleotide repeats. The workers evaluated a method using matrix-assisted laser desorption ionization-time of flight mass spectrometry (MALDI-TOF MS) as an alternative to Sanger sequencing to further differentiate the dominant mycobacterial interspersed repetitive-unit (MIRU)-variable-number tandem-repeat (VNTR) *M. avium* subsp. *paratuberculosis* type (n = 37) in Canadian dairy herds by targeting a highly discriminatory mononucleotide SSR locus. First, PCR-amplified DNA was digested with two restriction enzymes to yield a sufficiently small fragment containing the SSR locus. Second, MALDI-TOF MS was performed to identify the mass, and thus repeat length, of the target. Sufficiently intense, discriminating spectra were obtained to determine repeat lengths up to 15, an improvement over the limit of 11 using traditional sequencing techniques. Comparison to synthetic oligonucleotides and Sanger sequencing results confirmed a valid and reproducible assay that increased discrimination of the dominant *M. avium* subsp. *paratuberculosis* MIRU-VNTR type. Thus, MALDI-TOF MS was a reliable, fast, and automatable technique to accurately resolve *M. avium* subsp. *paratuberculosis* genotypes based on SSRs. Motiwala et al., (2006) studied the genetic diversity of *Mycobacterium avium* subsp. *Paratuberculosis*. Paratuberculosis is a chronic gastroenteritis mainly affecting ruminants like cattle, sheep and others. It is caused by *Mycobacterium avium subsp. paratuberculosis* (MAP) and is the etiological agent of Johne's disease (or paratuberculosis). MAP is also of concern in Crohn's disease in humans. The authors have studied the genesis of the disease in their work. Corbett et al., (2017) studied the faecal shedding and tissue infections transmission of *Mycobacterium avium* subsp. *paratuberculosis* in group-housed dairy calves. Current Johne's disease control programs are important as they decrease the transmission of *Mycobacterium avium* subsp. *paratuberculosis* (MAP) from infectious adult cows to susceptible calves. Transmission between calves of MAP infection in calves contact-exposed to infection was checked. It was found that new MAP infections occurred due to exposure of infectious penmates. Therefore, calf-to-calf transmission is a potential route of uncontrolled transmission on cattle farms. Nowotarska et al. (2017) studied the mechanisms of antimicrobial action of cinnamon and oregano oils, cinnamaldehyde, carvacrol, 2,5-dihydroxybenzaldehyde and 2-hydroxy-5-methoxybenzaldehyde against *Mycobacterium avium* subsp. *paratuberculosis* (Map) which infected food like milk, cheese and meat also, animals and humans. The incubation of Map cultures in the presence of all six compounds caused phosphate ions to leak into the extracellular environment in a time- and concentration-dependent manner. Cinnamon oil and cinnamaldehyde decreased the intracellular adenosine triphosphate (ATP) concentration of Map cells, whereas oregano oil and carvacrol caused an initial decrease of intracellular ATP concentration that was restored gradually after incubation at 37 °C for 2 h. The compounds tested caused a leakage of ATP into the extracellular environment. Monolayer studies involving a Langmuir trough apparatus revealed that all anti-Map compounds, especially the essential oil compounds, altered the molecular packing characteristics of phospholipid molecules of model membranes, causing fluidization.

Derakhshani et al., (2016) studied the diagnostic tests for Johne's disease (JD). It is a chronic granulomatous inflammation of the gastrointestinal tract of ruminants which is caused due to the presence of *Mycobacterium avium* subspecies *paratuberculosis* (MAP). The work was done in order to have a potential biomarker for early detection of infected cattle. Glutathione metabolism and leucine and isoleucine degradation pathways in MAP-infected calves have clarified the contribution of ileal MAM in the development of intestinal inflammation. Finally, simultaneous over representation of families Planococcaceae and Paraprevotellaceae, as well as under representation of genera *Faecalibacterium* and *Akkermansia* in the faecal microbiota of infected cattle, served as potential biomarker for identifying infected cattle during subclinical stages of JD.

Rabbi et al., (2017) studied the reactivation of intestinal inflammation suppression by catestatin in a murine model of colitis via M1 macrophages and not the gut microbiota. They found that an hCTS-treated mouse develops less severe acute colitis. The scientists have also studied the implication of the macrophages and the release of pro-inflammatory cytokines. Eissa et al., (2017) have studied the amelioration of the progression of colitis by regulating activated macrophages. Ulcerative colitis (UC) is characterized by a functional dysregulation of alternatively activated macrophage (AAM) and intestinal epithelial cells (IECs) homeostasis. The scientists worked on the Chromogranin-A (CHGA) secreted by neuroendocrine cells in intestinal inflammation and immune dysregulation. They found the effects of the CHR in dextran sulfate sodium (DSS) colitis in mice and also on macrophages and human colonic epithelial cells. It was found that CHR treatment reduces the severity of colitis and the inflammatory process due to enhancing AAM functions and maintaining IECs homeostasis. CHR is involved in the pathogenesis of inflammation in experimental colitis. These findings provide insight into the mechanisms of colonic inflammation and could lead to new therapeutic strategies for UC. Pedersen (2015) explained the term Ulcerative colitis (UC) and Crohn's disease (CD), collectively as inflammatory bowel disease (IBD). These are chronic immune disorders that affect the gastrointestinal tract. IBD gets enhanced if there is a disturbance in the balance between gut commensal bacteria and host response in the intestinal mucosa. Tumour necrosis factor (TNF)-β and interferon (IFN)-γ are pro-inflammatory cytokines, present in high amounts in colonic mucosa after inflammation. Using the model, both cytokines were found directly to impair the viability of colonic epithelial cells and to induce secretion of IL-8 *in vitro*. TNF-α and IFN-γ may also be involved in regulation of intestinal inflammation through stimulation of MMP expression and proteolytic activity. They found that colonic epithelial cells, like myofibroblasts and immune cells can cause intestinal mucosal damage. They found that colonic epithelium may regulate immune responses to microbial antigens including commensal bacterial DNA through modulation of the TLR9 pathway. Recently it has been found that PPARγ is also involved in modulation of inflammatory processes in the colon as studied in rodent colitis models. Topical application of rosiglitazone in patients with active distal UC reduced clinical activity and mucosal inflammation proving that PPARγ may be a new potential therapeutic target in the treatment of UC. It has been observed that intestinal epithelial cells participate actively in the immunological processes in the colonic mucosa. These studies help in the knowledge on the pathogenesis of human inflammatory colonic diseases. Ingrid et al., (2008) studied the bacteriocins from plant pathogenic bacteria. Bacteriocins are antimicrobial substances produced by many bacteria that are either nonribosomally synthesized antibiotics or ribosomally synthesized proteinaceous compounds having regulatory roles also. These bacteriocins do not have side effects. These help in the plant disease control. The antimicrobial compounds are important as they affect bacterial population dynamics, survival and virulence (Riley and Wertz, 2002a,b; Gardner and Buckling., 2004). Some antimicrobial compounds can have regulatory functions (Hauge et al., 1998; Eijsink et al., 2002; Kodani et al., 2005; Linares et al., 2006). They play a vital role in combatting bacterial infections and in plant disease control (Montesinos, 2007). Antimicrobials from bacteria are of two types. The bacteriocins (ribosomally synthesized proteinaceous substances; usually narrow spectrum) and the antibiotics (secondary metabolites; usually broader spectrum). They have a vital role in the food fermentation industry (Cotter et al., 2005). However, little is known on the bacteriocin production by bacterial plant pathogens.

Regulatory roles of Bacteriocins

Literature shows a link between bacteriocins and antibiotics to functions like signalling, virulence and sporulation (Kuipers et al., 1995; Hauge et al., 1998; Kodani et al., 2005; Linares et al., 2006). Linares (2006) studied the regulation of virulence in *Pseudomonas aeruginosa*. Many of the class I and class II peptide bacteriocins are produced by Gram-positive bacteria. They act as an auto-inducing peptide pheromone that stimulates bacteriocin production (Kuipers et al., 1995; Brurberg et al., 1997; Hauge et al., 1998; Eijsink et al., 2002). Kodani (2005) has shown a lanthionine produced by *Streptomyces tendae* is involved in aerial hyphae formation and has (limited) antimicrobial activity confirming the regulatory function of the bacteriocin. Lee (2008) found the avirulence gene product AvrXa21 produced by a number of *X. oryzae* pv. oryzae strains. According to Dirix et al., (2004) it has to be seen whether the peptide is a pheromone or a bacteriocin or is multifunctional. Alleson et al., (2012) have studied if bacteriocin production is a probiotic trait. Traditionally, bacteriocin production is associated in the selection of probiotic strains. Studies have demonstrated the impact of bacteriocin production of a strain on the effect of host health. It seems that bacteriocins may function in different ways in the gastrointestinal tract. They could directly inhibit the invasion of competing strains or pathogens, or modulate the composition of the microbiota and influence the host immune system. Probiotic strengthens the intestinal barrier, modulates the host immune system and produces antimicrobial substances. Many probiotic bacteria produce a variety of antimicrobial compounds (short-chain fatty acids, hydrogen peroxide, nitric oxide and bacteriocins (Havenaar, et al., 1992). These enhance their ability to compete against other GI microbes and could potentially inhibit pathogenic bacteria (Cintas et al.,1995). Bacteriocins are bacterially produced peptides that are active against pathogenic bacteria and producer of a specific immunity mechanism (Cotter, 2005). They constitute a heterogeneous group of peptides with respect to size, structure, mode of action, antimicrobial potency, immunity mechanisms and target cell receptors (Guaní-Guerra, 2010).

An ecological perspective of bacteriocin function:

Scientists have found that majority of all bacteria and archaea produce at least one bacteriocin (Klaenhammer,1988). The apparent ubiquity of this trait implies that bacteriocins play an important role, despite the associated energy costs imposed by their production (Gillor et al., 2009). However, their exact ecological function needs attention. It is possible that bacteriocins could contribute to probiotic functionality in a number of ways like colonizing peptides, facilitating the introduction and/or dominance of a producer into an already occupied niche (Riley et al., 2002a,b). They can act as antimicrobial or killing peptides, directly inhibiting competing strains or pathogens (Majeed et al., 2011). They can also function as signaling peptides which could be signaling other bacteria through quorum sensing and bacterial cross talk within microbial communities or signaling cells of the host immune system (Veldhoen et al., 2012).

The high cell density is in the GI tract which might result in close cell to cell contact between members of the same or different species. This will be able to promote cooperative and antagonistic microbial interactions (Fuqua , 2006). Gillor et al. (2008) demonstrated that *E. coli* producing the bacteriocin colicin was able to persist in the large intestine of streptomycin-treated mice for an extended period of time relative to their non-colicin-producing counterparts. Over time, the density of the colicin-producing strains was much higher. In a similar study, Hillman et al. (1987) noted a strong correlation between the ability of a *Streptococcus mutans* strain to colonize the oral cavity and the production of the bacteriocin mutacin (Hillman et al.,1987). Work has been done to show that the production of BlpMN bacteriocins by the *S. pneumoniae* type 6A strain contributes to the ability of this strain to colonize and compete in the mouse nasopharynx. BlpMN bacteriocin contributes to the competitiveness of the associated strain within the complex microbial environment of the nasopharynx (Eijsink et al., 2002).

A five-strain probiotic mixture composed of *Lactobacillus murinus* DPC6002 and DPC6003, *Lactobacullus pentosus* DPC6004, *Lactobacillus salivarius* DPC6005, and *Pediococcus pentosaceus* DPC6006 has been shown to improve the clinical and microbiological outcome of *Salmonella* infection in pigs when bacteriocin producer, *L. salivarius* DPC6005, was coadministered in both the ileum digesta and mucosa of weaned pigs. High rate of ileal survival of the probiotic strain could be attributed to bacteriocin production (Casey et al., 2007).

Recently it has been found that the bacteriocin-producing strain *Enterococcus faecium* KH24 affected the faecal microbiota of mice to a large extent and that the probiotic might be helping in controlling the indigenous microbiota in a beneficial manner (Bhardwaj et al., 2010). Bacteriocin production can contribute to microbial survival in the human GI tract. Using intestinal isolate *Bifidobacterium longum* subsp. *longum* DJO10A had a significantly greater ability to compete against the intestinal isolates *Clostridium difficile* DJOcd1 and *E. coli* DJOec1 than did the nonproducing isogenic variant *B. longum* DJO10-JH1 (Lee et al., 2008). These studies indicate that bacteriocin production is an important trait with regard to microbial competition in the human intestine. The ability of bacteriocin-producing microorganisms to inhibit pathogens *in-vitro* has been studied well (Gillor et al., 2009; Le Blay et al., 2007). *Lactococci* sp. which produces the broad-spectrum antimicrobial peptide lacticin 3147 does not provide protection against *Listeria monocytogenes* infection in a mouse model, despite the efficacy of the bacteriocin against this pathogen *in vitro* (Ryan et al., 1996). It has also been observed that pediocin PA-1 production by *Pediococcus acidilactici* UL5 reduces *L. monocytogenes* viability by approximately 3 logs *in-vitro*, however the results were not found *in-vivo* (Dabour et al., 2009). Work has been done to show that the increase in pathogen numbers lowers the intestinal pH. It appears that the production of lactic acid by *Pediococcus* spp. and *Lactobacillus* spp. induces virulence gene production in the pathogen (O'Driscoll et al., 1996). There are some studies which demonstrated the ability of bacteriocin producers to inhibit pathogens in the GI tract. Most notably, Corr et al. (2009) found that *L. salivarius* UCC118 provides protection against *L. monocytogenes* infection in mice. According to the workers, the inhibition of the pathogen was shown to be the direct result of the production of the two-peptide bacteriocin Abp118, as it was demonstrated that a non-bacteriocin-producing isogenic derivative failed to protect mice from infection. (Millette et al., 2008) have shown the bacteriocin-producing human isolates *P. acidilactici* MM33 and *L. lactis* MM19 to reduce the vancomycin-resistant enterococci (VRE) populations *in vivo*. *P. acidilactici* MM33 produces the bacteriocin pediocin PA-1/AcH, while L. lactis MM19 produces the bacteriocin nisin Z.

According to Nandane et al., (2007) bacteriocins are a heterogeneous group of substances which consists of an active protein moiety either alone or is in conjugated form. These save the food from spoilage and preserve foods through natural fermentations and have commercial applications in dairy, baking, meat, vegetables and alcoholic beverages. Numerous bacteriocin producing lactic acid bacteria which are inhibitory to *Listeria monocytogenes* have been isolated from fermented products and ready to eat foods as well as raw meat products. A lot of emphasis is being given on the identification and properties of these proteins that includes production condition, purification procedures and mode of action. Studies at the genetic engineering level have facilitated the characterization of proteins at the genetic level. This has helped in providing information about protein synthesis, structure and cloning of bacteriocin genes into other organisms.

Célia et al., (2018) have looked into the application of bacteriocins in the prevention of dairy products. The use of chemical preservatives in foods has become a matter of concern by the consumers. The dairy industry has a lot of demand to extend the shelf-life and to prevent spoilage of dairy products. Bacteriocins are antimicrobial peptides and are safe in dairy industry. These are applied to dairy foods either in the purified/crude form or as a bacteriocin-producing LAB as a part of fermentation process or as an adjuvant culture. These control pathogens in milk, yogurt and cheeses. Incorporation of bacteriocins, directly as purified or semi-purified form or incorporation of bacteriocin-producing LAB into bioactive films and coatings have become common. These can be applied directly onto the food surfaces or on packaging. Bacteriocins also inhibit other related or unrelated microorganisms (Leroy and De Vuyst, 2004; Cotter et al., 2005). Bacteriocins have the ability to kill bacteria that are taxonomically close and also of a broad spectrum (Cotter et al., 2005; Mills et al., 2011). They have an added advantage as they can be easily digested by the human gastrointestinal tract (Mills et al., 2011). Sometimes, the application of bacteriocins as food additives becomes limited due to the lack of effectiveness of pathogen elimination and its high price (Chen and Hoover, 2003).

A large number of bacteriocins have been isolated and identified from Gram-positive and Gram-negative microorganisms. All the information and the databases have been created and compiled for the automated screening of bacteriocin gene clusters (Blin et al., 2013; van Heel et al., 2013).

Mechanisms of action of bacteriocins can be divided into those that promote a bactericidal effect, with or without cell lysis, or bacteriostatic in which the cell growth is inhibited (da Silva Sabo et al., 2014). Usually, the bacteriocins produced from LAB, particularly those inhibiting Gram-positive bacteria target the cell envelope-associated mechanisms (Cotter et al., 2013). Several lantibiotics and some class II bacteriocins target Lipid II, an intermediate in the peptidoglycan biosynthesis machinery within the bacterial cell envelope and, by this way they inhibit peptidoglycan synthesis (Breukink and de Kruijff, 2006). Other bacteriocins use Lipid II as a docking molecule to facilitate pore formation resulting in variation of the cytoplasm membrane potential and ultimately, cell death (Machaidze and Seelig, 2003). Nisin, the most studied lantibiotic, is capable of both mechanisms (Cotter et al., 2005). Some bacteriocins damage or kill target cells by binding to the cell envelope-associated mannose phosphotransferase system (Man-PTS) and subsequent formation of pores in the cell membrane (Cotter et al., 2013). Other bacteriocins can kill their target cells by inhibition of gene expression (Parks et al., 2007; Vincent and Morero, 2009) and protein production (Metlitskaya et al., 2006). Dawid et al.,(2007) isolated bacteriocin from *Streptococcus pneumoniae*.

Application of Purified/Semi-Purified Bacteriocins in Dairy Products

In the biopreservation of various foods, either alone or in combination with other methods of preservation bacteriocins has been used which is known as hurdle technology (De Vuyst and Leroy, 2007; Perez et al., 2014). The application of bacteriocins into foods needs to be tested to confirm their effectiveness. Bacteriocin-producing strains into foods mainly include meat, dairy products, fish, alcoholic beverages, salads, and fermented vegetables (O'Sullivan et al., 2003; Ramu et al., 2015). To date, only nisin (Nisaplin, Danisco) and pediocin PA1 (Microgard™, ALTA 2431, Quest) have been commercialized as food preservatives (Simha et al., 2012). The enterocin AS-48 (Sánchez-Hidalgo et al., 2011) or lacticin 3147 (Suda et al., 2012) have a lot of potential but are not proposed for industrial application. When applied to food the producing strains should be food grade (GRAS or QPS). They should exhibit a broad spectrum of inhibition, present high specific activity, have no associated health risks, present beneficial effects (e. g., improve safety, quality and flavor of foods), display heat and pH stability and optimal solubility and stability for a particular food (Cotter et al., 2005; Leroy and De Vuyst, 2010). The type of food matrix plays an important role in the inactivation of several food borne pathogens (Muñoz et al., 2007). It is therefore important to know the effectiveness of different bacteriocins to foodborne pathogens in all food systems. Mostly, the bacteriocins get adsorbed into food matrices which get degraded and results in a loss of antibacterial activity. An alternative method is the incorporation of bacteriocins into food packaging films/coatings has been found effective which improves their activity and stability in complex food systems (Salgado et al., 2015). Use of immobilized bacteriocins is common in the development of antimicrobial packaging films in order to control food-borne pathogenic bacteria, such as *L. monocytogenes* (Sánchez-González et al., 2013; Ibarguren et al., 2015; Narsaiah et al., 2015). In fermented foods, the contamination with *L. monocytogenes* is a major concern (Melo et al., 2015). Ready to use products like different varieties of cheese are conserved at refrigeration temperatures that allow the survival and growth of psychrotrophic bacteria *L. monocytogenes*. *Listeria* active bacteriocins emerge as an ideal solution for preventing the growth of this pathogen either after cooking or packaging (Cotter et al., 2005). These are also used to control non-starter lactic acid bacteria (NSLAB) in cheese and wine (Oumer et al., 2001; O'Sullivan et al., 2003). Bacteriocins also help in cheese ripening and to improve its flavor (Oumer et al., 2001). Bonade et al., (2011) partially characterized bacteriocin from *Lactobacillus helveticus*.

Purchasing safe food in the market

The precautions should be taken while purchasing vegetables and fruits. It has been found that most of the fresh vegetables and fruits retain their freshness for a short time under ideal conditions of storage. Therefore, the consumer usually purchases them depending on the harvesting time. It is advised that while buying vegetables and fruits the consumer must check for the freshness, firmness, crisp, bright in colour and without any visible bruises or signs of decay and wilting. Buying vegetables and fruits of the season helps in better quality to the consumer. In case of buying fish it is advisable to check for a firm flesh, a stiff body and tight scales. While pressing the body no indentation should be left. It should be fresh, which have red gills and bright eyes, fish which have been refrigerated or stored on ice and are not slippery or slimy to touch. There are many chemicals used as colourants or preservatives in fish and are associated with health risks. It has been found that the red colour indicates the carbon monoxide treatment to the fish. This makes the quality of the fish inferior. It is not permitted in countries like Singapore, Canada, the EU and Japan. However, it is common in some Asian countries. Some use formalin solution to preserve for a longer time. However, this gives a special smell. The use of formalin to preserve food is illegal as it comes under the category of a potent poison. Such treated fish with formalin are not fit for consumption. Seafood and fish usually get contaminated with pathogens like *Vibrio cholera, Salmonella, E. coli, Shigella and Listeria* due to human activity or poor hygienic conditions during food production and processing. Many outbreaks of foodborne illnesses and infections have occurred due to the consumption of contaminated fish and seafood. In an aquatic environment dumping of industrial mercury as well as natural sources of elemental mercury produces methyl mercury due to bacterial action. Recently, most of the importing countries demand the testing of fish against methyl mercury. Many a times some parasite larvae are present in fish. Therefore the cooking must be thorough.

Importance of food labelling

It is vital to label the food material properly which involves legal, technical and administrative reasons. Food labelling is the outside label in the written, printed or graphic matter present on the product outside. It also includes the function of the product after use. Labelling is different in each country depending on their national food control system and the existing legislation. The label should usually have; the name of the food, list of ingredients,net contents and dry weight, name and address of the company, country of origin, batch number, identification code, date marking, storage instructions, temperature of storage, nutritional value and shelf life.

Looking into the importance of food labelling for international trade, the Codex Alimentarius Commission has made standards and recommendations commodity specific that must be followed for food labelling as it is important for the protection of consumer interest. Labels on food products helps in checking the freshness of the food as can be checked with the expiry dates. It helps in 'food recall' or 'food withdrawal'.

Expiry date or use-by Date' of a food product is the end of the estimated period. After the given time the products do not have the quality attributes for consumption. Till the expiry date the food product maintains the microbiological, physical stability and the nutrient content declared on the label. This assures the consumer to use the food before the expiry date to get the most nutritional value from it and to be safe. The first things a consumer must do before purchasing any food product is to check the "best before" and "use-by date" labels on foods. It is compulsory in countries of the South-East Asia Region where food safety law enforcement is weak and the understanding of food safety issues for consumer's is not explained to a large extent. It has been observed that many super markets reduce the prices of food products that are near to their "expiry date". Therefore, while purchasing special attention should be paid before purchasing such items.

Generally, food producers and processors comply with regulatory food standards and recommended food safety practices. Compliance of food safety aspects depends on the rules any company sets. The senior management should be able to absorb the additional costs (FAO/WHO, 2008). The possibility of being sued for foodborne illness is low and if sued the damages are very low. However, most food companies comply with the regulations and implement additional measures to ensure for product safety. Presently, the progress of the industry's food safety programs involves monitoring the occurrence of foodborne diseases in the population primarily through the identification and reporting by clinicians. Additionally, science-based performance and microbiological standards for levels of pathogen contamination are applied under the assumption that if products are not meeting these production standards they should be removed from the market, so that the incidence of foodborne disease will decrease. These practices have reduced the number of outbreaks and illnesses for *Escherichia coli* O157:H7 by 50% during the past 15 years (WHO, 2019). *Campylobacter* infections have continued to increase after a modest decline in the late 1990s (Doyle, 2015), and *Salmonella* infections have neither substantially increased nor declined (Barat et al., 2012).

Current status of food safety initiatives that will impact the food industry

Foodborne illnesses are usually found to be infectious and toxic. These are mainly caused by bacteria, viruses, parasites or chemical substances which enter the body through contaminated food or water. Foodborne pathogens can cause severe diarrhoea or debilitating infections including meningitis. The food safety system requires a proactive approach to predict where problems might arise so that the detection is not to be done after they have occurred (Buzby, 2001). The food industries and regulatory authorities have adopted a food safety system focused on science-based prevention of food safety problems and on products and foodborne pathogens associated with a large number of outbreaks (Tauxe et al., 2010) like *E. coli* O157:H7, *Salmonella* and *Listeria monocytogenes* in leafy greens, sprouts, dry foods, and ground meat and poultry products. The Food Safety Modernization Act (FSMA) was signed and made into a law in January, 2011. However, the implementation is still awaited as specific regulations need to be adopted. When fully implemented, the food industry will be required to apply science- and risk-based preventive measures at all appropriate points across the farm-to-table spectrum to ensure the safety of foods. This new food safety system will have some links during its early stages of implementation. However, in long term, improved food safety system should lead to a safer food supply and reduced burden of foodborne illnesses. Even after the implementation of FSMA some level of risk of foodborne illnesses remains (Alexandra, 2013). Quantitative microbial risk assessment (QMRA) can help in defining the level of acceptable risk and the associated performance and microbiological standards. Using QMRA, the actual risk to public health (based on surveillance data) is related to the levels of a microbiological hazard ingested through food at consumption, and those levels are in turn dependent on the initial contamination levels and modifying influences during processing and distribution (FAO/WHO, 2003). The World Health Organization/Food and Agricultural Organization of the United Nations QMRA models for *L. monocytogenes* in ready-to-eat food have revealed that most cases of listeriosis are associated with consumption of food contaminated with higher *L. monocytogenes* numbers than the current standards of zero tolerance or 100 colony-forming units per gram (USFDA/FSIS [U. S. Food and Drug Administration/Food Safety and Inspection Agency (USDA; 2001). Many a times the data needed for QMRA are not available in the correct format. There is a need for epidemiologic studies and surveillance programs to fill the gap where data for these models are not available, and to provide independent assessment of the sources of illnesses. As new information becomes available (the ecology of foodborne pathogens, the efficiency of alternative pathogen control approaches, the major sources of illnesses), microbiological and performance standards should be changed to align them with the state-of-the-science knowledge of the food system. Collection of environmental and finished product testing data should be continued and possibly enhanced to identify weaknesses that may occur post-FSMA implementation. Assessment of illnesses, hospitalizations and deaths requires robust human surveillance systems for sporadic illnesses and outbreaks that supply data needed to estimate the incidence of illness caused by each pathogen, and the exposures that result from those infections.

The authors Cláudia et al., (2013) have evaluated flies as a vector for foodborne pathogens. Many flies were collected from different rural areas and were analyzed for the presence of *Enterobacteriaceae, Escherichia coli, Staphylococcus coagulase* and *Listeria monocytogenes.* They were also studied for the virulence factors in the collected pathogens. They also studied the resistance to antibiotics and the presence of enterotoxigenic *S. aureus.* The results indicated that the flies in the presence of animals had significantly higher number of pathogens than those in the kitchens. *Enterobacteriaceae* was the indicator organism with the highest microbial counts followed by *E. coli* and *S. aureus.* The workers did not find *Listeria monocytogenes* from the collected flies. The antimicrobial susceptibility test showed that the bacteria carried by the flies possessed multiantibiotic resistance profiles and enterotoxin A was produced by 17.9% of the confirmed *S. aureus* isolates. Therefore, it was found that flies could transmit foodborne pathogens and their associated toxin and resistance and the areas of higher risk are those in closer proximity to animal production sites. They are an important vector of human diseases around the globe (Olsen,1998 and Jones et al., 2008) as they move and are able to fly long distances. Flies are found mostly on decaying organic materials and where food is prepared and stored. They transport microbial pathogens from reservoirs (animal manure) where they present a minimal hazard to people in places where they pose a great risk (Olsen,1998). Many of them are blood sucking insects and important pests of domestic animals. Especially in food they cause a lot of economic losses in the animal industry (Campbell et al., 2001), they also play a role in the ecology of various bacteria (Rochon et al., 2005). Mainly insects carry foodborne pathogenic bacteria as *Escherichia coli, Salmonella* spp.,

Campylobacter and Shigella spp. (Gil et al., 2004; Shane et al., 1985). They are capable of transmitting *E. coli* O157:H7 to cattle (Ahmad et al., 2007). Fruit flies and vinegar fruit fly are potentially competent vectors for *E. coli* O157:H7 and contaminate fruits with the pathogen (Sela et al., 2005). It was Cardozo et al., (2009) who identified *E. coli* and *Salmonella* spp. from houseflies on milk and manufacturing of handmade Minas cheese. There is also a concern that flies may also contribute to the spread of avian influenza (Sawabe et al., 2006). Blunt et al., (2011) worked on the potential of the dominant *Musca domestica* flies to act as vectors for circovirus and many of the nonenveloped viruses. Faulde and Spiesberger (2013) showed moth fly to be a vector of bacterial pathogens associated with nosocomial infections. Cláudia et al., (2013) studied flies as a vector for foodborne pathogens. The results showed that flies in the presence of animals had higher prevalence of the pathogens than those collected in the kitchens. Highest microbial counts were of *Enterobacteriaceae* followed by *E. coli* and *S. aureus*. *Listeria monocytogenes* was not detected from any of the collected flies. The bacteria carried by the flies had resistance to multiantibiotic. Climate change can affect the microbiological safety of the US food supply and change environmental conditions in which pathogens survive (Faulde and Spiesberger, 2013; Anonymous et al., 2008b). For the safety of imported foods it needs a different strategy as compared to that for domestic foods. As the hygienic criteria in the United States are very high the export of food by other countries to the United States needs attention (ISO 16649-2,(2001) and ISO 11290-2, (1998). Faecal wastes serve as vehicles of a variety of foodborne pathogens. Many a times the production and processing establishments in developing countries lack hygienic controls which include cleaning and sanitization of equipment and quality assurance management systems (CLSI–Clinical and Laboratory Standards Institute, 2007). These safety issues include almost 114 000 foreign food processing facilities (Zhang et al., 2004) which are registered to export in USA (Løvseth et al., 2004). This has led towards formation of FSMA rules. It requires importers to verify that preventive control systems and produce safety standards have been applied to food brought into the United States, and the food is not adulterated or mislabeled (Løvseth et al., 2004). Educating consumers of their needs in handling and storing of food is very important. Another emerging issue is for the aging and immune deficient population that affects the number of outbreaks and severity of foodborne illnesses. It has been found out that the elderly are more infected by *Salmonella* or *L. monocytogenes* (Urban and Broce,(1998) and Alam and Zurek (2004). It is noted that increase in foodborne outbreaks and illnesses will be present inspite of efforts by the food industry in order to reduce foodborne pathogens in their products. Looking into this the hospitals are providing "extra-safe" food products, which are irradiated, sterilized, or pasteurized foods. Describing the problem and approaches to minimize it will be helpful if the educational tools are made available to the common man. As today the lifestyles of consumers are more diverse and both spouses go out for work the demand for ethnic foods and semiprepared ingredients has increased. The products like unpasteurized milk and cheeses made from unpasteurized milk as well as minimally processed natural foods with no or reduced levels of antimicrobial additives are risky and should be avoided (Graham et al., 2009; Davari et al., 2010 and Macovei and Zurek, 2006). There is a need for alternative interventions to be developed that will mitigate the risk.

Recently, with food recalls due to infections the safety awareness of foods has grown among the consumers. A lot of campaigns are being conducted. However, studies show that it does not have much effect on reducing the actual level of bacteria in a meal or risk on illness (Macovei and Zurek, 2007). Many a times the consumers are unable to understand the labels given for safety. If the consumers use insufficient temperatures or cooking time a warning in this regard could be displayed. Ensuring food safety of products is expensive and has to be borne by the consumer. The role of food safety and the risks of foodborne pathogens should be brought to the public for better health.

Biopreservation of Dairy Foods

One of the most potent methods to reduce the food pathogens is the use of bacteriocins. Food can be inoculated with LAB,a purified or semi-purified bacteriocins can be added to food as preservative, usage of a product previously fermented with a bacteriocin producing strain as an ingredient in food processing. Biopreservation of dairy products against *L. monocytogenes* (Fleming et al., 1985). Bacteriocin nisin has been found to be very effective against *L. monocytogenes* in dairy products. Ferreira and Lund (1996) have shown decrease in the number of L. *monocytogenes* if inoculated with nisin-resistant strain into cottage cheese. Davies et al.,(1997) determined the efficacy of nisin to control *L. monocytogenes* in ricotta type cheeses over 70 d at 6 to 8 °C. However, the control cheese had unsafe levels of the organism even after 1 to 2 wk of storage. Zottola et al., (1994) used nisin produced from *lactococci* into cheddar cheese. The shelf life of the nisin containing pasteurized process cheese (301 and 387 IU nisin/g) was much more as compared to the control cheese spread. In cold pack cheese spreads, nisin (100 and 300 IU/g) significantly reduced the numbers of *L. monocytogenes, S. aureus*, and heat-shocked spores of *C. sporogenes*. Another major pathogen is the *Clostridium* sp. associated butyric acid fermentation in cheese. Nisin is commonly added to pasteurize processed cheese spreads. This helps to prevent the outgrowth of clostridia spores, as *Clostridium tyrobutyricum* (Schillinger et al., 1996). Lacticin 3147 a bacteriocin produced by *L. lactis subsp. lactis* DPC 3147 is used to control the quality of cheddar cheese by reducing non-starter LAB populations during ripening (Ross et al., 1999). With the lacticin 3147 a transconjugant of *L. lactis* DPC4275 is produced. In cottage cheese the population of *L. monocytogenes* was reduced by 3-log10cycles over a 7 days ripening period when it was manufactured with *L. lactis* DPC4275; however, the number of *Listeria* in the control cheese, manufactured with a non-lacticin 3147-producing starter, remained unchanged (10^4 CFU/g). The lacticin 3147-producing transconjugant has also been used as a protective culture to inhibit *Listeria* on the surface of a mold-ripened cheese. Presence of the lacticin 3147 producer on the cheese surface reduced the number of *L. monocytogenes* by 3-log10 cycles (Ross et al. 1999).

Biopreservation of meat products

Characteristically, *L. monocytogenes* is a gram-positive; non-sporeforming facultative anaerobic rod widely distributed in the natural environment. It can grow over a pH range of 4. 1 to 9. 6 and a temperature range of 0 to 45°C. It is resistant to desiccation and can grow at aw values as low as 0. 90. As *L. monocytogenes* is hardy and has the ability to grow even at refrigeration temperatures and is anaerobic in nature it is a threat to the safety of foods. It is a big threat to the food products as it can cause serious illnesses and death. In the USA the government has set a zero tolerance level for *L. monocytogenes* in ready to-eat foods (Jay 1996 b). According to Jay, (1996 a) it has been the cause of many foodborne outbreaks. They also found that it is very common in slaughter house and meat-packing environments and has been isolated from raw meat and cooked and ready-to-eat meat products. Degnan et al., (1992) have shown the use of bacteriocinogenic cultures of *Pediococcus acidilactici* (pediocin AcH producer) to control the growth of *L. monocytogenes* in vacuum packaged all-beef wieners. They also found out that the bacteriocin production is in the early logarithmic phase of *pediococci* which continues upto the late logarithmic phase. In curing meat, the usage of nitrite is made in order to stabilize red meat color and to inhibit poisoning organism *C. botulinum*. However, nitrite can react with secondary amines in meats to form carcinogenic nitrosamines. Using bacteriocins has a potential as an alternative to nitrite. Rayman et al., (1981) reported that a combination of 3000 IU/g of nisin and 40 ppm of nitrite inhibited the growth of *Clostridium sporogenes* spores in meat slurries at 37 °C for 56 d. In a study (Rayman et al., 1983) it was found that up to 22000 IU/g of nisin in combination with 60 ppm of nitrite could not inhibit the growth of *C. botulinum* spores in meat slurries at pH 5. 8. If the pH was reduced it helped in increasing the activity of nisin. They found that (8000 IU/g) nisin in combination with (60 ppm) nitrite inhibited growth of the *C. botulinum* spores in pork slurries at pH 5. 5. Taylor et al., (1987) showed that combinations of nisin and nitrite could delay botulinal toxin formation in chicken (pH 5. 9 to 6. 2). It was also found out that the control samples (without preservatives) were toxic within 1wk; where as a combination of 4000 IU/g of nisin along with 120 ppm of nitrite could help in delaying the toxin formation upto 5 weeks.

Hurdle technology to enhance food safety

The bacteriocins have limitations to be used in foods due to their narrow activity spectra, antibacterial effects and are inactive against gram negativebacteria. To overcome these limitations, more and more researchers use the concept of hurdle technology to improve shelf life and enhance food safety. Gram-negative bacteria have a tendency to become sensitive to bacteriocins if the permeability barrier properties of their outer membrane are not perfect. According to Abee et al., (1995) chelating agents, like EDTA, can bind magnesium ions from the lipopolysaccharide layer and disrupt the outer membrane of gram-negative bacteria, thus allowing nisin to gain access to the cytoplasmic membrane. Nisin enhances the thermal inactivation of bacteria. This reduces the time of the treatment with better food qualities. Boziaris et al., (1998) found that the time of pasteurization could be reduced up to 35% with the addition of nisin(500 to 2500 IU/ml) in media, liquid whole egg or egg white. According to Budu-Amoako et al., (1999) nisin could reduce the heat resistance of *L. monocytogenes* in lobster meat and could help in cost savings. The synergistic effect between bacteriocins and other processing technologies on the inactivation of microorganisms has been reported by Schlyter et al., (1993). They have shown the synergistic effects between sodium diacetate and pediocin against *L. monocytogenes* in meat slurries. However, it was found that in the control samples, counts of *L. monocytogenes* increased from 4. 5 $\log^{10}$ to approximately 8 $\log^{10}$ CFU/ml within 1d at 25 °C and within 14 d at 4 °C. Zapico et al., (1998) have found out the synergistic effect of nisin (10 or 100 IU/ml) and the lactoperoxidase system to inactivate *L. monocytogenes* in skim milk. Addition of nisin and lactoperoxidase together reduced the counts of L. *monocytogenes* which was not observed with nisin alone. Combinations of bacteriocins have been found to increase the antibacterial activity (Hanlin et al., 1993; Mulet-Powell et al., 1998). Leucocin F10 along with nisin gives greater activity against *L. monocytogenes* (Parente et al. 1998). Use of nonthermal processing technologies as high hydrostatic pressure (HP) and pulsed electric field (PEF) are common in food preservation. Bacterial inactivation can be increased if bacteriocins are used along with these processing techniques. Gram-negative bacteria are usually insensitive to *LAB* bacteriocins, as has been observed in *E. coli* O157:H7 and *S. typhimurium*. However, they become sensitive following HP/PEF treatments that induce sublethal injury to bacterial cells (Kalchayanand et al. 1992). Roberts and Hoover 1996 and Stewart et al., (2000) have demonstrated that nisin enhances the pressure inactivation of spores of *Bacillus coagulans, Bacillus subtilis* and *C. sporogenes.*

According to Settanni and Corsetti (2008) bacteriocins are generally recognized as "natural" compounds able to influence the safety and quality of foods. Recently, much work has been done towards the detection, purification, characterisation of bacteriocins and in food preservation. Bacteriocins from lactic acid bacteria (LAB) can promote the microbial stability both for fermented and non-fermented vegetable food products using bacteriocinogenic strains as starter cultures, protective cultures or co-cultures and the employment of pure bacteriocins as food additives. According to Settanni and Corsetti (2008) bacteriocins from lactic acid bacteria (LAB) help in the microbial stability of both fermented and non-fermented vegetable food products. They utilize bacteriocinogenic strains as starter cultures, protective cultures or co-cultures and the employment of pure bacteriocins as food additives.

As it is possible that the fresh produce products can be contaminated with human pathogenic bacteria from different sources (manure, irrigation water, insects and during harvesting and other process operations) and result in a number of outbreaks (Lynch et al. 2009). Scientists have found out (Galvez et al. 2008; Abriouel et al. 2011) that many of the bacteriocin preparations (nisin, pediocin and enterocin AS-48) inactivate the foodborne pathogenic bacteria (*Listeria monocytogenes, Bacillus cereus and Bacillus weihenstephanensis, Escherichia coli, Salmonella* and other enterobacteria). Bacteriocin treatments are also helpful to decontaminate whole fruit surfaces and to avoid transmission of pathogenic bacteria from fruit surfaces to processed fruits (Ukuku et al. 2005; Silveira et al. 2008).

Demand for ready-to-eat (RTE) and minimally processed foods have increased to a large extent in the last decade. Consumers are more aware of the risks derived from foodborne pathogens and from artificial chemical preservatives which are used to control pathogens (Parada et al., 2007; Rodríguez, et al., 2003; Schuenzel and Harrison, 2002). This has brought many challenges (Settanni and Corsetti, 2008). In olden times the undesirable microorganisms were kept under control by simple methods like drying, salting, heating or fermentation. However, these classic preservation techniques are not suitable for fresh meats and RTE products (Gram et al., 2002; Quintavalla and Vicini, 2002; Rao et al., 2005). According to Singh (2018) it is the fermentation products along with beneficial bacteria which help in controlling the spoilage by making the pathogen inactive. The lactic acid bacteria (LAB) and their metabolites are capable to exhibit antimicrobial properties and helpful in imparting unique flavour and texture to the food products. The major compounds produced by LAB are bacteriocin, organic acids and hydrogen peroxide. Bacteriocin is a peptide with antimicrobial activity. As they are non-toxic, non-immunogenic, thermo-resistant and have broad bactericidal activity they act as good bio-preservative agents. Among these nisin has wider applications in food industry and has been approved by the Food and Drug Administration (FDA). It can be used in vegetables products, dairy and meat industries. Apart from LAB metabolites, bacteriophages and endolysins are found to be effective in food processing, preservation and foodsafety. Bacteriocins and endolysins have the property for DNA shuffling and protein engineering in order to generate highly potent variants with expanded activity spectrum. In future it is envisaged that the use of genetically modified bacteriophages may be helpful in bio-preservation.

Leyva et al., (2017) have written a good review article on microbial agents which could protect food items against fungi. For the food industry a major issue is spoilage. This leads to food waste, economic loss for manufacturers and consumers. Fungal contamination is found at different stages of the food chain (post-harvest, processing or storage). Fungal development spoils the product not only visually but also the odor, flavour and texture gets changed. It also gives negative health impacts via mycotoxin production by many molds. To overcome this, different treatments like that of fungicides and chemical preservatives are used. Public health authorities want the food industries not to use these chemical compounds and instead develop natural methods for food preservation. The consumers are looking for more natural, less severely processed and safer products. In this context, microbial agents corresponding to bioprotective cultures, fermentates, culture-free supernatant or purified molecules, exhibiting antifungal activities represent a growing interest as an alternative to chemical preservation. Sundh and Melin (2011) have studied the potential of utilising antagonistic yeasts like *Pichia anomala* and *Candida* spp., for post-harvest biological control of spoilage fungi. Many yeast species are safe. However, some cause opportunistic infections in humans. These include *P. anomala* and bakers' yeast, *Saccharomyces cerevisiae*. Scientists are studying the pathogenicity and virulence factors in opportunistic yeasts and their prevalence in yeast infections. Usage of yeasts has to go from regulations that are used in post-harvest biocontrol. It is complex and the products in the market are registered as biological pesticides. Recently, the European Food Safety Authority has launched by the qualified presumption of safety (QPS) approach for the safety assessments of microorganisms that are added to food or feed. *P. anomala* is one of several yeast species that have been given QPS status, although the status is restricted to use of this yeast for enzyme and metabolite production purposes. Chopra et al., (2015) studied the biofilms which cause infections along with antimicrobial resistance. In biofilm the bacteria are many folds more resistant to antibiotics than the same bacteria circulating in a planktonic state. The authors found sonorensin which comes under the subfamily of bacteriocins to be effectively killing active and non-multiplying cells of both Gram-positive and Gram-negative bacteria. It inhibited biofilm formation against *Staphylococcus aureus*. With fluorescence and electron microscopy the authors found that growth is inhibited due to the increased membrane permeability. When a low density polyethylene (LDP) film was coated with sonorensin it was found to control the growth of food spoilage bacteria like *Listeria monocytogenes* and *S. aureus* to a large extent. It has been shown in chicken meat and tomato samples. Rai et al., (2016) have shown an enhancement of shelf life in foods by using antimicrobial peptides (AMPs) present in animals, plants, insects and bacteria. These are peptides of natural proteins and are responsible for defense of host from pathogenic organisms. For the synthesis of antimicrobial peptides various methods are utilized which include chemical, enzymatic and recombinant techniques. These can be an alternative to the chemical preservatives. Presently, nisin is the only antimicrobial peptide, which is widely utilized in the preservation of food. However, antimicrobial peptides can be either used by itself or in combination with other antimicrobial, essential oils and polymeric nanoparticles which increases the shelf-life of the food. Authors have studied the mode of action and application in food preservation. Nithya et al., (2018) have studied the bio-preservative efficacy of a partially purified antibacterial peptide (ppABP) produced by *Bacillus licheniformis* Me1. The work has been done in an economical medium which is developed using agro-industry waste. It was evaluated by direct application in milk and milk-based food products. The addition of ppABP in milk samples stored at 4°C and 28°C resulted in the growth inhibition of pathogens like *Listeria monocytogenes* Scott A, *Micrococcus luteus* ATCC 9341, and *Staphylococcus aureus* FRI 722. The shelf life of milk with added ppABP was found to increase. Curdling and off-odor were found in control samples which were without ppABP. A milk sample with ppABP was found to be acceptable sensorily and had antilisterial effect and reduces the risk of food-borne diseases.

For biopreservation the packaging system is critical. A container system should have closure integrity, sample stability and ready access to the preserved material. In order to ensure integrity of the food product during processing, storage and distribution it must remain stable over long periods of time as many biobanked samples may be stored indefinitely. Closed access systems must be used to avoid contamination. Woods and Thirumala (2011) have worked on a new commercially available container with a blood bag style closure and access system with a hermetical seal. This remains stable through cryopreservation and biobanking procedures. The authors have studied the cell seal system for the durability, closure integrity through transportation and maintenance of functional viability of a cryopreserved mesenchymal stem cell model. They have found that the Cell Seal vials are highly suitable for biopreservation, biobanking and provide a suitable container system for clinical and commercial cell therapy products which can be frozen in small volumes. GÃmez-Sala et al.,(2016) have worked on the development of a biopreservation strategy for fresh fish based on the use of bacteriocinogenic LAB of marine origin. They have utilized two multibacteriocinogenic LAB strains, *Lactobacillus curvatus* BCS35 and *Enterococcus faecium* BNM58. These culture strains were isolated from fish and fish products. The strains were selected depending on their capability to inhibit the growth of several fish-spoilage and food-borne pathogenic bacteria. The work was done on two commercially important fish species namely young hake (*Merluccius merluccius*) and megrim (*Lepidorhombus boscii*). The biopreservation potential and the application strategies of these two LAB strains were first tested at a laboratory scale, where several batches of fresh fish were inoculated with: (i) the multibacteriocinogenic LAB culture(s) as protective culture(s); and/or (ii) their cell-free culture supernatant(s) as food ingredient(s), and (iii) the lyophilized bacteriocin preparation(s) as lyophilized food ingredient(s). These were stored in polystyrene boxes, filled with ice at 0-2°C, for 14 days. Microbiological analyses, as well as sensorial analyses, were carried out during the biopreservation trials. Subsequently, *Lb. curvatus* BCS35 was selected to up-scale the trials and combinations of the three application methods were assayed. The microbiological analyses showed that the strain had protective properties in food ingredient. Chen et al., (2013) have studied a novel botanic biopreservative which was prepared from the combination of the bamboo leaf extract and ebony extracts. This was designated as ebony-bamboo leaves complex extracts (EBLCE). The scientists found that EBLCE had antimicrobial activity. It was found to be edible, safe and economical. They also studied the effect in prolonging the shelf life and improving the quality of peeled *Penaeus vannamei* during storage at 4°C. It was found to prevent the spoilage of peeled *P. vannamei* efficiently. In pre-treated peeled shrimps with shelf life increased to 16 days in comparison to 6 days in the control group. Therefore, EBLCE can be used as a promising alternative in the preservation of food. Wayah and Philip (2018) studied the major food-borne pathogenic and spoilage bacteria *Micrococcus luteus, Listeria monocytogenes,* and *Bacillus cereus* in food items. The workers evaluated the biopreservative potential of pentocin from *Lactobacillus pentosus* CS2. Pentocin MQ1 is a novel bacteriocin isolated from *L. pentosus* CS2 from coconut shake. It was found to be potent against *M. luteus, B. cereus* and *L. monocytogenes* even at low concentrations. The pathogen killing was mediated by membrane permeabilization as a mechanism of action of the pentocin from the study against Gram-positive bacteria. Pentocin MQ1 is a cell wall-associated bacteriocin which has been shown to improve the microbiological quality and shelf life of fresh banana. da Silva Sabo et al., (2017) have worked on cheese whey which is the main byproduct of the dairy industry. It is an industrial waste which causes environmental pollution. It can serve as a raw material for the production of biocomposites. The authors' cultured *Lactobacillus plantarum* ST16Pa in hydrolyzed fresh cheese whey. Supplementation with soybean flour was found to be helpful. The cell-free supernatant showed high biopreservative efficiency in chicken breast fillets artificially contaminated with *Enterococcus faecium* 711 during 7 days of refrigerated storage, thus indicating the potential use of this BLIS as a biopreservative in the food industry. Semeraro et al., (2017) studied the biopreservation by sugar and/or polymeric matrixes with technological relevance. Samples of myoglobin were embedded in amorphous gelatin along with trehalose + gelatin matrixes at different hydrations which were compared with amorphous myoglobin-only and also myoglobin-trehalose samples. It was observed that the gelatin appears to be more effective than trehalose against massive denaturation in the long time range. Whereas, the mixed trehalose + collagen matrix is very effective in preserving protein functionality as compared to gelatin-only or trehalose-only matrixes.

Zudaire et al., (2018) studied the biological preservation with probiotic bacteria in order to control the growth of foodborne pathogens in food. Biopreservation with *Lactobacillus rhamnosus* GG was studied on the quality and bioaccessibility of total phenolic content and antioxidant activity in fresh-cut pears. The immersion of whole pears in a calcium chloride solution did not provide added value. Despite the increase in observed activity of PME and PPO enzymes in fresh-cut pears during storage, the browning index and firmness values were constant for all samples. It appears that for fresh-cut pears a postharvest treatment of calcium immersed in a solution of antioxidant agents and probiotic bacteria could be a suitable alternative to dairy products for maintaining the overall quality of fruit during storage. Francis et al., (2014) studied the hybrid food preservation program. Kellie et al., (2016) have associated obesity with food.

Therdtatha et al., (2016) isolated *Lactobacillus salivarius* KL-D4 from the intestine of the duck which produced bacteriocin. This was found to be stable at high temperatures and also to a wide pH range of 3-10. The cell free supernatant at pH 5. 5 had inhibitory effect against both G+ and G- bacteria. The purified product was tested against *Bacillus cereus, Enterococcus faecalis, Pseudomonas stutzeri, Staphylococcus sp.* and *Stenotrophomonas* sp. which resulted in the inhibition of the pathogens. With the work it was proved that salivaricin KLD can be a tentative biopreservative for food industry in the future. According to Oliveira et al., (2015) fruits and vegetables can become contaminated by foodborne pathogens such as *Escherichia coli* O157:H7,

Salmonella and *Listeria monocytogenes*. Current industrial sanitizing treatments do not eliminate the pathogens. Therefore, the industries are forced to use chemical control. However, biological control may be a better solution by using the native microbiota which is present on the fresh produce. The scientists have isolated native microbiota from whole and fresh-cut produce and determined the antagonistic effects on foodborne pathogens, to inhibit the growth of *Salmonella enterica* on lettuce disks. The cell count of the pathogens was found to reduce. Bacteriophages and nisin are used commercially against *Salmonella* and *L. monocytogenes* on fresh-cut lettuce. The study highlighted the potential of biocontrol, but the combination with other technologies may be required to improve their application on fresh-cut lettuce. Claudia et al.,(2017) studied the microbial spoilage of bread and the consequent waste problem causes large economic losses for both the bakery industry and the consumer. Mycotoxins due to fungal contamination in cereals and cereal products are an issue to be seen. Sourdough fermented with antifungal strains of lactic acid bacteria (LAB) produces gluten-containing and gluten-free breads with improved nutritional value, quality and safety due to shelf-life extension. Other alternative of biopreservation techniques include the utilization of antifungal peptides, ethanol and plant extracts. These can be added either to the bread dough or incorporated in antimicrobial films for active packaging (AP) of bread.

Andrade et al., (2014) worked on the ability of the osmotolerant yeast *Debaryomyces hansenii* which could inhibit *Penicillium nordicum*, the most common ochratoxigenic mould. It is commonly found in dry-cured meat products. They found that most of the *D. hansenii* strains were able to delay *P. nordicum* spore germination when co-cultured in malt extract broth with maximum activity by *D. hansenii* FHSCC 253H. Ochratoxin A (OTA) production in ham samples was detected by HPLC-MS. Co-culturing of *P. nordicum* with *D. hansenii* FHSCC 253H resulted in lower OTA levels compared with control samples without *D. hansenii*. Therefore, *D. hansenii* can be used as a biopreservative agent for preventing ochratoxigenic mould growth and OTA accumulation in dry-cured meat products. Claudia et al., (2016) worked on the use of sourdough fermented with specific strains of antifungal lactic acid bacteria which could reduce chemical preservatives in bakery products. They studied the production of antifungal carboxylic acids after sourdough fermentation of quinoa and rice flour using the antifungal strains *Lactobacillus reuteri* R29 and *Lactobacillus brevis* R2Î as bioprotective cultures and the non-antifungal *L. brevis* L1105 as a negative control strain. Durability tests were used to study the effects against environmental moulds for biopreservative potential and prolonging the shelf life of the bread. The sourdough fermentation of the different flour substrates generated a complex and significantly different profile of carboxylic acids. Extracted quinoa sourdough detected the greatest number of carboxylic acids with high concentration than from rice sourdough. Comparing the lactic acid bacteria strains, *L. reuteri* R29 fermented sourdoughs contained generally higher concentrations of acetic and lactic acid but also the carboxylic acids. Among them, 3-phenyllactic acid and 2-hydroxyisocaproic acid were present at a significant concentration. This was correlated with the superior protein content of quinoa flour and its high protease activity. With the addition of *L. reuteri* R29 inoculated sourdough, the shelf life was extended by 2 days for quinoa (+100%) and rice bread (+67%) when compared to the non-acidified controls. Francis et al., (2014) studied the hybrid food preservation program. Kellie et al., (2016) have associated obesity with food.

Alvarez et al.,(2016) studied the combined effects of bioactive agents (tea tree essential oil, propolis extract and gallic acid) and storage temperature on the microbiological and sensory quality of fresh-cut mixed vegetables for soup (celery, leek and butternut squash). Biopreservatives applied on mixed vegetables were effective only when combined with optimal refrigeration temperature (5°C). Bioactive compounds showed effectiveness in controlling the microbiota present in mixed vegetables, coliforms were found to reduce with gallic acid and propolis treatmentsduring storage. Antimicrobial activity was also found against endogenous *Escherichia coli* and inoculated *E. coli* O157:H7 at 10 days of refrigerated storage. The use of natural agents as propolis extract towards the preservation of the quality and safety of mixed vegetables for soup addresses the concerns of the consumer about the use of synthetic chemical antimicrobials potentially harmful to health. A study was done by Sparo et al., (2013) on the bio-preservation of ground beef meat by *Enterococcus faecalis* CECT7121. Meat and particularly ground beef is frequently associated with Food Poisoning in Food Safety. The bactericidal effect of the probiotic *Enterococcus faecalis* CECT7121 was studied against different pathogens like *Escherichia coli* O157:H7, *Staphylococcus aureus, Clostridium perfringens* and *Listeria monocytogenes.* when the probiotic was added 24 h before and 24 h after the pathogen *E. coli* O157:H7, viable cells were not detected at 24 h and 48 h post-treatment, respectively. The work confirmed the bactericidal activity of *E. faecalis* CECT7121 strain on bacterial pathogens in ground beef meat. Zbrun et al., (2016) studied the use of indigenous lactic acid bacteria (LAB) with specific additives as a Biopreservation System (BS) for poultry blood during its storage in slaughterhouses. The BS consisted of two LAB (*Enterococcus faecalis* DSPV 008SA or *Lactobacillus salivarius* DSPV 032SA. They found that after 24 h storage at 30°C the counts of *Enterobacteria, Coliforms, Pseudomonas* spp. and *Staphylococcus aureus* were found to be less in blood. The ability of LAB to prevent haemolysis was evident. Smaoui et al., (2016) studied the major compounds in *Mentha piperita* essential oil (EOMP). They found them to be menthol (33. 59%) and iso-menthone (33%). It was found that the combination of EOMP and BacTN635 decreased the pathogenic microflora proliferation and enhanced the sensory acceptability. The shelf life of meat beef was enhanced by 7 days.

Hintz et al., (2015) studied the use of plant antimicrobial compounds for food preservation. The workers studied the structure, modes of action, stability and resistance to the plant compounds and their application in food industries. They have also worked out the technologies of delivery. Nithya et al., (2013) studied the effectiveness of partially purified antibacterial peptide (ppABP) produced by *Bacillus licheniformis* Me1 for food preservation by means of active packaging. The active packaging films containing ppABP were developed using two different packing materials [low-density polyethylene (LDPE) and cellulose films] by soaking and spread coating. The activated films showed antibacterial activity against pathogens. The activated films demonstrated its biopreservative efficacy in controlling the growth of pathogens in cheese and paneer. The ppABP-activated films were found to be effective for biopreservation. The ppABP from active films got diffused into the food matrix and reduced the growth rate and maximum growth population of the target micro-organism. Both types of ppABP-activated films can be used as a packaging material to control spoilage and pathogenic organisms in food, thereby extending the shelf life of foods. Ahmad et al., (2013) have isolated and characterized a proteinaceous antifungal compound from *Lactobacillus plantarum* YML007 for its application as a food preservative. Korean kimchi is known for its myriad of lactic acid bacteria (LAB) with diverse bioactive compounds which were screened against *Aspergillus niger*. The strain exhibiting the highest antifungal activity was identified as *Lactobacillus plantarum* YML007 by 16S rRNA sequencing and biochemical assays using API 50 CHL kit. The biopreservative activity was evaluated using dried soybeans. Muhialdin et al., (2011) have worked on looking for new preservatives from natural resources to replace or to reduce the use of chemical preservatives using lactic acid bacteria (LAB) for their antifungal activity on selected foods. The supernatants of the selected strains delayed the growth of fungi in tomato puree, processed cheese and commercial bread with enhanced shelf. Rodgers (2016) studied the technological and operational pathways in minimally processed functional foods and foods with fresh-like qualities commanding premium prices. According to the workers these will provide new exciting opportunities for small and medium size enterprises (SMEs) specializing in fresh produce to play an active role in the health market. Supporting SMEs, governments could benefit from savings in healthcare costs and value creation in the economy. Consumers could benefit from novel FF formats such as refrigerated RTE (ready-to-eat) meals, a variety of fresh-like meat-, fish-, and egg-based products, fresh-cut fruits and vegetables, cereal-based fermented foods and beverages. In the operational sense, moving from nourishment to health improvement demands a shift from defensive market-oriented to offensive market-developing strategies including collaborative networks with research organizations. Alvarado et al.,(2006) conducted the identification of indigenous LAB for antimicrobial activity in traditional Mexican foods towards natural preservation of food. *Lactobacillus* and *Lactococcus* strains were found to enhance the safety of food products. Cell free cultures of *Leuconostoc mesenteroides* reduced the number of viable cells of enteropathogenic *E. coli. Lb. plantarum* CC10 showed high inhibitory activity against *S. aureus* 8943. *E. faecium* QPII (from panela cheese) produced a bacteriocin which had activity against *L. monocytogenes*. Selected LAB from traditional Mexican foods showed good potential as bio-preservatives.

Wang et al., (2016) studied the high hydrostatic pressure as non-thermal technology that can achieve the same standards of food safety as those of heat pasteurization and keeps the food items fresher. High-pressure processing inactivates the pathogens and spoil causing microorganisms and the enzymes. It also modifies the structures with little or no effects on the nutritional and sensory quality of foods. The U. S. Food and Drug Administration (FDA) and the U. S. Department of Agriculture (USDA) have approved the use of high-pressure processing (HPP), which is a reliable technological alternative to conventional heat pasteurization in food-processing procedures.

PÃrez Pulido et al., (2016) have studied the application of bacteriophages towards the application in food biopreservation. Lytic bacteriophages have been tested in different food systems in order to inactivate the main food-borne pathogens like *Listeria monocytogenes, Staphylococcus aureus, Escherichia coli O157:H7, Salmonella enterica, Shigella spp., Campylobacter jejuni, Cronobacter sakazkii* and also for control of spoilage bacteria. It can control the pathogens without interfering with the remaining food microbiota. Bacteriophages could also be applied for inactivation of bacteria attached to food contact surfaces or grown as biofilms. Phage preparations specific for *L. monocytogenes, E. coli* O157:H7 and *S. enterica* serotypes have been commercialized and approved for application in foods. Phage endolysins have a broader host specificity compared to lytic bacteriophages. Cloned endolysins could be used as natural preservatives, singly or in combination with other antimicrobials such as bacteriocins. Gray et al., (2018) studied the possibility of reducing *Listeria monocytogenes* as a foodborne pathogen to improve public health. Novel biocontrol methods may offer effective means to help improve control of *L. monocytogenes* and decrease cross contamination of food. Essential oils are natural antimicrobials from the plants have been used as food additives and preservatives for many years to prevent food spoilage by microorganisms. Competitive exclusion of food borne pathogens has potential as a biocontrol application. Lee et al., (2016) have written a good review which discusses the status, antimicrobial mechanisms, application and regulation of natural preservatives in livestock food systems. Earlier preservatives used were nitrates/nitrites, sulfites, sodium benzoate, propyl gallate, and potassium sorbate. Due to the side effects of chemicals new natural antimicrobial compounds of various origins are being developed, including plant-derived products (polyphenolics, essential oils, plant antimicrobial peptides (pAMPs)), animal-derived products (lysozymes, lactoperoxidase, lactoferrin, ovotransferrin, antimicrobial peptide (AMP), chitosan and others), and microbial metabolites (nisin, natamycin, pullulan, Îμ-polylysine, organic acid, and others). The action of natural preservatives is usually by inhibiting microbial cell walls/membranes, DNA/RNA replication and transcription, protein synthesis, and metabolism. However, they can affect color, smell and toxicity in large amounts. Therefore, many trials are being performed to check the same. Natamycin and nisin are currently being used on commercial scale. McLeod et al., (2010) have discussed the role of *Lactobacillus sakei* as an important food-associated lactic acid bacterium. It is commonly used as a starter culture for industrial meat fermentation and with great potential as a biopreservative in meat and fish products. It is important to know the metabolic mechanisms underlying the growth performance of a strain to be used for food fermentations. This helps in bringing out high-quality and safe products. On proteomic analysis it was found that the expression varied depending on the carbon source. D-ribose pyranase was directly involved in ribose catabolism and enzymes were involved in the phosphoketolase and glycolytic pathways. With the change in carbon source it was found that the expression of enzymes involved in pyruvate and glycerol/glycerolipid metabolism were also affected. A commercial starter culture and a protective culture strain down-regulated the glycolytic pathway more efficiently than the rest of the strains when grown on ribose. *Lactobacillus sakei* is used as a starter culture for industrial meat fermentation and also a biopreservative in meat and fish products. The strain was isolated from fermented fish which showed a higher expression of stress related proteins growing on both carbon sources. Grande Burgos et al., (2014) isolated a cyclic antibacterial peptide Enterocin AS-48 by *Enterococcus* which is a circular bacteriocin. It contains a 70 amino acid-residue chain circularized by a head-to-tail peptide bond. The confirmation of enterocin AS-48 is arranged into five alpha-helices with a compact globular structure. Enterocin AS-48 has a wide inhibitory spectrum on Gram-positive bacteria. The cationic peptide gets into the bacterial membrane and causes membrane permeabilization which leads to cell death. With microarray analysis up-regulation and down-regulation of genes was observed in *Bacillus cereus* cells after treating them with sublethal bacteriocin concentration. This has wide applications as a food biopreservative against many foodborne pathogens like *Listeria monocytogenes, Bacillus cereus, Staphylococcus aureus, Escherichia coli, Salmonella enterica* and spoilage bacteria like *Alicyclobacillus acidoterrestris, Bacillus* spp., *Paenibacillus* spp., *Geobacillus stearothermophilus, Brochothrix thermosphacta, Staphylococcus carnosus and Lactobacillus sakei.* Morgan et al., (2000) developed a novel method for the isolation of microbially-derived inhibitory substances from food sources. This involves an enrichment step along with a killing assay. It has been tested to identify the antimicrobial potential of a number of Kefir grains which are incubated in 10% reconstituted skim milk for 20 h at 32 degrees ^{0}C for the production of potential biopreservatives to inhibit *Listeria innocua* DPC1770 and *Escherichia coli* O157:H45. Das and Goyal (2014) isolated a bacteriocin produced by probiotic *Lactobacillus plantarum* DM5 from indigenous fermented beverage Marcha from India. *Lb. plantarum* DM5 possessed probiotic properties along with bacteriocin activity against several major food borne pathogens. Cytotoxicity analysis of plantaricin DM5 on human embryonic kidney 293 (HEK 293) and human cervical cancer (HeLa) cell lines revealed it to be nontoxic and biocompatible in nature. Pellegrini et al., (2018) studied the essential oils (EOs) of some plants from Abruzzo territory (Italy) for their antimicrobial, antioxidant activities along with their volatile fraction after chemical characterization. The extracts used were of *Rosmarinus officinalis, Origanum vulgare, Salvia officinalis, Mentha piperita, Allium sativum, Foeniculum vulgare, Satureja montana, Thymus vulgaris* and *Coriandrum sativum* seeds. The results allowed the study to establish that these EO's are good candidates for potential application as biopreservatives in foods and/or food manufacture environments. MaÃtre et al., (2014) studied the food allergies and if the immune mechanism is involved or not. The first category includes immune mediated reactions like IgE mediated food allergy, eosinophilic oesophagitis, food protein-induced enterocolitis syndrome and celiac disease. The second category implies non-immune mediated adverse food reactions called as food intolerances. The workers have studied the adverse food reactions based on the pathophysiologic mechanism that can be useful for both diagnostic approach and management.

Iweala and Choudhary (2018) have studied the relation between the history of major childhood and adult food allergies and their treatments. Peanut/tree nut allergies generally appear in adulthood. Adults can develop new IgE-mediated food allergies which is a very common oral allergy syndrome. The prevalence of food allergy is increasing. However, newer approaches of treatment are being actively pursued which will be helpful in future days to come. In a survey done by Grief (2016) food allergies are common and seem to be increasing in prevalence. Preventive measures have become far more evident in the public arena (schools, camps, sports venues, and so forth). Evaluation and management of food allergies have evolved in such a way that primary care practitioners may choose to provide initial diagnostic and treatment care or refer to allergists for similar care. Food allergies, once considered incurable, are now being diminished in intensity by new strategies. Spencer (1974) has discussed food additives from the food technology point of view. These can be used to protect food from chemical and microbiological attack, to even out seasonal supplies, to improve their eating quality and to improve their nutritional value. The various types of food additives include colours, flavours, emulsifiers, bread and flour additives, preservatives and nutritional additives. The author has also explained the reasoning when the addditives can be used. Staden et al., (2007) studied the reasons of food allergy in children. Werfel et al., (2016) studied food allergy in adulthood.

Capps and Park (2003) have studied food retailing and food service sector which is an important component of the food marketing channel and is also vital to the United States economy. In recent times the business of food retailing and food services has undergone a lot of change due to the consumer. The authors have covered farm-to-retail prices for red meat and dairy industries. Schwenk (1991) has written an overall perspective on the trends in food consumption. Nutrition awareness is at an all-time high. The consumption is usually influenced by changes in disposable income, availability of convenience foods, smaller household size and an increasing proportion of ethnic minorities in the population. Martel et al., (1995) made a survey on the antimicrobial resistance in bacteria from diseased cattle in France.

McBride and Anne (2010) have written an article on food porn. The terminology typically refers in watching others cook on television or gazing at unattainable dishes in glossy magazines without actually cooking oneself. The term is used mainly by commentators rather than by people actively engaged in the world of cooking. Practitioners (chefs and a food television producer) and academics address whether or not food porn exists, what shape it could take, what purpose it might serve, and/or what usefulness it might have, showing that these contentious issues are more complex than the ease with which the term is used. Forsyth et al., (2011) studied the challenges in access to food using Geographic Information Systems (GIS). This is important both for food during travel and eating behavior. The authors have made a survey and collected data available from fieldwork, land use, parcel data, licensing information, commercial listings, taxation data and online street-level photographs. Various methods have been suggested to classify different kinds of food sales to deliver healthy food. According to Berglund (1978) the use of additives into food satisfies many purposes, as shown by the index issued by the Codex Committee on Food Additives. These include acids, bases and salts; preservatives, antioxidants and antioxidant synergists; anticaking agents; colours; emulsifiers; thickening agents; flour-treatment agents; extraction solvents; carrier solvents; flavours (synthetic); flavour enhancers; non-nutritive sweeteners; processing aids and enzyme preparations. Many additives occur naturally in foods, but that does not exclude toxicity at higher levels. Some food additives are nutrients or even essential nutrients like NaCl. There are examples of food additives which cause toxicity in man even when used according to regulations (cobalt in beer). Sometimes, the poisoning is due to carry-over (nitrate in cheese whey) when used for artificial feed for infants. Sometimes, poisoning occurs due to permitted substance being added at very high levels. This could happen due to accident or carelessness (nitrite in fish). Many a times hypersensitivity is noticed against food additives as found in tartrazine and other food colours. The toxicological evaluation, based on animal feeding studies, may be complicated by impurities. This is seen in orthotoluene-sulfonamide in saccharin, transformation or disappearance of the additive in food processing in storage as seen in bisulfite in raisins, by reaction products with food constituents like formation of ethylurethane from diethyl pyrocarbonate, by metabolic transformation products like formation in the gut of cyclohexylamine from cyclamate. Metabolic end products may differ in experimental animals and in man. It has been observed that guanylic acid and inosinic acid are metabolized to allantoin in the rat but to uric acid in case of man. The magnitude of the safety margin in man of the Acceptable Daily Intake (ADI) is not identical to the "safety factor" used when calculating the ADI. Sicherer et al., (2014) have studied the food allergy reactions to milk, eggs, peanuts, soy, wheat, tree nuts (such as walnuts and cashews) and shellfish. It has been observed that about 10% of kids suffer with these allergies and therefore have to be given specific foods. Jordi et al., (2006) have written an article on allergy. The food products must display the nutrition information.

With young children it is important to encourage the children to have a clean plate after food. The children must be taught to eat only when they are hungry and stop when they are full. Children should be allowed to choose foods based on their likes and dislikes. According to Borchers et al., (2010) food can never be safe completely. Many pathogens threaten food safety that cause a number of foodborne diseases. Algal toxins cause acute disease and fungal toxins can be highly toxic. The toxicity can be teratogenic, immunotoxic, nephrotoxic or estrogenic. The industrial activities have resulted in massive increases in our exposure to toxic metals such as lead, cadmium, mercury and arsenic. These are present in the entire food chain and exhibit various toxicities. The various processes of the industries release chemicals that persist in the environment and contaminate the food. These include organochlorine compounds, such as; 1, 1, 1-trichloro-2, 2-bis (p-chlorophenyl) ethane (dichlorodiphenyl dichloroethene)

(DDT), other pesticides, dioxins, and dioxin-like compounds. DDT and its breakdown product dichlorophenyl dichloroethylene affect the developing male and female reproductive organs. They also have neurodevelopmental toxicities in human infants and children. Other food contaminants can arise from the treatment of animals with veterinary drugs or the spraying of food crops, which may leave residues. Among the pesticides applied to food crops, the organophosphates have been the focus of much regulatory attention because there is growing evidence that they, too, affect the developing brain. Numerous chemical contaminants and carcinogens are formed during the processing and cooking of foods. Other food contaminants leach out from the packaging or storage containers. These include phthalates and bisphenol A which induce malformations in the male reproductive system.

Food labelling with regard to allergies is also important. It has been observed that the manufacturers usually do not label the food product against allergies. According to Dennis O'Brien. (2019) USDA provides links to relevant agricultural research data from multiple sources, enhances the transparency about the sources of nutrition information, provides data that is based on the latest scientific research, and is representative of the marketplace. Food Data Central contains in one place five distinct types of food and nutrient composition data. It incorporates the data USDA has provided for years, but adds two new types of data that are grouped under new topics. "Foundation Foods" provides expanded nutrient information and extensive underlying metadata that will help users understand the variability in the nutrient values of foods. "Experimental Foods" links to data about foods produced by agricultural researchers that will allow users to see for themselves how factors such as climate, soils and agricultural practices can affect a food's nutritional profile. The new system is designed to strengthen the capacity for rigorous research and policy applications through its search capabilities, downloadable datasets, and detailed documentation. Application developers will be able to incorporate the information into their applications and web sites.

The constantly changing and expanding food supply is a challenge to those who are interested in using food and nutrient data. Including diverse types of data in one data system will provide researchers, policymakers, and other audiences a key resource for addressing vital nutrition and health issues.

When Fast Foods are discussed there seems to be no standard definition of fast food. Fast foods are usually eaten without cutlery. Failure to have a standardized definition makes it difficult to compare studies. Foods available outside the home tend to be high in energy. It will be helpful if restaurants offer information about the the nutritive values of such food. Samadi et al., (2012) studied the methods for the biopreservation of hamburgers. Hamburgers with high nutrient supply and loosely-packed structure that provide favourable conditions for microbial growth. The authors have studied the chemical composition and antimicrobial activity of the essential oil of *Zataria multiflora* and its potential application as a natural preservative have been studied in order to inhibit the indigenous microbial population in hamburgers. It was found that carvacrol, thymol and linalool were the constituents of the essential oil. The essential oil exhibited strong antibacterial activity against Gram-positive and Gram-negative bacteria. Addition of *Z. multiflora* essential oil in concentrations higher than MIC values influenced the microbial population of hamburgers stored at 25°C, 4°C and -12°C. Chauliac et al., (1987) have looked into the projects which produce weaning foods in developing countries. They have also described and discussed the global epidemiology of malnutrition and have offered guidelines to reduce malnutrition.

Siegris et al., (2008) have studied the use of nanotechnology which has the potential to generate new food products and food packaging. They conducted a survey in the German speaking part of Switzerland, lay people's (N=337) perceptions of 19 nanotechnology applications which were examined. The goal was to identify food applications that were more likely, and food applications, that were less likely to be accepted by the public. The psychometric paradigm was employed, and applications were described in short scenarios. Results suggested that perceived controls are important factors which influence risk and benefit perception. Nanotechnology food packagings have been assessed to be less problematic than nanotechnology foods. Analyses of individual data showed that the importance of naturalness in food products and trust were significant factors influencing the perceived risks and the benefits of nanotechnology foods and nanotechnology food packagin. Astonishing growth in the market for nanofoods is predicted in the future, from the current market of $2. 6 billion to $20. 4 billion in 2010. The market for nanotechnology in food packaging is increasing every day.

In a study done by Wüthrich, (2009) food allergy is a parameter which is widely used for all different symptoms and diseases caused by food. Food allergy (FA) is an adverse reaction to food (food hypersensitivity) that is usually found in susceptible individuals. The term "food allergy" is widely misused for all sorts of symptoms and diseases caused by food. Food allergy (FA) is an adverse reaction to food (food hypersensitivity) occurring in susceptible individuals, which is mediated by a classical immune mechanism specific for the food itself. The best established mechanism in FA is due to the presence of IgE antibodies against the offending food. Food intolerance (FI) are all non-immune-mediated adverse reactions to food. The subgroups of FI are enzymatic (e. g. lactose intolerance due to lactase deficiency), pharmacological (reactions against biogenic amines, histamine intolerance), and undefined food intolerance (e. g. against some food additives). The diagnosis of an IgE-mediated FA is made by a carefully taken case history, supported by the demonstration of an IgE sensitization either by skin prick tests or by *in-vitro* tests, and confirmed by positive oral provocation. For scientific purposes the only accepted test for the confirmation of FA/FI is a properly performed double-blind, placebo-controlled food challenge (DBPCFC). A panel of recombinant allergens, produced as single allergenic molecules, may in future improve the diagnosis of IgE-mediated FA. Due to a lack of causal treatment possibilities, the elimination of the culprit "food allergen" from the diet is the only therapeutic option for patients with real food allergy. The established mechanism in FA is due to the presence of IgE antibodies against the offending food. Food intolerance (FI) is all nonimmune mediated adverse reactions to food. The subgroups of FI are enzymatic (lactose intolerance due to lactase deficiency), pharmacological (reactions against biogenic amines, histamine intolerance) and undefined food intolerance (against some food additives). The diagnosis of an IgE-mediated FA is usually done after taking the case history. This is supported by the demonstration of an IgE sensitization either by skin prick tests or by *in vitro* tests and then it is confirmed by positive oral provocation. However, for the confirmation of FA/FI a properly performed double-blind, placebo-controlled food challenge (DBPCFC) is necessary. According to Gupta et al., (2018) food allergy diagnosis remains challenging. Most standard methods are unable to differentiate sensitization from clinical allergy. Sometimes, faulty interpretation of tests leads to overdiagnosis and unnecessary food avoidances. Highly predictive models have been established for major food allergens based on skin prick testing and food-specific immunoglobulin E. Although many newer diagnostic techniques are improving the accuracy of food allergy diagnostics, an oral food challenge remains the only definitive method of confirming a food allergy. Natalie et al., (2013) studied a relation between chocolate towards the mood. Higher depression was found to be associated with greater chocolate consumption. It has been observed that usually it is the local food that protects the human health and the environment, enhances revitalization, increases access to healthy foods and creates economic opportunities by promoting local foods. According to Rosalie (2005) Agricultural Research Service (ARS) scientists have launched a searchable database on the World Wide Web where users can view a 60-nutrient profile for each of more than 13,000 foods. The new resource is called "What's in the Foods You Eat—Search Tool. "It has been found that for the processing of foods electrical energy is utilized. In conventional food processing plants, electricity drives mechanical devices and controls the degree of process. Recently, many processing technologies are being developed for the processing of foods directly with electricity.

Sathe and Sharma (2009) studied the increase of food allergies especially in the western countries. In the US and EU, the labeling for any food allergen present in the food product is a must. There is an interest to study the the role of food processing in food allergens. Foods are processed in various ways at home, in institutional settings and in the industries. Many a times it depends on the processing method and the food, partial or complete removal of the allergen. Scientists have found the reduction of peanut allergen *in-vitro* IgE immunoreactivity upon soaking and blanching treatments. When the allergen is discretely located in a food, one may physically separate and remove it from the food. Some of the examples include lye peeling to produce hypoallergenic peach nectar, protein denaturation and/or hydrolysis during food processing. These can be used to produce hypoallergenic products. Allen and Wilson (2006) have studied the recent concern about food availability arising from a general survival security motivation during childhood. When materialists set a high goal of food security reported more food insecurity. When the materialists stored/ hoarded more food at home higher obesity was observed. In another case in which the survival security were low led to greater endorsement of materialism, food security as goal and using food for emotional comfort. The results imply that materialists overcame the food insecurity of their childhood by making food security a top life goal. Golem et al., (2015) studied the limited food supply which could be paired with reduced access to food during emergency disasters and could lead to malnutrition. The workers have looked into the household calorie availability and nutrient density in a normal situation and the changes that could occur during emergencies like natural disasters. When there is replenishment of food supplies with disrupted water and/or energy required for food preparation and storage or in case of loss of power for greater than five days could reduce the availability of household calories by 27%.

Ft. Lee (1971) has described the ways on the prevention of food poisoning. The author has come out with a course which provides a programed text of a single lesson with a duration of four-hours on the prevention of food poisoning. It covers the following areas: a definition of food poisoning; chemical food poisoning; biological food poisoning; causes and prevention of trichinosis; six factors controlling bacteria growth; bacterial infection and their prevention. Winter (2016) have brought out the need of presenting food science information effectively which is a critical competency for food scientists by the Institute of Food Technologists. Many a times the food scientists do not have adequate training in this area which needs to be taken and therefore the author has brought it out effectively. AraÃjo et al., (2018) studied the association between food insecurity and food intake in a study conducted by users of a primary health care service in Belo Horizonte, Brazil. In their survey the authors found that the food insecurity did not affect the consumption of ultra-processed foods, were independent of age, sex, marital status, educational level and employed status. It was noticed that the consumption of beans was higher in households with children and adolescents whereas, the consumption of tubers was higher in households without children and adolescents. Halvorsen et al., (1995) studied the reaction to foods. Adverse reactions to food occur in about 1-2% of the population as reported by patients. Most reactions to food are not caused by allergy. IgE-mediated food reactions are well known and are of major clinical significance owing to their potentially dangerous life-threatening character. Enzyme deficiencies, active pharmacological substances in food and psychological mechanisms also cause adverse reactions to food. SzÃpfalusi (2012) studied the various allergies which appear during childhood. Food allergies are life-threatening and reduce the quality of life. The prevalence of food allergies is found around the globe. The food allergens cover the majority of food-related reactions (milk, egg, wheat, soy, fish, crustacean, nuts and peanut). Immunological mechanisms include IgE-mediated (most common) and non-Ig E-mediated. During the treatment of food allergy trigger food should be strictly avoided. Currently, there is no cure for food allergy.

Sobal and Bisogni (2009) studied the food choice decision among people. Choice decisions of food leads to food behaviors where people acquire, prepare, serve, give away, store, eat and clean up. Many disciplines and fields examine decision making. The food choice process model includes; life course events and experiences that establishes a food choice trajectory through transitions, turning points, timing and contexts, influences on food choices that include cultural ideals, personal factors, resources, social factors and present contexts and a personal system that develops food choice values, negotiates and balances values, classifies foods and situations and forms/revises food choice strategies, scripts and routines. The parts of the model dynamically interact to make food choice decisions leading to food behaviors. No single theory can fully explain decision making in food behavior. Multiple perspectives are needed, including constructionist thinking.

Doyle et al., (2005) attributed illness to food. They Identified and prioritized the effective food safety interventions. This requires an understanding of the relationship between food and pathogen from farm to consumption. Critical is the capacity to attribute cases of foodborne disease to the food vehicle or other source responsible for illness. It includes the analysis of outbreak data, case-control studies, microbial subtyping, source tracking methods and expert judgment. The Food Safety Research Consortium sponsored the Food Attribution Data Workshop in October 2003 to discuss the virtues and limitations of these approaches and to identify future options for collecting food attribution data in the United States. The authors have identified various challenges that affect progress in this critical component of a risk-based approach to improving food safety.

Kong and Epstein (2016) studied the food reinforcement during infancy. The motivation for eating is cross-sectionally related to obesity. Research has shown avid sucking for milk in early infancy which predicts adiposity later in life. Relationship between food reinforcement and excess body weight has been observed in infants as young as 9 months of age. New methodological developments in studying food reinforcement in infants and young children have provided opportunity to study the origin of food reinforcement. Olszyna-Marzys (1990) have looked into the relationship between radioactivity with food specifically in food irradiation for preservation purposes and food contamination from radioactive substances. Most of the times food is irradiated by using electromagnetic energy (X and gamma rays) emitted by radioactive substances or produced by machine in order to destroy the insects and microorganisms present in order to prevent their germination. Under strictly controlled conditions no undesirable changes take place in food that has been irradiated nor is radioactivity induced. The paper has also discussed the steps taken by different international organizations to set limits on acceptable radioactivity in food.

Garber and Lustig (2011) have looked into the studies of food addiction by highly palatable foods. The review examines the nutrients present in fast food, the characteristics of fast food consumers or the presentation and packaging of fast food can encourage substance dependence, as defined by the American Psychiatric Association. Mostly, the fast foods are accompanied with soda. This increases the sugar content to ten times. Sugar addiction, including tolerance and withdrawal, has been demonstrated in rodents but not humans. Caffeine is a substance of dependence and therefore there is an upsurge in the fast food sales with caffeine. It is usually the obese adults who eat more of fast food than those of normal weight. Obesity is characterized by resistance to insulin, leptin and other hormonal signals that would normally control appetite and limit reward. Neuroimaging studies in obese subjects provide evidence of altered reward and tolerance. Fast food advertisements, restaurants and menus all provide environmental cues that may trigger addictive overeating. While the concept of fast food addiction remains to be proven, these findings support the role of fast food as a potentially addictive substance that is most likely to create dependence in vulnerable populations. According to Blanco-Padilla et al., (2014) natural food antimicrobials are bioactive compounds that inhibits the growth of microorganisms involved in food spoilage or food-borne illnesses. However, stability issues result in degradation and loss of antimicrobial activity. Nanoencapsulation allows protection of antimicrobial food agents from unfavorable environmental conditions and incompatibilities. Encapsulation of food antimicrobials control delivery increasing the concentration of the antimicrobials in specific areas and the improvement of passive cellular absorption mechanisms resulted in higher antimicrobial activity. The review article has covered the present state of the art of the nanostructures used as food antimicrobial carriers including nanoemulsions, nanoliposomes, nanoparticles and nanofibers.

Lindeman (2002) has described the need for functional foods for the good health of women. Foods or food ingredients that could provide health benefits beyond basic nutrition. This explains the need of both whole and modified foods to be included as functional foods. He has discussed the history, regulation, and promotion of functional foods and the consumer interest in functional foods. Smith (1977) has explained the energy flow in food webs using candy bars as food sources.

Norovirus illness (www. cdc. gov/norovirus, 2017) is commonly found in nursing homes, hospitals, restaurants, cruise ships, schools, banquet halls, summer camps and the outbreaks occur in the food service settings like restaurants. Infected food workers are frequently the source of contamination of pathogenic microorganisms. Diet-busting foods taste good, but are low in nutrition and high in calories. Many of these foods make the person feel very hungry because they are low in fiber or protein.

Arsenic in Food is found in water, air, food, and soil in organic and inorganic forms. It is important to be careful of these sources. According to Asthma and Food Allergies (2015) sulfites are used during processing of some foods. These cause allergies and needs caution. According to Healthy food trends (J Nutr., 2015) flax seeds and linseeds come under healthy food trends. They are also functional foods and keep blood pressure in control. Kerr and David (1993) studied the heat content of a food at a given temperature by the thermodynamic property of enthalpy. They have shown a method of a simple calorimeter to measure the enthalpy changes of different foods during freezing. According to Falvey (1997) intensive agriculture is restricted to suitable land which will be required in the future due to global population growth, declining food prices and extreme poverty. The challenge of balancing environmental care with food production has been brought out clearly. Guandalini and Newland (2011) have differentiated the food allergies from food intolerance. Adverse reaction to foods are very common that are attributed to allergies. Recently, the epidemiological data in North America has shown that the prevalence of food allergy in children has increased. Common food allergens in the United States include egg, milk, peanut, tree nuts, wheat, crustacean shellfish and soy. The authors have also looked into the various forms of food intolerances (immunoglobulin E [IgE] and non-IgE mediated), including celiac disease and gluten sensitivity. Immune mediated reactions can be either IgE mediated or non-IgE mediated. IgE-mediated food allergies are entities such as eosinophilic, esophagitis and eosinophilic gastroenteropathy. Non IgE-mediated immune mediated food reactions include celiac disease and gluten sensitivity, two increasingly recognized disorders. Finally, non-immune mediated reactions encompass different categories such as disorders of digestion and absorption, inborn errors of metabolism, as well as pharmacological and toxic reactions. In a study done by Chapman and Gunter (2018) the foodborne diseases cause an estimated 48 million illnesses and 3,000 deaths annually (Scallan et al., 2011), with U. S. economic costs estimated at $152 billion to $1. 4 trillion annually (Roberts,2007 and Scharff, 2010). An increasing number of these illnesses are associated with fresh fruits and vegetables. An analysis of outbreaks from 1990 to 2003 found that 12% of outbreaks and 20% of outbreak-related illnesses were associated with produce (Klein and DeWaal , 2008; Lynch et al., 2009). These food safety problems have resulted in the recommendation of the shift to a more preventative and risk-based food safety system. A modern risk-based food safety system takes a farm-to-fork preventative approach to food safety and relies on the proactive collection and analysis of data to better understand potential hazards and risk factors, to design and evaluate interventions in order to prioritize prevention efforts. Due to the coming up of shared kitchens, food hubs and local food systems in the United States the foodborne illness outbreaks have become common at these locations. As the food safety methods are rare, it gives rise to a growing number of foodborne illness outbreaks. Food risk analysis is a holistic approach to food safety because it considers all aspects of the problem. Risk assessment modeling is the foundation of food risk analysis. Proper design and simulation of the risk assessment model is important to properly predict and control risk. Microbial food-borne illnesses pose a significant health problem in Japan. In 1996 the world's largest outbreak of *Escherichia coli* food illness occurred in Japan. Afterwards, a new regulatory measure was established, including strict hygiene practices in meat and food processes. Wilsey and David (2014) have worked on the personal food system mapping which is a practical means to engage community participants and educators in individualized and shared learning about food systems, decisions and behaviors. It is a useful approach for introducing the food system concept.

Wright and Nault (2013) have discussed on the participatory approaches as to how to engage youth into the food systems. The authors observed that the youth had valuable knowledge in order to enhance the understanding of food environments. According to a survey done by DeVoe (2008) the incidence of food allergies in children has grown fivefold, from 1 in 100 children to 1 in 20 children since 1960 according to the Food Allergy Initiative. Food allergies cause anaphylactic shock, the most severe type of allergic reaction, which can lead to death within minutes if left untreated. Frenkel et al., (2017) have surveyed the literature published in 2016 and early 2017 which are related to food processing waste treatment and can be utilized for industrial applications. The review has covered the food processing industries and applications which include meat and poultry, fruits and vegetables, dairy and beverage and miscellaneous treatment of food wastes. Food composition is a critical bridge between nutrition, health promotion, disease prevention and food production. Therefore, compilation of data into useable databases is very important for the development of dietary guidance.

Brown et al., (2012) have looked into the improper food cooling practices which are the major cause of foodborne illnesses. Not much is known about restaurant food cooling practices. The authors have examined the food cooling practices in restaurants for food safety. In the study they have assessed the frequency with which restaurants meet U. S. Food and Drug Administration (FDA) recommendations aimed at reducing pathogen proliferation during food cooling. Members of the Centers for Disease Control and Prevention's Environmental Health Specialists Network collected data on food cooling practices in 420 restaurants. The data collected indicated that many restaurants are not meeting FDA recommendations concerning the cooling system. Although most restaurant kitchen managers report that they have formal cooling processes (86%) and provide training to food workers on proper cooling (91%), however, many managers do not test and verify the cooling processes (39%). They neither monitored the time or temperature during the cooling processes (41%), nor did they calibrate the thermometers used for monitoring temperatures (15%). It was the food managers (86%) who reported that the cooling processes did not incorporate all FDA recommended components. Additionally, restaurants do not always follow recommendations concerning specific cooling methods, such as refrigerating cooling food at shallow depths, ventilating cooling food, providing open-air space around the tops and sides of cooling food containers, and refraining from stacking cooling food containers on top of each other. Data of this study can be helpful in food safety programs and the restaurant industry to target training and intervention efforts concerning cooling practices. Radio frequency (RF) heating is a commonly used food processing technology that has been applied for drying and baking as well as thawing of frozen foods. Its use in pasteurization, sterilization and disinfection of foods is limited. EPA sets limits on how much of a pesticide may be used on food during growing and processing, and how much can remain on the food.

Bonner (2004) in his book has discussed on the kitchen chemistry, post-eating food metabolism, origins of different foods (from crop breeding to evolution) and ecological and environmental impacts of farming and harvesting practices.

Wild and Lehrer (2001) have explained the important cause of life-threatening hypersensitivity reactions as food allergy. Avoidance of allergenic foods is the only method of prevention presently available to the sensitized patients which is often impossible. With better characterization of allergens and better understanding of the immunologic mechanism, investigators have developed several therapeutic modalities that are potentially applicable to the treatment and prevention of food allergy. The methods include peptide immunotherapy, DNA immunization, and immunization with immunostimulatory sequences, anti-IgE therapy and genetic modification of foods. These exciting developments hold promise for the safe and effective treatment and prevention of food allergy. Jackson and Viehoff (2016) have written an excellent review on the recent researches been done on the consumption of 'convenience' foods. It highlights the nature of the term and explores the implications for public health and environment. It distinguishes between different types of convenience food which includes ready-meals, tracing the structure and growth of the market for such foods. They have highlighted the significance of convenience foods in time-saving and time-shifting, the importance of recent changes in domestic labour and family life and the way the consumption of convenience food is frequently moralized. They have also discussed the current levels of public understanding about the links between convenience food, environmental sustainability and food waste.

Rice et al., (2007) have studied if the fast food can be friendly. Usually the fast foods are linked towards obesity and are therefore excluded from professional dietary recommendations. However, several sections of society including senior citizens, low-income adult and children, minority and homeless children, or those pressed for time appear to rely on fast food as an important source of meals. Considering the dependence of these nutritionally vulnerable population groups on fast food, the authors have demonstrated the possibility to design a fast food menu which could provide a reasonable level of essential nutrients without exceeding the caloric recommendations. Therefore, the authors have emphasised that the fast foods should not be forbidden. According to the authors such foods should be looked as foods which will enable its inclusion in meal for those who depend on it out of necessity. Joseph (2000) has looked into the WebQuest frameworks that are developed to help students investigate the topic of nutrition. These include food labels; the Food Guide Pyramid; three levels of inquiry related to nutrition and ingredients in foods; how food choices affect health; historical background of food and food companies and online grocery shopping. According to Novembre et al., (1988) many foods induce respiratory symptoms by both reaginic and nonreaginic mechanisms. Asthma is one of the most common respiratory manifestations in children, and it is well known that many factors may provoke an attack. Food allergy may coexist with an inhalant allergy and with nonallergens pollutants like smoke or additives may modify bronchial reactivity and thus favor the food allergen action. The authors demonstrated how the respiratory symptoms were related to food allergy especially asthma in children which are always associated with other clinical manifestations. The recognition of food dependent IgE-mediated respiratory symptoms is essentially limited to those cases characterized by food allergy with asthmatic expression. In many cases foods might have a nonspecific role in the determination of asthma. Singh, (2014) studied food as a potent natural reward and food intake as a complex process. Reward and gratification associated with food consumption leads to dopamine (DA) production, it activates reward and pleasure centers in the brain. An individual will repeatedly eat a particular food to experience the positive feeling. This type of repetitive behavior of food intake leads to the activation of brain reward pathways that eventually overrides other signals of satiety and hunger. It has been observed that overeating and obesity arise from many biological factors that involves both central and peripheral systems in a bi-directional manner that involve mood and emotions and ultimately leads to overeating and obesity. Schmidt, and Kim (2002) studied the interactions between food and drugs which might reduce or increase the drug effect. The majority of clinically relevant food-drug interactions are caused by food-induced changes in the bioavailability of the drug. The bioavailability and clinical effect of most drugs are correlated. In healthcare systems, healthcare providers are increasingly developing methods to identify and prevent adverse food-drug interactions. The most important interactions reduced the bioavailability in the fed state. These are caused due to the chelation with components in food (as occurs with alendronic acid, clodronic acid, didanosine, etidronic acid, penicillamine and tetracycline) or dairy products (ciprofloxacin and norfloxacin) or by other direct interactions between the drug and certain food components (avitriptan, indinavir, itraconazole solution, levodopa, melphalan, mercaptopurine and perindopril). The gastric acid secretion can also reduce the bioavailability of many drugs like ampicillin, azithromycin capsules, didanosine, erythromycin stearate or enteric coated and isoniazid. Some foods may increase the drug bioavailability either because of a food-induced increase in drug solubility. Jedermann et al., (2014) have looked into the ways for reducing food losses by logistics. The need to feed an ever-increasing world population makes it obligatory to reduce the millions of tons of avoidable perishable waste along the food supply chain. Much of these losses are caused by non-optimal cold chain processes and management. The authors have brought focus on the technologies, models and applications in order to monitor changes in the product shelf life. They have studied three cases on the cold chain for berries, bananas and meat and an overview of different post-harvest treatments. Further contributions focus on the required technical solutions, such as the wireless sensor and communication system for remote quality supervision, gas sensors to detect ethylene as an indicator of unwanted ripening and volatile components to indicate mould infections. Naturally occurring antimicrobial compounds could be applied as food preservatives to protect food quality and extend the shelf life of foods and beverages (Juneja et al., 2012). These compounds are naturally produced and isolated from various sources like plants, animals and microorganisms, in which they constitute part of host defense systems. Many naturally occurring compounds, like nisin, plant essential oils and natamycin have been reported to be effective antimicrobial agents against spoilage and pathogenic microorganisms. Many natural antimicrobials are available commercially and utilized in food processing. However, their efficacy, consumer acceptance and regulations have not been well defined. With this in view and the mechanism of action, factors affecting their antimicrobial activities and future prospects for use of natural antimicrobials in the food industry have been discussed. Stanton (2015) studied the relation between food retailers and obesity. According to the scientist we live in an 'obesogenic environment' where we are constantly bombarded with choices that encourage us to move less and eat more. Many factors influence our dietary choices, including the expert marketers who advise manufacturers on ways to encourage the population to buy more, especially profitable, palatable 'ultra-processed' foods. Supermarkets themselves have become skilled in manipulating buying behaviour, using their layout and specific product placement as well as advertising to maximise purchases of particular foods. The energy intake from foods and drinks has increased than that of energy output which has decreased. Obesity has a range of relevant factors and there is a role of supermarkets and fast food retailers in it. Heine (2015) has worked on the gastrointestinal food allergies present during early childhood with many symptoms. Cow's milk, soy and wheat are the three most common gastrointestinal food allergens. The article deals with the food protein-induced enteropathy, proctocolitis and enterocolitis. In IgE-mediated food allergies the gastrointestinal symptoms are delayed by 1-2 hours after ingestion in non-IgE-mediated allergic disorders. The pathophysiologies of these non-IgE-mediated allergic disorders are not understood and there are no useful *in-vitro* markers. Diagnosis is usually difficult as the results of the skin prick test / measurement of the food-specific serum IgE level is found negative. There is also a clinical overlap between non-IgE-mediated food allergy and several common paediatric gastroenterological conditions. The treatment of gastrointestinal food allergies requires the strict elimination of fending food allergens until tolerance has developed. These details have been worked out well by the author. Brown et al., (2017) have studied the food cooling practices in the restaurants. Improper food cooling practices are a significant cause of foodborne illness. Not much is known about food cooling practices in the restaurants. The study has assessed the frequency with which various restaurants meet U. S. Food and Drug Administration (FDA) recommendations. The worker has looked into the ways to reduce pathogen proliferation during food cooling. Members of the Centers for Disease Control and Prevention Environmental Health Specialists Network collected data on food cooling practices in 420 restaurants. The data collected indicate that many restaurants are not meeting FDA recommendations concerning cooling. Although most restaurant kitchen managers report that they have formal cooling processes (86%) and provide training to food workers on proper cooling (91%), many managers said that they do not have tested and verified cooling processes (39%), do not monitor time or temperature during cooling processes (41%), or do not calibrate thermometers used for monitoring temperatures (15%). Indeed, 86% of managers reported cooling processes that did not incorporate all FDA-recommended components. Additionally, restaurants do not always follow recommendations concerning specific cooling methods, such as refrigerating cooling food at shallow depths, ventilating cooling food, providing open-air space around the tops and sides of cooling food containers, and refraining from stacking cooling food containers on top of each other. Allen and Prosperi (2016) have looked into the processes which underlie environmental, economic and social unsustainability which are derived from the food system. Building sustainable food systems is the need of the day which aims in redirecting our food systems and policies towards better-adjusted goals and improved societal welfare. Policies should be interdependent and interacting which encourages the public perception of humanity and nature. This is a stepwise approach with a logical application. Buluswar(2018) has discussed the issue of urban food deserts and malnutrition in American inner cities. According to Ganzin (1975) food for all, requires an interdisciplinary approach. It needs an evaluation of all sectors of the economy and the formulation of objectives which include proper nutrition in overall development plans which otherwise will continue to give low priority to agriculture and social reform and the objective of food for all will not be met. Tivadar and Luthar (2005) have analysed a data obtained from a random sample of the Slovenian population within a research project entitled 'Lifestyles in a Mediated Society. 'They have aimed on the exploration of the role of socio-demographic variables in food practices and to discover the inherent logic that motivates each particular set of food practices. A cluster analysis revealed six food cultures lying along a continuum where traditionalism occupies one end and post-traditionalism the other. The authors conclude that although two out of six food cultures crosscut socio-demographic affiliations and transform food consumption into a constituent part of a lifestyle as an identity project, there is still a significant influence of socio-demographic characteristics on food practices in contemporary Slovenia.

A study done by Burks et al., (2012) revealed that the food allergies can result in life-threatening reactions and can diminish the quality of life. In the last several decades, the prevalence of food allergies has increased in several regions throughout the world. Although more than 170 foods have been identified as being potentially allergenic, a minority of these foods cause the majority of reactions, and common food allergens vary between geographic regions. In the treatment of food allergy the trigger foods should be avoided. The American Academy of Allergy, Asthma & Immunology; European Academy of Allergy and Clinical Immunology; World Allergy Organization; and American College of Allergy along with Asthma and Immunology have come together to increase the information on allergies and asthma at a global level. Within the framework of this collaboration, termed the International Collaboration in Asthma, Allergy and Immunology, a series of consensus documents called International Consensus ON (ICON) have been developed that will serve as an important resource and support physicians in managing different allergic diseases. According to Wagner et al., (2014) most of the times people seek out their own idiosyncratic comfort foods when in negative moods. They believe that these foods improve their mood. They have studied to know if these comfort foods could actually provide psychological benefits or not. Do they really improve mood better than comparison foods. Although people believe that comfort foods provide them with mood benefits, comfort foods do not provide comfort beyond that of other foods (or no food). Ott (1988) has discussed on the developments in food packaging, processing and preservation techniques in terms of packaging materials, technologies, consumer benefits in current and potential food product applications. Willet and Stampfer (2003) have discussed the old food guide pyramid which was released in 1992 by the U. S. Department of Agriculture. It contradicts the message that fat is bad, which was presented to the public by nutritionists and the effects of plant oils on cholesterol. BlÃzquez and Berin (2017) tried to coorelate the microbiome and the food allergy. Food allergy is a common disease affecting approximately 8% of the children and 5% of adults. The prevalence has increased over the last two decades, suggesting an important environmental contribution to susceptibility. Scientists have looked into the human cohorts support as dysbiosis in food allergy. It has suggested that it occurs early in life, preceding the onset of sensitization. In animal models it has been found that the intestinal microbiota and food allergy are interdependent. There are organisms like *Clostridia* species which provide protection in the development of food allergy. Work should be done to identify commensal-derived microorganisms that could be used therapeutically to prevent food allergy. Tordesillas et al., (2017) have reviewed the recent understanding of basic mechanisms which consists of IgE-mediated food allergies and novel therapeutic approaches under investigation for both the prevention and treatment of IgE-mediated food allergies. Respiratory allergy that includes asthma and allergic rhinitis is plaguing westernized countries to the extent of up to 8% in young children and 2%-3% of adults in the United States. They have been affected by hypersensitivity reactions to various foods. In the past decade, there have been great strides in our understanding of the underlying immunopathogenesis of these disorders, which have led to improved diagnostic techniques, management strategies and therapeutic approaches. Murkovic (2004) has discussed the affects of acrylamide which is known for its potential health hazards. Recently acrylamide was found in starch containing heated foods in high concentrations which leads to the assumption that a cancer risk could be associated with the uptake of foods containing high amounts of acrylamide. In the work scientists found highest concentrations in potato crisps with concentrations of above 1500 ng/g (median: 499 ng/g). However, it was less in cookies with a median of 99 ng/g; crisp bread with a median of 69 ng/g; breakfast cereals with a median of 0 ng/g; popcorn and rice products with a median of 97 ng/g; potato chips with a median of 161 ng/g and coffee with a median of 169 ng/g. Carrard et al., (2015) have written an update on food allergies. According to them cure for food allergy has not yet been achieved and patients are forced to alter their eating habits and social engagements. New technologies and improved *in-vitro* and in *-vivo* models have enhanced the knowledge of the pathogenesis of food allergies. It will elucidate interactions between genetic background, lifestyle and environmental factors. Infrared(IR) processing of foods has been gaining popularity over conventional processing in several unit operations, including drying, peeling, baking, roasting, blanching, pasteurization, sterilization, disinfection, disinfestation, cooking and popping. The Murphy Elementary School District (School Business Affairs, 1983) in Phoenix, Arizona has cut food service costs and improved community relations by cooking and baking from "scratch" and utilizing the staff's ethnic cooking skills. (MLF). Wender (1977) has reviewed the hypothesis of food additives with hyperkinesis and learning disabilities in children and found out the possibility of the interaction. Werfel (2016) studied the allergies in adulthood. It has been found that food allergies can suddenly appear in adulthood. The prevalence of primary food allergy is basically higher in children than in adults. However, recently, food allergies in adulthood are appearing increasingly. Allergy symptoms of the immediate type can be observed as well as symptoms occurring after a delay, such as indigestion, triggering of hematogenous contact eczema or flares of atopic dermatitis. Skin tests, molecular allergy diagnostics and *in-vitro* tests are employed to eliminate diets. The most frequent food allergies in adults are nuts, fruit and vegetables, which can cross-react with pollen as well as wheat, shellfish and crustaceans. Valenta et al., (2015) have looked into the basics of food allergies. According to their survey the IgE-associated food allergy affects approximately 3% of the population and has severe effects on the daily life of patients. The manifestations occur in the gastrointestinal tract and also affect many other organ systems. Birth cohort studies have shown that allergic sensitization to food allergens starts in early childhood. Mechanisms of pathogenesis include cross-linking of mast cells and basophil-bound IgE and immediate release of inflammatory mediators, as well as late-phase and chronic allergic inflammation, resulting from T-cell, basophil and eosinophil activation. Characterization of the molecular features of food allergens and the development of chip-based assays for multiple allergens are very helpful. Importantly, learning about the structure of disease-causing food allergens has allowed researchers to engineer synthetic and recombinant vaccines. Verhoeckx et al., (2015) have discussed the effects of food processing. Processing can alter the allergenic properties of food proteins. A wide variety of processing methods are available and their use depends largely on the food to be processed. The scientists have reviewed the impact of processing (heat and non-heat treatment) on the allergenic potential of proteins and on the antigenic (IgG-binding) and allergenic (IgE-binding) properties of proteins. It is known that fermentation and hydrolysis may have the potential to reduce allergenicity to a large extent. Speer (1975) has looked into the multiple food allergy. This could be the sensitivity due to three or more foods. As per the studies in the literature, multiple food allergies occur in both sexes and at all ages. However, it is more common in boys than in girls and more common in women then in men. Most of the patients reacted to air-borne allergens as molds, pollens, house dust and animal epithelials. This indicates that food allergy and inhalant allergy are fundamentally the same. The common food allergens were found to be everyday foods which include; milk, chocolate, corn, egg, tomato, peanut and citrus fruits. Andre (1980) has discussed a study that was done at the Nottingham University Consumer Study Group. It determined the market operation for popular convenience foods in England. He has provided information on the distribution of purchases, brand loyalties of respondents to a questionnaire regarding convenience foods and market fluctuation due to inflation. According to Walker (2005) 60% of the high school children eat food in cafeterias. This should provide 33% of the students' energy intake but the data shows that it provides only 19% because of the sale of competitive foods. This is alarming and needs inputs from the scientists. Dolan et al., (2010) have studied the toxins in food that occur naturally. The adverse effects of foods are low. This is because of the result of the solutions given by the US Food and Drug Administration (FDA) and other regulatory agencies. They have used specifications, action levels, tolerances, warning labels and prohibitions. Role of manufacturers by setting limits on certain substances and developing mitigation procedures for process-induced toxins is also noteworthy. This brings in a possibility of toxicity due to contamination, overconsumption or allergy response. Pectin, a complex polysaccharide, is a major component of non-lignified cell walls of dicotyledonous and some monocotyledonous plants. Its food-related technological functions are numerous and mirrors many of its biological functions (USDA-ARS?s Scientific Manuscript database. Pectin in foods). Kolata (1982) has brought the interrelation between food and nutrients on human behavior. He has reviewed the various research studies on this topic including the effects of food on brain biochemistry (particularly sleep) and effects of tryptophan as a pain reducer. Czerwiecki (2004) have collected data on food authenticity and has discussed it in their work. The most known are adulteration of vegetable and fruit products, adulteration of wine, honey, olive oil etc. The modern analytical techniques for detection of food adulteration have been discussed. Among physicochemical methods isotopic techniques (SCIRA, IRMS and SNIF-NMR) are discussed. The main spectral methods used are: IACPAES, PyMs, FTIR and NIR. The chromatographic techniques (GC, HPLC, HPAEC and HPTLC) with several kinds of detectors have been described and the ELISA and PCR techniques have been mentioned. There is a necessity of rigorous control of food in order to fight fraud in food industry. Peleg, (1993) have looked into the fractal geometry and related concepts which had only a very minor impact on food research. The very few reported food applications deal mainly with the characterization of the contours of agglomerated instant coffee particles, the surface morphology of treated starch particles, the microstructure of casein gels viewed as a product limited diffusion aggregation, and the jagged mechanical signatures of crunchy dry foods. Fractal geometry describes objects having morphological features that are scale invariant. A demonstration of the self-similarity of fractal objects can be found in the familiar morphology of cauliflower and broccoli, both foods. Processes regulated by nonlinear dynamics can exhibit a chaotic behavior that has fractal characteristics. Examples are mixing of viscous fluids, turbulence, crystallization, agglomeration, diffusion and possibly food spoilage. Ring et al., (2001) studied the adverse allergic reactions to foods. These represent a prominent, actual and increasing problem in clinical medicine. Among the many symptoms of food allergy includes skin reactions (urticaria, angioedema, eczema) respiratory (bronchoconstriction, rhinitis), gastrointestinal (cramping, diarrhea) and cardiovascular symptoms with the maximal manifestation of anaphylactic shock. With little amounts of allergens these symptoms get elicited. Most of the times the food allergy is confirmed by looking into the history, skin test, *in- vitro* allergy diagnosis and oral provocation tests. The numbers of cases are increasing every day. These include fatalities, due to inadvertent intake of food allergens. Therefore, it is a must to improve declaration laws and develop methods for allergen detection in foods. Allergens can be detected by serological methods (enzyme immunoassays, *in-vitro* basophil histamine release or *in-vivo* skin test procedures in sensitized individuals).

However, the diagnosis of food allergy becomes very complicated due to the cross-reactivity between allergens in foods and aeroallergens (pollen, animal epithelia and latex). Elicitors of pseudo-allergic reactions consist of low-molecular-mass chemicals (preservatives, colorings and flavor substances). For some of them (sulfites) detection assays are available. In some patients classic allergic contact eczema can be elicited systemically after oral intake of low-molecular-mass contact allergens such as nickel sulfate or flavorings such as vanillin in foods. The role of xenobiotic components in foods (pesticides) is not known at the moment. In order to improve the situation of the food allergic patient, research programs should be undertaken to elucidate the pathophysiology and improve allergen detection strategies which can be implemented together. In a study by Bushra et al., (2011) the drug taken by any person interacts with another drug the person is already taking (drug-drug interaction). It also interacts with the food, beverages, dietary supplements the person is consuming (drug-nutrient/food interaction) or another disease the person has (drug-disease interaction). It was found that in case of food and drug when taken simultaneously can also be responsible to alter the body's ability to utilize a particular food or drug or may also cause serious side effects. Francis (2014) has discussed on the food preservation and food safety knowledge. Due to the increased demand for food preservation by the public new comprehensive food preservation programs have been developed. This is known as Preserve the Taste of Summer (PTTS). This PTTS programme is a comprehensive hybrid food preservation programme and will help the consumers to a large extent.

12

Safety of Food in Space

Josephson (1977) studied the effects of food irradiation and sterilization. In radiation sterilization of food (radappertization) the food is exposed in sealed containers to ionizing radiation at absorbed doses which are high enough (25-70 kGy) to kill all organisms of food spoilage and public health significance. Radappertization is analogous to thermal canning for enhancing the shelf stability. In dry products the radappertization process requires the food to be heated to an internal temperature of 70-80Â°C (bacon to 53Â°C) to inactivate autolytic enzymes which catalyze spoilage during storage without refrigeration. In order to reduce off-flavors and odors, undesirable color changes, and textural and nutritional losses from exposure to the high doses of irradiation the foods are vacuum sealed and irradiated frozen. Radappertozed foods have the characteristic of fresh foods prepared for eating. Irradiation can be a good substitute for some chemical food additives such as ethylene oxide and nitrites which are toxic, carcinogenic, mutagenic or teratogenic either wholly or in part. Till now, there is no evidence of any adverse effects of radappertization on the foods. Extensive research has taken place towards the space food. During the planning of Apollo missions, Space Food NASA Technical Reports Server (Space Food NASA Technical Reports Server, NTRS, 1994) conducted trials including freeze drying. Action Products commercialized this technique, concentrating on snack food including the first freeze-dried ice cream. The foods include instant, quickly frozen and then slowly heated in a vacuum chamber to remove the ice crystals formed by the freezing process. The final product retains 98 percent of its nutrition and weighs only 20 percent of its original weight (Food packets, NASA Image and Video Library, 2012-08-15. ISS032-E-019031). The library has covered various food items which were utilized for an Expedition with 32 crew members. It has shown to prepare a meal in the Unity node of the International Space Station. Galvez et al., (2016) have studied the food applications in NASA and the regulations on the working of various food stuffs. Brynjolfsson (1985), Josephson et al., (1997, 1983) studied the safety of irradiated foods. They have brought out the guidelines and regulations for processing irradiated foods. The radiolytic products formed in food when it is irradiated and its wholesomeness has also been discussed. They found that the food irradiation processing is not a panacea for all problems in food processing but when properly used will serve the space station well. Perchonok and Dane (2001) studied the food and its package in Space Food Systems Laboratory (SFSL). It is a multipurpose laboratory responsible for space research. It develops the flight food, menus, packaging and food related hardware for Shuttle, International Space Station, and Advanced Life Support food systems. All foods used to support NASA ground tests and/or missions must meet the highest standards before they are 'accepted' for use on actual space flights. The foods are evaluated for nutritional content, sensory acceptability, safety, storage and shelf life, and suitability for use in micro-gravity. The food packaging is also tested to determine its functionality and suitability for use in space. According to Weber (1988) for food scientists to take up food as a business is a recent option. However, it should be known that food business entrepreneurship is a difficult career in which one needs to work for long hours with extensive decision making. The task requires knowledge beyond food science. According to the report NASA Technical Reports Server (NTRS,1992) the 3M food service system 2 employs a "cook/chill" concept for serving food in hospitals. The system allows staff to prepare food well in advance, maintain heat, visual appeal and nutritional value as well as reducing operating costs. The integral heating method, which keeps hot foods hot and cold foods cold, was developed by 3M for the Apollo Program. Nielson et al., (2016) have written an introduction to food analysis in NASA. Urquía-Fernández (2013) in an overview has presented an article on the food security and nutrition in Mexico. It is based on the analysis of the four pillars of food security as availability, access, utilization of food and stability of the food supply. They have analysed malnutrition in Mexico based on obesity and undernourishment. The author has collected the data from the food security indicators of the United Nations's Food and Agriculture Organization, from the Mexican Scale of Food Security, and from the National Health and Nutrition Survey. According to the data Mexico is affected by a double burden of malnutrition. Children and adults are affected with problems (14% of stunt and 30% of the adult population is obese). Also, more than 18% of the population cannot afford the basic food (food poverty). Using perception surveys, people report important levels of food insecurity, which concentrates in seven states of the Mexican Federation. The production structure underlying these indicators shows a very heterogeneous landscape, which translates into a low productivity growth. Food security being a multidimensional concept, to ensure food security for the Mexican population requires a revision and redesign of public productive and social policies, placing a particular focus on strengthening the mechanisms of institutional governance. The authors Galvez et al.,(2008) have discussed the food applications of bacteriocins and their regulatory issues for the incorporation of bacteriocins as bioprotectants. Specific applications of bacteriocins or bacteriocin-producing strains are described for main food categories like milk and dairy products, raw meats, ready-to-eat meat and poultry products, fermented meats, fish and fish products or fermented fish. They have also dealt with applications in foods and beverages derived from plant materials which include raw vegetable foods, fruits and fruit juices, cooked food products, fermented vegetable foods and fermented beverages. Results obtained for application of bacteriocins are discussed with a special emphasis on novel food packaging and food-processing technologies. These include irradiation, pulsed electric field treatments or high hydrostatic pressure treatment. Smith et al.,(1975) have studied the large improvements and advances in space food systems achieved during the Apollo food program. Modifications of the Apollo food system were directed primarily towards improving the delivery of adequate nutrition to the astronaut. Individual food items and flight menus were modified as nutritional counter measures to the effects of weightlessness. Unique food items were developed. They provided nutritional completeness, high acceptability, ready-to-eat, shelf-stable and were found to be convenient. Specialized food packages were also developed. The Apollo program showed that future space food systems will need well-directed efforts to achieve the optimum potential of food systems in support of the physiological and psychological well-being of astronauts and crews. According to King et al., (2003) due to fungal infection in food there has been a great loss in revenue worth billions of dollars per annum along with a lot of health problems both to animals and humans. Using the modern food sorting techniques which are based on colour or other physical characteristics is helpful in sorting out the diseased food stuffs from healthy food stuffs. This also brings a change in their texture, taste and colour. Till now, only visible wavelengths and colour identification have been used to bulk-sort food, but there has been little research in the ultra violet regions of the spectrum to identify fungus or toxin infections. This research specifically concentrated on the ultra violet (UV) spectral characteristics of food in an attempt to identify the spectral changes that occurs when healthy food items like peanuts become infected with toxin-producing fungi. Ultimately, the goal is to design, build and construct an optical detection system to use these 'spectral fingerprints' so as to detect toxically infected food items rapidly and efficiently. Thurmond et al., (1986) have looked into the research on the food system of space station. This has given rise to a set of specifications which are used to produce concepts. Concept verification includes testing of prototypes, both in 1^{-g} and microgravity. The end-item specification provides an overall guide for assembling a functional food system for Space Station. MikÅ¡-Krajnik et al., (2015) have studied the ways of food security through increasing the microbial food safety. Both food safety and food security are interrelated and have a great impact on the quality of life. Food safety deals with the handling, preparation and storage of foods in order to prevent foodborne illnesses and ensure food security through improving food safety. It is important to deal with innovative thermal and non-thermal technologies in the area of food processing. This helps in the reduction and control of microbiological risks to extend the shelf life of food product. It is the Food table (ISS. NASA Image and Video Library,2015) which gives a view of the food table located in the Russian Zvezda service module on the International Space Station which was taken by Expedition 43 Flight Engineer Scott Kelly. Assorted food, drink and condiment packets are visible. Kelly tweeted this image along with the comment: ""Looks messy, but it's functional. Our #food table on the @space station. What's for breakfast? #YearInSpace". Ahmed et al., (1989) worked to obtain basic data on ethnic foods by studying dietary patterns and multicultural foods, and to determine nutritional status of multicultural space explorers by evaluating dietary, clinical, biochemical, and socioeconomic factors. The study had planned a significant role in providing nutritional research for space explorers of different ethnic backgrounds. They felt that the results will provide data which will help in the development of future food plans for long duration flights involving manned exploration to Mars and lunar base colonies.

Survival Kit - Food KitNASA Image and Video Library1962-10-01S62-08743 (1962) has looked into the food kit which is used by Mercury astronauts. Some are dehydrated and need water, other packets are ready to eat. Size is measured relative to a ruler. Included are packets of mushroom soup, orange-grapefruit juice, cocoa beverage, pineapple juice, chicken with gravy, pears, strawberries, beef and vegetables and other assorted food containers. Ehlermann (2016) has discussed on the safety of treated food by ionizing radiation that has a great concern. The author has covered many questions which have been of concern. They have elucidated their scientific relevance. There has never been any other method of food processing studied in such depth as food irradiation and concluded that consumption of any food treated at any high dose is safe, as long as the food remains palatable. This conclusion has been adopted by WHO, also by international and national bodies. Finally, this finding has also been adopted by Codex Alimentarius in 2003, the international standard for food. However, this conclusion has not been adopted and included at its full extent in most national regulations. A study done by Rappole et al., (1985) was conducted to design a food service system using current technology to serve a small scale Space Station. The psychological, sociological and nutritional factors affecting feeding in microgravity conditions were investigated. The logistics of the food service system was defined. Analysis of food is an important aspect in knowing the quality of any food. Nielsen (2016) has looked into the analytical methods used by the food industry, governmental agencies or universities for the determination of food composition and characteristics. He has also looked into the demands of the consumers, the food industry and national and international regulations to monitor the composition of food and to ensure the quality and safety of the food supply. All food products need to be analysed as part of a quality management program during the development process (including raw ingredients), through production and also after any product comes into the market. The characteristics of foods (i.e., chemical composition, physical properties and sensory properties) are responsible for the regulatory purposes and quality control which includes speed, precision, accuracy and ruggedness. Validation of the method for the specific food matrix being analyzed is necessary to ensure usefulness of the method. Such official methods are critical in the analysis of foods to ensure that they meet the desired quality. Allen and Prosperi (2016) have studied the modelling system of sustainable foods. These processes underlie the environmental, economical and social unsustainability from the food system. Building sustainable food systems is very vital towards improved societal welfare. Food systems are complex social-ecological systems that involve many interactions between human and natural components. The scientists have proposed a conceptual model that has articulated crucial vulnerability and resilience factors to global environmental and socio-economic changes, postulating specific food and nutrition security issues as priority outcomes of food systems. In a report of NASA Technical Reports Server (NTRS),(1997), when NASA manned space travel in 1959 the challenge was of sustaining life in space and what to feed an astronaut? It was Pillsbury Company who solved the problem by coating bite-size foods to prevent crumbling. They developed the hazard analysis and critical control point (HACCP) concept to ensure against bacterial contamination. Hazard analysis is a systematic study of product, its ingredients, processing conditions, handling, storage, packing, distribution and directions for consumer use to identify sensitive areas that might prove hazardous. Hazard analysis provides a basis for blueprinting the Critical Control Points (CCPs) to be monitored. Later incorporation of HACCP in the FDA's Low Acid Canned Foods Regulations set down in the mid-1970s to ensure the safety of all canned food products in the U. S. NASA Technical Reports Server (NTRS,1981) have reported the development of a universal closure lid for the space shuttle food package. The revised lid needed a folded configuration which, when unfolded, fully conforms to the interior surfaces of the food cup. Experimental thermoform molds were fabricated and the test lids were formed. The lid material was made in such a way that it did not touch the food in the cup interior permitting full nesting of the cups. The final lid design was established and thermoform tooling was designed and fabricated. Lids formed on these molds were tested for strength. The heating elements were replaced and repositioned in order to eliminate the hot spots if any as it could cause warpage.

Popov (1975) has discussed the supply of adequate food and water to the astronauts on a short and also on a long-term space flights. They have looked into the various factors and parameters of study which included the food consumption, energy requirements and suitability of the foodstuffs in the space flights. Physicochemical and biological methods of food production and regeneration of water from astronaut metabolic wastes, as well as wastes produced in a closed ecological system or as a result of technical processes taking place in various spacecraft systems were studied. Fohey et al., (1978) have written an excellent review on the food packaging techniques that are used in the space flight missions and for the Space Shuttle. Attention is directed to bite-size food cubes used in Gemini, Gemini rehydratable food packages, Apollo spoon-bowl rehydratable packages, thermostabilized flex pouch for Apollo, tear-top commercial food cans used in skylab, polyethylene beverage containers, skylab rehydratable food package, space shuttle food package configuration, duck-bill septum rehydration device and a drinking/dispensing nozzle for Space Shuttle liquids. Constraints and testing of packaging is considered, a comparison of food package materials is presented and typical shuttle foods and beverages are listed.

Sirmons et al., (2016) has looked into the assessment of the stability of vitamin content, sensory acceptability and color variation in fortified spaceflight foods over a period of 2 years. The work has been done to identify optimal formulation, processing and storage conditions so as to maintain stability and acceptability of commercially available fortified nutrients. Changes in food quality are being monitored to indicate whether fortification affects quality over time (compared to the unfortified control), thus indicating their potential for use on long-duration missions.

According to a NASA Image and Video Library (2017) it was Bryan Onate, as the advanced Plant Habitat project manager, with the Exploration Research and Technology Directorate who did the brainstorming using innovative approaches towards the food production with industry representatives inside a laboratory at the Space Station Processing Facility at NASA's Kennedy Space Center in Florida. The scientists at NASA (NASA Image and Video Library,2017) at the Exploration Research and Technology Directorate brainstormed the innovative approaches to food production with industry representatives at the Space Station Processing Facility at NASA's Kennedy Space Center in Florida.

In a technical report of NASA by Bush (1974) the Skylab food program was a major effort involving a complex spectrum of activities which are necessary for the preparation of a crew feeding system. Approximately 17,000 individual food packages and support items, weighing more than 1225 kgs, were launched into space as a single unit on board. The Skylab crew was given nourishing food and beverage for a total of 156 days, as well as with eating utensils and accessory items. Additionally, provisions for 5 days (15 man-days) were provided in each of the three command and service modules in a manner similar to that of the Apollo flights. The Skylab food system not only provided the crew with a palatable balanced diet in a familiar and acceptable manner but also supported the formidable mineral balance medical experiment series (M070). According to Neilson (1985) psychobiology is a scientific discipline which encompasses the phenomena known to be important as regards nutrition and food consumption in space. It includes those areas of biology which are clearly related to behavior, human subjective experience and problems of coping up and also adapting to stress. Brown et al., (2015) have discussed on the climate change and global food security. This report has mainly focused on the impact of climate change on global food security which will provide sufficient, safe and nutritious food to meet their dietary needs. Bourland (1999) has studied the food systems during space travel. Space food systems have evolved from tubes and cubes to earth-like food being planned for the International Space Station. As there are many constraints in the space like the weight, volume and oxygen it is aital to know the type of food to be consumed. However, with food systems being better, now the spacecraft conditions, have become more habitable. Nutritional requirements for long-duration missions in microgravity and problems associated with meeting these requirements have been discussed. Sirmons et al., (2017) have studied the stability of vitamin content, sensory acceptability and color variation in fortified spaceflight foods over a period of two years. The work has looked into the optimal formulation, processing and storage conditions which will help in maintaining the stability and acceptability of commercially available fortification nutrients. A monitoring was done to check if fortification affects the product quality over time. This has helped to know the use in long-duration missions.

13

Food Regulation and Food Security Bill

Henderson et al., (2010) have studied the different aspects of food regulation. The authors have looked into the perceptions of Australian consumers' who is and should be responsible for food safety. Forty-seven participants were interviewed as part of a larger study on trust in the food system. Participants associate food governance with government, industry and the individual. While few participants can name the national food regulator, there is a strong belief that the government is responsible for regulating the quality and safety of food. Personal responsibility for food safety practices was also identified. Little evidence is found in the politicisation of food, reflecting a level of trust in the Australian food governance system that may arise from a lack of exposure to major food scares. According to Wood (1990) the marketing of food service program in an Ohio district is directed toward the students and also at the community, school administrators, teachers, and employees. Students are encouraged to follow a healthier way of eating. Jones (1998) studied the role of food safety in the food chain from farm to fork. In response to growing threat of food-borne illnesses the federal government had launched the Food Safety Initiative. A key element is the Hazard Analysis Critical Control Point's system (HACCP). Under this it has been designed to make everyone in the food-delivery chain to be responsible for ensuring a safe food supply.

Heinonen (2014) studied the use of additives which are used for improving the structure of the food and in the prevention of food spoilage. In case the high concentration of any additive causes adverse health effects to the humans in that case a limit of acceptable daily intake (ADI) is set for it. In case the ADI becomes high then there is a risk. The nutrient density decides the healthiness of food. De Decker et al., (2017) has studied the importance of the home food environment on unhealthy food consumption in children. Parents reported on their children's fast-food consumption frequency, food cue responsiveness (by means of the subscale 'food responsiveness') and on the home availability of unhealthy foods. The findings point at the importance of the home food environment in children and suggested in limiting the home availability of unhealthy foods. Lebel et al., (2016) have looked into the rural food deserts. Food insecurity is an important public health issue and affects many households. It happens due to poor accessibility to fresh, diverse and affordable food products. The work involves evaluation of 25 food products for the food stores located in rural counties in Quebec. The quality of food products was estimated using four indices: freshness, affordability, diversity and the relative availability. The result was compared with the food parameters proposed by the US Department of Agriculture (USDA), as well as with the perceptions of a group of regional stakeholders. Food stores' characteristics are different in rural areas and requires an in-store estimation to identify potential rural food needs to develop relevant interventions. Wight and Killham (2014) have studied the food mapping. This is a new, participatory, interdisciplinary pedagogical approach for learning about the modern food systems. Patten et al.,(2017) have looked into the determination of food safety practices at Summer Food Service Program (SFSP) sites, to identify the different types of food served at the sites and collect associated temperatures and to establish recommendations for food safety training in the SFSP. According to Anderson (2013) the right to food is widely accepted by nations, with the notable exception of the United States (US) and four other countries. The US government deals with domestic food insecurity through an array of need based food assistance programs instead of right based approaches. Ghosh (2014) has discussed the food safety regulations and food standards in Australia and New Zealand as they recognise it as a major global issue. Food security also affects Australia and New Zealand's status as premier food exporting nations and the health and wellbeing of the Australasian population. Australia is building up a resilient food value chain and support programs aimed at addressing existing and emerging food security challenges.

Adams et al., (2011) have studied the portrail of foods via advertisements. It has been seen that exposure to food promotion influences the preferences to food and diet. It has been found out that the food advertisements usually promote 'less healthy' products. In order to increase regulation the food manufacturers are engaging in a variety of health-promoting marketing initiatives. Towards this, television advertisements is the major mass communication system. In a survey it was found that only two-thirds of food advertisements tended to be towards healthy food.

Swaminathan and Bhavani (2013) have looked into the production of food and its availability as an essential prerequisite for sustainable food security. Food and nutrition security are interconnected as only a food based approach can help in overcoming malnutrition in an economically and socially sustainable manner. Food production provides the base for food security as it is a key determinant of food availability. According to Watthanakulpanich (2015) the natural foods which are obtained from soil or water environments might contain the infective stages of parasites endemic to these environments. Through the infected animal hosts it is possible that infection can enter the human food supply. Therefore, it is important to have awareness on the impact of parasites during the food supply. To prevent or reduce the risk posed by contaminated foodstuffs it is vital to handle the food in a safe manner along with good kitchen hygiene. It has been found out that parasites are not able to cause a health problem if the food is cooked thoroughly. Oexle et al., (2015) has surveyed on the neighborhood fast food availability and its consumption. The research has focused on how the availability of various types of food in a person's immediate area or neighborhood could influence his or her food choices and eating habits. According to them availability might not be the dominant factor influencing fast-food consumption. They have recommended subjective availability measures and individual characteristics. These could influence both perceived availability of fast food and its impact on fast-food consumption. Hartman et al., (2017) have described a general model of food webs within tidal wetlands and have looked into the parameters as to how physical features of the wetland affect the structure and function of the food web. The conceptual model given by the authors has focused on the technique of food web. This model is part of a suite of conceptual models designed to guide monitoring of restoration sites throughout the San Francisco Estuary (SFE), but particularly within the Sacramento-San Joaquin Delta (Delta) and Suisun Marsh. The conceptual models have been developed based on the Delta Regional Ecosystem Restoration Implementation Plan (DRERIP) models, and are designed to aid in the identification and evaluation of monitoring metrics for tidal wetland restoration projects. Intertidal wetlands and shallow subtidal habitat can be highly productive. Restoring areas of tidal wetlands may result in an increase in productivity that will provide food web support for the fish species. Recently DÃaz and Ãngel(2015) have looked into the radical new approaches in the EU Commission to ensure the highest standards of food safety for EU citizens. This is based on a more coordinated and integrated organization. The basic principle is to extend the application of control measures at all stages of the production chain. Beef and cow's milk, two basic components of the European diet needs attention towards the quality from the source. Koppel (2014) has looked into the sensory aspects of pet foods. Pet food palatability depends first and foremost on the pet and is related to the pet food sensory properties such as aroma, texture and flavor. Sensory analysis of pet foods provides additional information on palatability and unpalatable foods as pets lack linguistic capabilities.

DeJesus et al., (2018) have done an investigation of maternal food intake and maternal food talk as predictors of child food intake. Research has been done to examine maternal food intake and maternal food talk as independent predictors of children's food intake. The present study has looked into the maternal food talk during a structured eating protocol, in which mothers and their children had the opportunity to eat a series of familiar and unfamiliar vegetables and desserts. In the study it has been found that higher maternal body mass index (BMI) predicted lower amounts of food talk, pronoun use, and questions. However, maternal food talk did not predict children's food intake. Heikkila et al., (2016) have studied the elements that affects the food waste in the food service sector. According to them avoidable food waste is mostly produced in the food service sector which has significant ecological and economical impacts. In a study to find out the reasons for the generation of wastes in restaurants and catering businesses it was found that eight elements influenced the production and reduction of food wastes. Results revealed the diversity of managing food waste in the food service sector and a holistic approach to prevent and reduce it. It is important to know that food waste is manageable. The model of eight factors provides a framework for recognition and management of food waste in the food service sector. According to Koivisto (1999) the food habits are not stable in a lifetime. Usually the base for healthy food habits is created in early childhood. Children's food habits can be assumed to be influenced by their parents' food habits and choices. The author has reviewed the various factors which influences the choice of food in children as well as in adults. The results demonstrate that the development of children's food habits is influenced by a multitude of factors with parents playing an important role in the formation of food habits and preferences of young children. They also get afraid of new foods and are not ready to eat them. Parents should therefore be encouraged to make healthy foods available to the child and serve these foods in positive mealtime situations in order to help their children to develop healthy food habits. Martindale (2017) has studied the potential of food preservation to reduce food waste. The food waste can help us to determine the sustainability of diets. The work has shown that there is a need to consider the role of food preservation in order to reduce the waste produced from foods. There is a need for new marketing terms associated with sustainability actions that can be used to stimulate changes in consumption behaviours. Use of frozen foods has shown less waste as compared to preparation of fresh good. Sobrino et al., (2014) have studied the connection between peptides and food intake. The controlling of food intake involves an interaction between the major organs which involve gut, brain and adipose tissue (AT). For the communication between the brain and the satiety center, it is the gut, AT parasympathetic, sympathetic and other systems which are required. The neuronal circuits consists of peptides and hormones. Nutrients generated by food digestion have been proposed to activate G-protein-coupled receptors on the luminal side of enteroendocrine cells, e. g., the L-cells. It stimulates the release of gut hormones into the circulation. These include glucagon-like peptide-

1 (GLP-1), oxyntomodulin, pancreatic polypeptides, peptide tyrosine and cholecystokinin. Ghrelin is a peptide secreted from the stomach and in contrast to other gut hormones, plasma levels decrease after a meal and stimulate the intake of food. Other circulating factors like insulin and leptin relay information for long-term energy storage. Both hormones circulate at proportional levels to body fat content, enter the CNS proportionally to the plasma levels reducing the food intake. These can also influence the activity of the arcuate nucleus (ARC) neurons of the hypothalamus, after passing across the median eminence. Gut hormones being the circulating factors can influence the nucleus of the tractus solitarius (NTS) through the adjacent circumventricular organ. From the NTS it is the neural projections that carry signals to the hypothalamus. The ARC acts as an integrative center, with two major subpopulations of neurons influencing appetite, one of them coexpressing neuropeptide Y and agouti-related protein (AgRP) that increases food intake. According to a study by Hartel (2013) crystals play an important role in the quality and shelf life of any food product. In order to obtain the desired crystal content, size distribution, shape and polymorph in manufacturing products decides the functionality and shelf life of the product. The work done by Hartel (2013) has helped in the understanding of the physico-chemical phenomena that governs crystallization and improves the ability to control it during processing and storage. Recently much work has been done to analyse the allergies caused due to environment (Huffman, 2015). Immunoglobulin E-mediated allergic responses to food and environmental allergens cause mild allergic rhinitis and rashes to gastrointestinal distress and also anaphylaxis. It has been observed that many times the diagnosis becomes difficult, as it relies on complex interplay between patient history and diagnostic tests with low specificity. Adding to the difficulty in confirming the diagnosis is an increased public interest in food intolerances, which can be inappropriately attributed to an allergic response. In case the treatment is given in time it can improve the quality of life and can control the chronic conditions, like asthma and eczema. According to Jordan (1998) it has been observed that many adolescents turn to vegetarianism. This is manily found among females. They do not eat meat, have concern for the environment and animal welfare. The paper tells that this can be a precursor to eating disorders.

Kibler et al., (2018) have not only discussed the food waste but have looked into the food-energy-water nexus (FEW). The review has also discussed the food waste management alternatives. A lot of food which is produced is wasted all around the world. However, little is known about the energy and water consumed in managing the wasted food. The various food waste management options, which include waste prevention, landfilling, composting, anaerobic digestion and incineration, provide variable pathways for FEW impacts and opportunities. Characterization of FEW nexus impacts of wasted food, including descriptions of dynamic feedback behaviors, presents a significant research gap and a priority for future work. Large-scale decision making requires more complete understanding of food waste and its management within the FEW nexus, particularly regarding post-disposal impacts related to water. The food security bill is important as it makes food our legal right. It has been passed in India recently that will provide food to all. By the FAO (Food and Agricultural Organization) food security is defined to meet the dietary diversity requirements in line with the cultural preferences taking millets, proteins and oil to eradicate malnutrition. A connect is required between the nutritional needs, agricultural policy and food programmes at the national and international level. Using newer technologies, biometric systems, truck tracking systems, painting PDS transport vehicle with identifiable distinct color and SMS to panchayats at the village level when truck leaves the godown are some of the ways to enhance the efficacy. With more items of consumption being owned by overseas enterprises every purchase gives value out of the country to the foreign owners. To keep our own profits it is therefore, important to align corporates along the lines of national and international energies which may be a source of wealth.

Today, there is a large change in the pattern of food production, processing, distribution, storage that have created new challenges to food safety. Use of pesticides, veterinary drugs, genetic modifications and pollutants in sources of irrigations has enhanced increased production of food products. However, the risk of food contaminants has also increased which create new risks. Food safety is a basic human right which gives the right to healthy and safe environment. Consumers are concerned about the health risks associated with unsafe and unhealthy food. The public is aware about the potential dangers in food by microbial contaminations, heavy metals, hazardous chemicals and toxins. With the changing life styles and food habits and the outbreaks of food borne diseases, adulterations, the consumer has become very aware on the nutrition, quality and food safety issues. This has resulted to form new regulations. Looking into this, a new law, FSSAI (Food safety and standards authority of India) was constituted in 2008 and operationalized in 2011. This allows uniform rules, regulations and standards throughout the country.

It is important to have a connect between the nutritional needs, agricultural policy and the food programmes. The legislative authority, FSSAI, takes up the responsibility for national leadership on food safety along with concerned ministries to ensure farm to fork continuum. It has representatives from food industry, consumer organizations, food technical scientists, union territories, farmer's organization and retailer's organization. For functioning it includes central licensing authority, state licensing authority and registering by involving panchayat and municipal bodies. FSSAI is a simple reference point for all matters relating to food safety and standards. For food businesses it is important to establish food safety management system by NABL or any other body. The food testing system must be for good management techniques and improvement in test competencies.

The nations should have mutual benefits. For networking it is important to exchange information, development, implement joint projects, exchange expertise and use best practices in the field of food safety. These could be in the form of training and capacity building, food surveillance and scientific activities of the authority. For safe and wholesome food as per international standards it is important to understand food borne diseases and infections which would link them to food contamination on the basis of risk. Building requires insight, continuous nurturing of R&D, brand building capability, cutting edge manufacturing, an extensive trade marketing and distribution network. Research agencies should provide the grant with deliverables of international standards. Standards must reflect policy decisions about acceptable risks and costs of risk avoidance. For many food-safety hazards, consumers cannot detect the hazard at the time of purchase and producers may also be unable to measure or guarantee a particular level of safety. It is difficult for the producers to always supply as per the demand of the consumers. Due to the adverse health effects from the presence of chemical additives in foods the consumers are drawn to natural and "fresher" foods with no chemical preservatives added. There is an increasing demand for minimally processed foods with long shelf life and convenience, which has enhanced research in finding natural but effective preservatives. Bacteriocins, produced by LAB, may be considered natural preservatives or biopreservatives that fulfil these requirements. Biopreservation refers to the use of antagonistic microorganisms or their metabolic products to inhibit or destroy undesired microorganisms in foods to enhance food safety and extend the shelf life (Schillinger et al. 1996). For the biopreservation of foods application of bacteriocins is usually considered with multiple approaches (Schillinger et al. 1996). It is the FDA which conducts the environmental assessments to learn the probable cause(s) of an outbreak of foodborne illness or a food contamination event and uses that information to identify preventive controls to prevent reoccurrence of an outbreak or contamination event. The Agency's first priority should be to minimize the number of people who become ill in case of any outbreak or food contamination event. It must be checked to know the probable cause and to identify preventive controls for achieving the long-term goal of preventing outbreaks of foodborne illness and food contamination events that may be applicable to the industry.

14

Challenges in Food Safety

Food safety scientifically describes various methods of handling, preparation and storage of food to prevent any foodborne illnesses. However, it is important to check the safety between industry, market and the consumer. From the food industry to the market the practices relating to food labeling, food hygiene, food additives and pesticide residues are important to be looked into and the policies on biotechnology. The governmental imports and exports needs to be given proper guidelines. Certification systems and Inspection for foods need to be installed which should be clean and corruption free. When all this is properly considered the food could be safe in the market. This may allow for safe delivery and safe preparation of the food for the consumers. Many diseases are transmitted by food which serves as a growth medium for bacteria and cause food poisoning. In the developed countries it has been observed that there are many intricate standards for food preparation. However, in the developing countries the availability of adequate safe water itself is a question. It is possible to prevent food poisoning to 100%. An ordinance and code regulation in US (1943) was given.

The economists look into the socio-economic issues which are associated with the safety in food supply. This might reflect the real incidence of food-borne illnesses world-wide, and in part consumer concerns about the safety of the food they eat. The impact of food safety regulations goes on the global trade in agriculture and also on food products.

Considering the food products which directly go into the gut system of human has two aspects, health and safety which are very important. The supply and demand are connected with the economics and consumption of the food. It has been found out that if the diseases are more with the final products then the costs of curing increases. This reduces the market of the product. This is common in the developing countries. The high rates of food-borne illnesses are associated with low levels of economic development. There are close inter-relationships between food safety issues with environmental health, like sanitation, water quality and housing conditions (Sacramento, 2004). *E. coli* 0157 has been found to be carried by cattle and heavy rains with contamination of surface water which contributes to the outbreaks (Cooley et al., 2003). Jarvenpaa,. and Todd,(1992) studied the consumer study of electronic media. In Vietnam, 30 to 57 % of students in the university hostels in Hanoi were found to suffer from diarrhoea (1984-88). This was envisaged due to poorly prepared and/or stored food. The illnesses were associated with *Salmonella, E. coli* or *S. aureus*. Buzby (2001) studied the product liability and microbial food-borne illness. In order to bridge the gap in developing countries a broader group of studies have examined the levels of microbiological and/or chemical contamination in foods without any attempt to relate them to human diseases. Maximum contamination has been found through street foods and the products of the informal sector. Gran et al. (2002) investigated the occurrence of pathogenic bacteria in raw milk and fermented milk products in Zimbabwe. In certain cases high levels of contamination have been identified which gives significant health hazards to the consumers. Bonfoh et al., (2003) have found that the microbiological quality of milk in Bamako, Mali had high levels of bacterial contamination. Levels of aflatoxin B1 in milk in Argentina (Lopez et al., 2003); raw vegetables in Malaysia were found to be infested with *Salmonella* (Salleh et al., 2003) and polycyclic aromatic hydrocarbons (PAHs) were found in fruits and vegetables in Brazil (Camargo and Toledo, 2003). In an attempt to bridge gaps in the available data, the WHO has developed a new approach to estimate the incidence of food-borne diarrhoea using a risk assessment based approach (WHO, 2002). This approach uses the estimated annual incidence of food-borne non-typhi and paratyphi *Salmonella* as an indicator of the overall incidence of food-borne diarrhoea. Further, it employs the transfer rate (TR) risk assessment approach based on the link between exposure to a microorganism in food and a disease. Over the period 1984-89, there were 721 food-borne outbreaks and 1,199 sporadic cases of food-borne disease were recorded in the cities of Hyderabad and Secunderabad, India (Shekhar et al., 1992). The carriers of infection were stale food, rice dishes, sweets and curry. Stale food was most generally left over from a previous meal and stored at room temperature, usually overnight (Rao, 2005). According to Walker (2005) cafeteria food is a big challenge.

According to Teufel et al., (1992) in Pakistan many sweet dishes are found to be infested with *S. aureus*. Some samples of khoa obtained from manufacturers in a large city contained appreciable levels of *S. aureus*. Presence of *Salmonella* was seen in khoa and cheese-based confectionery. pulses, ground meat dishes and chick peas vended at bus and train stations in the same city contained high levels of *C. perfringens* at inadequate temperature (Bryan et al., 1992a). Bryan et al., (1992b) studied the food prepared at home and found pathogens like *S. aureus, C. perfringens* and *B. cereus* (In Latin America) which carry many disease systems (Todd 1997). Diarrhoeal diseases account for 967 deaths per

100,000 of the population amongst children who are less than one year. Usually, four to seven bouts of diarrhoea are experienced each year by children aged less than five years. In Argentina, *E. coli* 0157 (Lopez et al., 1992) have been recorded as responsible for significant rates of food-borne illness. In years 1986 and 1990, there were 35 outbreaks of *S. enteritidis* which affected 3,500 people, due to less cooked poultry / eggs in mayonnaise.

Over the past three decades the world has produced more grain per capita. But still several million people have died of hunger. What and where is it that we are going wrong? We need to work out mechanisms together. We need to check the global grain markets, international trade and price control for chronic food deficits which is impairing the lives of the people. This is directly related to the global market. Populist schemes are a boon for the country's economy if they reach out to the needy. We need an "International outlook" to take the country's progress to the next level. We need to change the way we do things. We need to bring in new technologies, innovation and change the earlier set processes. In 1972, there was a severe "World food crisis". But it was good weather and strong incentives given to the farmers which improved the conditions in 1978. The world food economy is intimately interdependent. The linkages among different countries and commodities are to be provided through international trade. Trade in agriculture, petroleum, fertilizers and pesticides connect the agricultural and industrial economies of most countries. The foreign exchange transactions and rate directly affects the food policy. Everything is globally interdependent. Interdependence will enhance specialization and promote higher productivity by better understanding and utilizing each other's potential at the international level. More than half of the world's population lives in Asia and most of the people are very poor, which creates 2/3rd of the world hunger. Solving hunger involves, expanding the available choices like income, food prices, food supplies and foremost the consumer knowledge. Practical processes for agricultural commodities are required at the farm level which would transform food for the marketing sector and provide better nutrition to the consumers. Solving malnutrition is the priority to all concerns in all Asian countries. The processes which make up the food system can improve by collaborations amongst all countries which could be from agri inputs to nutritional supplementation. A food system that has many hungry people is a failure. Not only that, it is important to produce enough food to eat, buy and to generate enough income. Also proper education should be given to people what they should eat to combat malnutrition. These factors in the policy will help in eradicating the hunger. The vulnerability is extreme for subsistence due to climate, droughts, floods, pests, failed crops and economic circumstances. A good food policy can protect this and can improve for good food products in the markets giving more economic opportunities among countries for trade. The food policy is a collective effort of the government, intergovernments, food producers, consumers and marketing agents to ultimately reach the goal. Food Policy is a process of research and a great thinking to design and discover the trade identifying government initiatives to the projects, programmes and policy arenas which should give intercountry benefits. Each domestic food policy maker must link their country's food system to the Asian and world commodity markets. Food price policy is an important element both for trade and price. A major role of the food policy should be to bridge the problems among nations which will lead to more rapid growth bridging towards rural development that reach the small producers, using food price effectively and carefully targeting the food subsidies.

Globally we all have same Objectives:

1. Efficient growth in food and agricultural sectors.

2. Improved income distribution and employment creation.

3. Satisfying nutritional status for the entire population.

4. Adequate food security to ensure against bad harvest, natural disasters, uncertain world food supplies and prices.

Therefore, it is very important to strategize minutely a consistent and workable food policy which should include land, income transfers and technical changes. It should be focused on job creation, incentives to enhance food production, investment in agricultural infrastructure, economic efficiency and rising productivity. It is impossible to afford mismanagement, waste or inefficiencies by bad policies. It is very important to keep in mind the intergovernmental trade benefit along with bilateral cooperation.

15

Conclusions and Future Prospectives

For looking into the biosafety in research laboratories, agriculture and other food producing industries it is important to look into the numerous strains of lactic acid bacteria that are associated with food systems and are capable of producing bacteriocin or antibacterial proteins with activity against foodborne pathogens such as *Listeria monocytogens, Staphylococcus aureus* and *Clostridium botulinum*. Recently considerable attention has been given on the identification of these proteins and their biochemical properties including spectrum of activity, production conditions, purification procedures, aminoacid composition and mode of action. Advances in genetic engineering have facilitated the characterization of proteins at genetic level, providing information about protein synthesis, structure and cloning of bacteriocin genes into other organism. One problem in the use of bacteriocin in their applications as bio-preservatives is the emergence of resistant population. The understanding on the mechanism of resistance to bacteriocins is limited which limits the application in bio-preservation. Much of the work on bacteriocin has been done under controlled laboratory conditions without the presence of adventitious microbial population. More research is needed to determine the efficacy of bacteriocin producing organisms and bacteriocins which are stable under commercial conditions. Food-borne illness is a major international problem and an important cause of reduced economic growth. The contamination of the food supply with the pathogens and its persistence, growth, multiplication and/or toxin production has emerged as an important public health concern. Most of these problems could be controlled with the efforts on the part of the food handlers, whether in a processing plant, a restaurant, and others. In contrast with most chemical hazardous compounds, the concentration of food pathogens changes during the processing, storage, and meal preparation, making it difficult to estimate the number of the microorganisms or the concentration of their toxins at the time of ingestion by the consumer. Food safety in the early twenty-first century is going to be an international challenge which requires close cooperation between countries in agreeing standards and in setting up transnational surveillance systems. The lessons of the past two decades are plain to those engaged in the food industry.

No longer can farmers grow just what they want or use technical aids to farming without taking into account the effect on the quality of the food produced. Food will always present some risk, and it is the task of the food industry to keep the level of risk to the minimum, which is practicable and technologically feasible. It should be the role of official bodies and the food industry to use risk analysis to determine realistic and achievable risk levels for foodborne hazards and to base food-safety practices on the practical application of the results of these analyses, thus continuously improving the safety of food and thereby lowering the disease risk. By providing effective food-control, countries not only provide public health benefits but enable themselves to participate in international food trade with greater confidence. The lasting solution to the problems, which have shaken the food industry in recent years, can only be overcome if all those involved – farmers, food processors, wholesalers, transporters, retailers, caterers, scientists, regulators and government work together towards common goals. Since the first antibiotic penicillin was discovered in 1928 by Alexander Fleming, many discovered antibiotics are applied to treat pathogens. Antibiotics were first approved by USFDA in 1951, and used in animal feed, which significantly reduced the number of deaths due to bacterial infection. However, the problem with multiple drug resistance pathogens has become increasingly serious, due to concerns regarding the abuse of antibiotics (Joerger, 2003). Bacteriocins are reported to inhibit important animal and plant pathogens, such as Shiga toxin-producing *E. coli* (STEC), enterotoxigenic *E. coli* (ETEC), methicillin-resistant *Staphylococcus aureus* (MRSA), VRE, *Agrobacterium*, and *Brenneria* spp. (Grinter et al., 2012; Cotter et al., 2013). The bactericidal mechanism of bacteriocins are mainly located in the receptor-binding of bacteria surfaces, and then through the membrane, which causes bacteria cytotoxicity. In addition, bacteriocins are low-toxic peptides or proteins sensitive to proteases, such as trypsin and pepsin (Cleveland et al., 2001). Jordi et al., (2001) have found that 20 kinds of *E. coli* could express colicin, which inhibited five kinds of Shiga toxin-producing *E. coli* (O26, O111, O128, O145, and O157: H7). These *E. coli* can cause diarrhea and hemolytic uremic syndrome in humans. In a simulated cattle rumen environment, colicin E1, E4, E8-J, K, and S4 producing *E. coli* can significantly inhibit the growth of STEC. Stahl et al., (2004) utilized purified colicin E1 and colicin N for effective activity against enterotoxigenic *E. coli* pathogens F4 (K88) and F18 in *in-vitro*, which caused post-weaning diarrhea in piglets. Furthermore, purified coilicin E1 proteins were mixed with the dietary intake of young pigs. The results showed a reduction of the incidence of post-weaning diarrhea by F-18 positive *E. coli* (Cutler et al., 2007). The growth of the piglets was thus ameliorated. Józefiak et al., (2013) used the nisin-supplemented bird diet to feed broiler chickens, and found a reduced number of Bacteroides and Enterobacteriacae in ileal digesta of nisin-supplemented chickens. The action of nisin was similar to that of salinomycin. After a 35-day growth, the average body weight gain of nisin-supplemented (2700 IU nisin/g) chickens is 1918 g/bird, which was higher than the 1729 g of non-nisin supplemented or the 1763 g of salinomycin-supplemented chickens. Stern et al., (2006) purified class II low molecule mass bacteriocin OR-7 from *Lactobacillus salivarius* strain NRRL B-30514 with an ability to kill human gastroenteritis pathogen *Campylobacter jejuni*. OR-7 was stable when treated with lysozyme, lipase, and heat to 90°C, or at pH ranges from 3.0 to 9.1. The purified OR-7 was encapsulated in polyvinylpyrrolidone for chicken feed. The populations of *C. jejuni* were reduced at least one million fold over that of non-OR-7 supplemented chickens in cecal material of OR-7-treated chickens and had potential to replace antibiotics in poultry and other animal feeds.

In the recent past cancer has become a threat to human health. Major being lung cancer (2.7%, including trachea and bronchus cancer) which caused 1.5 million (2.7%) deaths in 2011. This was higher than 1.2 million (2.2%, 9th) deaths in 2000. In the US around 1,660,290 new cancer cases, with 580,350 cancer deaths occured in 2013. This has shown cancer to be the second leading cause of death, exceeded by heart diseases (Center, 2013). Bacteriocins may be suitable as a potential anti-tumor drug. Some bacteriocins, such as pore-forming colicin A and E1 inhibited the growth of one human standard fibroblast line MRC5 and 11 human tumor-cell lines (Chumchalová and Smarda, 2003). The recent works (Fuska et al., 1979; Lancaster et al., 2007) have shown that pore-forming colicin U and with RNAase activity, colicin E3, had no growth inhibition capability. Colicin D, E2, E3, and pore-forming colicin A could inhibit the viability of murine leukaemia cells P388, while pore-forming colicin E1 and colicin E3 suppressed v-myb-transformed chicken monoblasts. In a study done by Bures et al., (1986) isolated *E. coli* from the faeces of 77 colorectal carcinoma patients, where 32 patients (41.6%) had barteriocins-producing *E. coli*. In the faeces of 160 healthy people, 102 people (63.8%) had barteriocins-producing *E. coli*, which also showed that colicins from bacteria in the intestine may be one of the factors in reducing human colorectal carcinoma. Colicins could act as an anti-cancer drug of moderate potential. Supplements of bacteriocin-producing probiotics may be another way to prevent cancer occurrence. Recently, Joo et al., (2012) found that nisin had capabilities to prevent cancer cell growth. Three head and neck squamous cell carcinoma (HNSCC) 17B, HSC, and 14A were treated by nisin at concentrations of 40 and 80 ig/ml. After 24 h nisin increased DNA fragmentation or apoptosis, arrested cell cycle and reduced cell proliferation of HNSCC. The floor-of-mouth oral cancer xenograft mouse model was used to test the anti- HNSCC function of nisin. A 150-mg/kg dose of nisin was administered orally every day for 3 weeks, and tumor volumes were significantly reduced in nisin-treated mice, as compared with the control mice, which received only water. These results indicated that nisin provides a safe and novel therapy for treating HNSCC. In general, the genes and immune genes of bacteriocins are encoded on the same plasmid or in adjacent regions of a chromosome. The bactericins genes can enter into other bacterial cells via conjugation (Ito et al., 2009; Burton et al., 2013). Therefore, it is by conjugation or transposon the bacterial pathogen can obtain some immune genes of bacteriocins. The use of any particular bacteriocin may not eliminate all bacterial pathogens. It can be a good strategy to mix varieties of bacteriocin proteins which may help to kill many pathogens. Bacteriocin can solve some of the most challenging problems in disease control caused due to multi-drug resistant pathogens. Using the recombinant PCR techniques, Acuña et al. (2012) have integrated enterocin CRL35 and microcin V genes to obtain a bacteriocin called as Ent35-MccV. This has activity against clinically isolated enterohemorrhagic *E. coli* and *Listeria monocytogenes*, as well as some Gram-positive and Gram-negative bacteria, which are not clinically isolated. Such a technique can crease new or multi-functional bacteriocins, which are more powerful in functionality and germicidal range. As a result, they can be widely used in food, animal husbandry and medicine.

Conclusion

Bacteriocins either as such or produced by live bacteria, can be successfully incorporated into dairy products to assure safety. It also provides extended shelf-life with better quality. The preferred method is the use of purified and concentrated bacteriocins. This provides more efficacy, compared to direct application of bacteriocinogenic cultures. The effectiveness of bacteriocins in food systems is usually found to be low due to several factors like adsorption to food components, enzymatic degradation, poor solubility, or uneven distribution in the food matrix. The bacteriocinogenic LAB added to dairy products such as yogurts and cheeses will ensure continuous production of bacteriocins throughout maturation and storage, and may be included as starter/adjunct cultures in fermentation. The lack of compatibility between the bacteriocin-producing strain and other cultures required in the fermentation of dairy foods makes the application little difficult. *Listeria monocytogenes* which is a foodborne pathogen, is of great concern in traditional cheeses made from raw milk and from pasteurized milk. Listeriosis is linked with the consumption of contaminated cheeses. It has been reported worldwide and there has been a significant research interest in recent years. In contrast, very limited research has been focused on the application of bacteriocins to the preservation of milk, cream, yogurts and other dairy products. More studies are necessary for effective application of bacteriocins in dairy foods to understand the action of bacteriocin in the complex environment of food matrices. Bacteriocins may be combined with other protection tools for the biopreservation and shelf life extension of dairy foods, or other preservation techniques, or in the incorporation in biofilms and active packaging as these compounds are able to influence the safety and quality of foods. In the past years, a lot of work has been done on the detection, purification and characterisation of bacteriocins and their use in food preservation strategies. The biopreservatives can be used in a number of ways in food systems both in fermented and non-fermented food products. However, there is still a considerable burden of foodborne illness, in which microorganisms play a prominent role. The microbes can enter the food chain at different steps, are highly versatile and can adapt to the different environment which allows their survival, growth and production of toxic compounds. This sets them apart from chemical agents and thus their study from food toxicology. A conference was organized by the Dutch Food and Consumer Products Safety Authority and the European Food Safety Authority in which the scientists discussed on the new challenges in food safety that are caused by micro-organisms as well as strategies and methodologies to counter these.

As food safety is required on different scales and in different time frames development of new risk assessment approaches is important. This requires active governmental policy setting and anticipation by food industries. Monitoring of contamination in the food chain, combined with surveillance of human illness and epidemiological investigations of outbreaks and sporadic cases are important to have sources of information. New approaches are the use of molecular markers for improved outbreak detection and source attribution, sero-epidemiology and disease burden estimation. This is possible with the current developments in molecular techniques where information can be obtained on the genome of various isolates of microbial infection. This also helps to develop new tracking and tracing methods and to investigate the behavior of micro-organisms under environmentally relevant stress conditions. These novel tools and insight needs to be applied to objectives for food safety strategies, as well as to models that predict microbial behavior. With the increasing complexity of the global food systems it is very vital that the knowledge is communicated between all parties involved: scientists, risk assessors and risk managers and the consumers.

At present HACCP (Hazard Analysis Critical Control Point) programs and GMP (good manufacturing practice) are mainly used to manage microbial hazards in foods. While these systems have proven very effective for the control of food safety (Van Der Spiegel et al., 2004; Arvanitoyannis et al., 2004), it must be realized that they are designed on the basis of known hazards, and do not necessarily take potential future developments in consideration. With any innovations, new validations and verifications should be made must. Furthermore, it should be realized that the implications of microbial adaptability are not sufficiently taken into account (McMeekin and Ross, 2002, McMeekin et al., 2006,2007). The validity of the effectiveness of the written guidelines and the adherence to them is a very important aspect of HACCP and GMP procedures. Producers and handlers of foodstuffs are more likely to adhere to the prescribed HACCP procedures if they recognize these as useful and implementable (Taylor and Taylor, 2004). Climate change is considered to be one of the greatest current challenges to mankind, affecting all sectors of the society, including nutrition, food security and food safety (Anonymous, 2008b). Due to climate change various risks may change, as a result of changed ecological conditions on various places on the earth. Changing ecology is expected to affect the distribution of plant and animal diseases. Water shortages may lead to limited quantities or quality problems with irrigation water, process water or ingredient water. This may lead to shifts in production areas and cultured crops, as well as an increased use of agrochemicals. This trend may be increased by competition for land-use, e. g. for biofuels or for settlements. Flooding can contaminate crops in the field. If the temperatures rise it may affect the cold chains. It has been found out that with increase in humidity the levels of mycotoxins may increase and certain foodborne pathogens may thrive better under warm conditions. The effect of climate change on food safety is very complex and interrelated with many other societal factors. With recent increase in international travels, pathogenic microorganisms can be spread very easily and quickly around the globe. People from various continents may come in contact with organisms and to the specific strains to which they have not been exposed earlier. This may increase the risk of illness.

In recent times the global trade has also increased which has a longer transit distance and duration in the food chain. This would also increase the contamination risk further. On the other hand more powerful stakeholders (trade companies, supermarkets) will have the intention to influence these complex chains in order to guarantee food safety. Several large retailers that operate internationally have organized the Global G. A. P. quality control system (www. globalgap. org) that aims to supervise the primary production process of all agricultural products. Sourcing food from various climate areas means more variation in hazards. Border controls are effective in regard to the control of only a small proportion of imported foods. The Codex Alimentarius Commission and the European Union have been useful towards a positive trade but may result in very different health risks in different parts of the world, due to differences in demography, immune status, food preparation , consumption habits and relevance of various routes of infection (Käferstein et al., 1997). Furthermore, we prefer to eat the food as fresh as possible (Doyle and Erickson, 2008). This trend increases the need for worldwide food safety systems along with collaboration and communication between all players in the food chain. As food chains extend or expand from local chains into worldwide chains, more and different factors may affect food safety. The food safety management knowledge or information from different scientific disciplines needs to be combined and the information made available should be standardized. Food safety starts at primary production. For foods to be eaten raw (fresh produce and shellfish) knowledge about contamination routes and preventive measures is of great importance. Sometimes, the interventions may be effective early in the primary production phase, including the production environment. For fresh produce, grown in the open field, *E. coli* (in particular the verocytotoxin producing strains VTEC) is, amongst others, a food safety hazard. This is underlined by a massive foodborne infection outbreak in the USA in 2006 (Anonymous, 2007e) caused by baby spinach eaten raw. It appears that this outbreak was either due to irrigation water or faeces from cattle or wild boar. Cattle are a known source of this pathogenic bacterium (Hussein and Sakuma, 2005), yet cow manure continues to be used as a main soil fertilizer in organic farming (Anonymous, 2000b). Preventive measures could be a change of feeding diet in order to reduce the numbers of VTEC shed by cattle (Diez-Gonzalez et al., 1998; Synge, 2000) or to reduce their survival in manure-amended soils (Franz et al., 2005). Thus, produce safety can be increased in this case by combining agricultural science and (food) microbiology. In 2008, a *C. jejuni* outbreak in Alaska was linked to the consumption of raw peas contaminated on the field by Sandhill crane faeces. The outbreak investigation identified a lack of chlorine residual in pea processing water, which could have been easily prevented (Gardner and McLaughlin, 2008). Introducing buffer zones, set-back distances and fences to restrict wildlife access to the production environment of produce such as leafy greens may prevent problems with feral swine or deer (Atwill, 2008). Although these solutions seem sometimes fairly simple, they could only be taken due to a proper outbreak investigation that elucidated the (probable) source of the food contamination. In order to stop an ongoing outbreak, such outbreak investigations should be carried out quickly, which relies on close collaboration between different (scientific) disciplines such as microbiologists, epidemiologists, wildlife control specialists, risk communicators, (Wijnands et al., 2006). A simple, spreadsheet-based tool is being developed to assess consumer risks associated with such products using available data (Evers and Chardon, 2008). These examples clearly show the benefits of inter- and intrascientific collaboration. Close personal collaboration may not always be necessary when data can be shared by other means. In the scientific literature, many data are published that can be used by others. However, translation of these data to a uniform data set is time consuming. In order to improve sharing of data on microbial growth and inactivation, the ComBase Initiative was established, a collaboration between the Food Standards Agency and the Institute of Food Research from the United Kingdom, the USDA Agricultural Research Service and its Eastern Regional Research Center from the United States and the Australian Food Safety Centre of Excellence (Combase Consortium, 2008). ComBase is a combined database of microbial responses to food environments and data can be used for predictive modeling. Recently, Combase started a collaboration with the Journal of Food Protection, which request authors to submit their data to the Combase database. A clearinghouse of interdisciplinary data is offered by Foodrisk. org, an initiative of the Joint Institute for Food Safety and Applied Nutrition (JIFSAN), a collaboration between the University of Maryland (UM) and the Food and Drug Administration (FDA). The clearing house provides data and methodology on food safety risk analysis offered by the private sector, trade associations, federal and state agencies, and international sources (www. foodrisk. org). In conclusion, it is the interdisciplinary collaboration which is promising. Along with this, extra efforts should be made to achieve more progress.

References

Abee, T., Krockel, L., and Hill, C. (1995). Bacteriocins: modes of action and potentials in food preservation and control of food poisoning. Int. J. Fd. Microbiol. 28, 169–185.

Abriouel, H., Franz, C. M., Omar, N. B., and Gálvez, A. (2011). Diversity and applications of *Bacillus* bacteriocins. FEMS. Microbiol. Rev. 35(1):201–232. doi:10. 1111/j. 1574-6976. 2010. 00244. x. PMID:20695901.

Achemchem, F., Abrini, J., Martinez-Bueno, M., Valdivia, E., and Maqueda, M. (2006). Control of *Listeria monocytogenes* in goat's milk and goat's Jben by the bacteriocinogenic *Enterococcus faecium* F58 strain. J. Fd. Prot. 69, 2370–2376.

Acuña, L., Corbalan, N. S., Fernandez-No, I. C., Morero, R. D., Barros-Velazquez, J., and Bellomio, A. (2015). Inhibitory effect of the hybrid bacteriocin Ent35-MccV on the growth of *Escherichia coli* and *Listeria monocytogenes* in model and food systems. Fd. Bioproc. Technol. 8, 1063–1075.

Acuña, L., Picariello, G., Sesma, F., Morero, R. D., and Bellomio, A. (2012). A new hybrid bacteriocin, Ent35–MccV, displays antimicrobial activity against pathogenic Gram-positive and Gram-negative bacteria. FEBS Open Bio. 2, 12–19.

Adams, J., Tyrrell, R., and White, M. (2011). Do television food advertisements portray advertised foods in a 'healthy' food context? The Brit. J. nutr. 105(6):810-5.

Adriana, D. T. F., and Guy A. C. (2016). A review of the impact of preparation and cooking on the nutritional quality of vegetables and legumes. Intern. J. Gastronomy Fd. Sci. 3, 2-11.

Adrienne, K., Christine, P., Diana, P., Margaret, R., and Robert, D. (2016). The Phenotypic and Genotypic Landscape of Colicin Resistance. Caister Academic Press. U. K. 141-154. DOI: https://doi. org/10. 21775/9781910190371. 08.

Aguayo, M. D. C. L., Burgos, M. J. G., Pulido, R. P., Gálvez, A., and López, R. L. (2016). Effect of different activated coatings containing enterocin AS-48 against *Listeria monocytogenes* on apple cubes. Innov. Fd. Sci. Emerg. Technol. 35, 177–183.

Ahlstrom, C., Barkema, H. W., and DeBuck. (2014). Improved short-sequence-repeat genotyping of *Mycobacterium avium* subsp. *paratuberculosis* by using matrix-assisted laser desorption ionization-time of flight mass spectrometry. J. Appl. Environ. Microbiol. 80(2):534-9.

Ahmad, A., Nagaraja, T. G., and Zurek L. (2007). Transmission of *Escherichia coli* O157:H7 to cattle by house flies. Prev. Veterin. Med. 80, 1, 74–81.

Ahmad, R. I., Seo, B. J., Rejish Kumar, V. J., Choi, U. H., Choi, K. H., Lim, J. H., and Park, Y. H. (2013). Isolation and characterization of a proteinaceous antifungal compound from *Lactobacillus plantarum* YML007 and its application as a food preservative. Lett. Appl. Microbiol. 57, (1), 69-76.

Ahmed, S., Cox, A., and Cornish, P. V. (1989). International food patterns for space food. NASA Technical Reports Server (NTRS).

Al Kassaa, I., Hober, D., Hamze, M., Chihib, NE and Drider, D. (2014). Antiviral potential of lactic acid bacteria and their bacteriocins. Probiotics and antimicrobial proteins. Probiotics Antimicrob. Proteins. 6 (3-4): 177–85.

Aladjadjiyan, A. (2005). In book: Safety in the Agri-Food Chain, Publisher: Wageningen Academic Publishers, Editors: P. A. Luning, F. Devlieghere, R. Verhe. 209-222.

Alam, M. J. and Zurek, L. (2004). Association of *Escherichia coli* O157:H7 with houseflies on a cattle farm. Appl. Environ. Microbiol. 70, (12), 7578–7580.

Alauddin, S. (2012). Food adulteration and society. Global research analysis international. 1(7), 3-5.

Alexandra, S. (2013). Food Safety: CDC Report shows rates of foodborne illnesses remain largely unchanged. *Morbidity and Mortality Weekly Report*. Alexandra Sifferlin @acsifferlin

Allen, K. J., Hill, D. J., Heine, R. G. (2006). Food allergy in childhood. Med. J. Aust. 2, 185(7):394-400.

Allen, T., and Prosperi, P. (2016). Modeling Sustainable Food Systems. Environ Manage. 57(5):956-75. doi: 10. 1007/s00267-016-0664-8.

Alleson, D., Paul, D., Cotter, R., Paul R., and Colin, H. (2012). Bacteriocin Production:a ProbioticTrait?0099-2240/12/$12. 00. Appl. Environ. Microbiol. 78(1):1–6. DOI: 10. 1128/AEM. 05576-11.

Almeida, J., Domingues, P., and Sampaio, P. (2014). Different perspectives on management systems integration. Total Quality Management & Business Excellence. 25. 10. 1080/14783363. 2013. 867098.

Aluminum Assn. (2006). North America aluminum industry—a quick review. Arlington, Va. AluminumAssn. http://www. aluminum. org/Content/NavigationMenu/News_and_Stats/Statistics_Reports/Facts_At_A_Glance/factsataglance05. pdf. Accessed 2007 Jan 10.

Alvarado, C., GarcÃ-a AlmendÃ¡rez, B. E., Martin, S. E., and Regalado, C. (2006). Food-associated lactic acid bacteria with antimicrobial potential from traditional Mexican foods. Rev. Latinoam Microbiol. 48 (3-4): 260-268.

Alvarez, M., Ponce, A. G., Goyeneche, R., and Moreira, Maria. (2016). Physical Treatments and Propolis Extract to Enhance Quality Attributes of Fresh-Cut Mixed Vegetables. J. Fd. Proc. Preserv. 10. 1111/jfpp. 13127.

Amanda, B. (2019). What are the pros and cons of GMO foods? Med. News Today. Reviewed by Debra Rose Wilson.

American Plastics Council. (2004). National post-consumer plastic recycling report. Arlington , Va . : American Plastics Council. p 12.

American Plastics Council. (2006a). The many uses of plastics. Arlington, Va, American Plastics Council. http://www. americanplasticscouncil. org/s_apc/sec. asp? TRACKID=& SID=6&VID=86&CID=312&DID=930. Accessed 2006 Oct 23.

American Plastics Council. (2006b). Resin identification codes—plastic recycling codes. Arlington, Va. :AmericanPlasticsCouncil. http://americanplasticscouncil. org/s_apc/sec. asp?TRACKID=&CID=313&DID=931. Accessed 2006.

Anderson, M. D. (2013). Beyond food security to realizing food rights in the US. ERIC Educational Resources Information Center.

Andrade, M. J., Thorsen, L., RodrÃ-guez, A., CÃrdoba, J. J., and Jespersen, L. (2014). Inhibition of ochratoxigenic moulds by *Debaryomyces hansenii* strains for biopreservation of dry-cured meat products. Int. J. Fd. Microbiol. 17:170:70-7. doi: 10. 1016/j. ijfoodmicro. 2013. 11. 004.

Andre, G. (1980). Pricing a Convenience Food. ERIC Educational Resources Information Center.

Anil, S. N., Tapre, A. R., and Ranveer, R. C. (2017). Applications of bacteriocins as bio preservative in foods. A review article. Curr. sci. 44: 110-117.

Anita, G., and Neetu, S. (2013). Hazards of new technologies in promoting food adulteration. J. Env. Sci. Tox. F. Sci. 5, (1), 08-10.

Anne, de Jong., Sacha, A. F. T. van Hijum., Jetta, J. E. Bijlsma., Jan, K., and Oscar, P. K. (2006). BAGEL: a web-based bacteriocin genome mining tool. Nucleic Acids Res. 1, 34(Web Server issue):W273–W279.

Anonymous. (2000b). National Organic Program. American Society for Microbiology. Available at http, //www. asm. org/Policy/index. asp?bid=3585. Accessed 9 December, 2008.

Anonymous. (2007e). Investigation of an *Escherichia coli* O157, H7 outbreak associated with Dole pre-packaged spinach. California Food Emergency Response Team Sacramento/ Alameda, CA, USA. Available at http, //www. marlerclark. com/ 2006_Spinach_ Report_Final_01. pdf. Accessed 22 December 2008.

Anonymous. (2008b). Climate change, implications for food safety. Food and Agriculture Organization of the United Nations, Rome.

Appendini, P., and Hotchkiss, J. H. (2002). Review of Antimicrobial Food Packaging. Innov. Fd. Sci. Emerg. Technol. 3, 113-126.

AraÃjo, M. L. de., MendonÃ§a, R. de. D., Lopes, F, J. D., and Lopes, A. C. S. (2018). Association between food insecurity and food intake. Nutri. 54 , 54–59.

Armistead C. et al., (1999). "Strategic Business Process Management for Organisational Effectiveness", Long Range Planning, 32, 1, 96 -106.

Arnison, P. G., Bibb, M. J., Bierbaum, G., Bowers, A. A., Bugni, T. S., Bulaj, G., et al. (2013). Ribosomally synthesized and post-translationally modified peptide natural products: overview and recommendations for a universal nomenclature. Nat. Prod. Rep. 30, 108–160.

Arqués, J. L., Rodríguez, E., Nuñez, M., and Medina, M. (2011). Combined effect of reuterin and lactic acid bacteria bacteriocins on the inactivation of food-borne pathogens in milk. Fd Cntrl. 22, 457–461.

Arvanitoyannis, I. S., and Bosnea, L. (2004). Migration of substances from food packaging materials to foods. Crit. Rev. Fd. Sci. Nutr. 44(3):63–76.

Asthma and Food Allergies. Last Updated (11/21/2015). Source:Nutrition: NIAID-Sponsored Expert Panel. Guidelines for the diagnosis and management of food allergy in the United States: Report of the NIAID-sponsored expert panel. J. Allergy Clin. Immunol. 2010: 126(6):S1- 58.

Atwill, E. R. (2008). Implications of wildlife in *E. coli* outbreaks associated with leafy green produce, Proceedings of the 23rd Vertebrate Pest Conference, San Diego, California, USA, published at Univ. Calif, Davis, p. 5–6.

Auras, R., Harte, B., and Selke, S. (2004). An overview of polyactides as packaging materials. Macromol. Biosci. 16, 4 (9), 835–64.

Awasthi, S., Jain, K., Das, A., Alam, R., Surti, G., and Kishan, N. (2014). Analysis of food quality and food adulterants from different departmental & local grocery stores by qualitative analysis for food safety. IOSRJESTFT: 8, (2), 22-26.

Awasthi, V., Bahman, S., Thakur, L. K., Singh, S. K., Dua, A., and Ganguly, S. (2012). Contaminants in milk and impact of heating: an assessment study. Ind. J. Pub. Health. 56, 95–99. doi:10. 4103/0019-557X. 96985.

Ayalew, H., Amare, B., and B, Asrade. (2013). Review on food safety system: Ethiopian perspective. Afri. J. Fd. Sci. 7(12), 431-440. DOI: 10. 5897/AJFS2013. 1064 ISSN 1996-0794 ©2013 Academic Journals http://www. academicjournals. org/AJFS.

Balciunas, E. M., Martinez, F. A. C., Todorov, S. D., De Melo Franco, B. D. G., Converti, A., and De Souza Oliveira, R. P. (2013). Novel biotechnological applications of bacteriocins: a review. Fd. Cntrl. 32, 134–142.

Barat Somedutta , Steeb Benjamin , Mazé Alain , and Bumann Dirk. (2012). Extensive *In Vivo* Resilience of Persistent *Salmonella*. PLoS One. 7(7):e42007. Published online 2012 Jul 24. doi: 10. 1371/journal. pone. 0042007. Ed. Michael Hensel.

Barbosa, M. S., Todorov, S. D., Jurkiewicz, C. H., and Franco, B. D. (2015). Bacteriocin production by *Lactobacillus curvatus* MBSa2 entrapped in calcium alginate during ripening of salami for control of *Listeria monocytogenes*. Fd. Cntrl. 47, 147–153.

Barker, J., Humphrey, T. J., and Brown, M. W. (1999). Survival of *Escherichia coli* O157 in a soil protozoan: implications for disease. FEMS Microbiol. Lett. 173: 291–295.

Bastioli., (2005). Handbook of biodegradable polymers. Toronto-Scarborough, Ontario, Canada : Chem. Tec. Publishing. 533.

Bastos, M. C. F., Coutinho, B. G., and Coelho, M. L. V. (2010). Lysostaphin: A Staphylococcal bacteriolysin with potential clinical applications. Pharmaceuticals. 3(4):1139-1161.

Batish, V. K., Roy, U., Lal, R., and Grover, S. (1997). Antifungal attributes of lactic acid bacteria – A review. Critical Rev. Biotechnol. 17:2009–2225.

Bauman, H. (1974). The HACCP concept and microbiological hazard categories. Fd. Technol. 28 (papers in issues 30 to 34).

Benech, R. O., Kheadr, E., Laridi, R., Lacroix, C., and Fliss, I. (2002). Inhibition of *Listeria innocua* in cheddar cheese by addition of nisin Z in liposomes or by *in situ* production in mixed culture. Appl. Environ. Microbiol. 68, 3683–3690.

Benjakul, S., and Visessanguan, W. (2013). Antibacterial activities of bacteriocins: application in foods and pharmaceuticals. Anti-listeria activity of poly(lactic acid)/sawdust particle biocomposite film impregnated with pediocin PA-1/AcH and its use in raw sliced pork. Int. J. Fd. Microbiol. 167(2):229-35. doi:10. 1016/j. ijfoodmicro. 2013. 09. 009. Epub 2013.

Benkerroum, N., Ghouati, Y., Ghalfi, H., Elmejdoub, T., Roblain, D., Jacques, P., et al. (2002). Biocontrol of *Listeria monocytogenes* in a model cultured milk (lben) by *in situ* bacteriocin production from *Lactococcus lactis* ssp. lactis. Int. J. Dairy Technol. 55, 145–151.

Bennet, W. L. and Steed, L. L. (1999). An integrated approach to food safety. Quality Progress. 32: 37-42.

Berglund, F. (1978). Food Additives. In: Leonard B. J. (eds.) Toxicological Aspects of Food Safety. Archives of Toxicology (Supplement 1), vol 1. Springer, Berlin, Heidelberg

Beshkova, D., and Frengova, G. (2012). Bacteriocins from lactic acid bacteria: microorganisms of potential biotechnological importance for the dairy industry. *Eng. Life Sci.*12, 419–432.

Bhardwaj A, et al. 2010. Safety assessment and evaluation of probiotic potential of bacteriocinogenic Enterococcus faecium KH 24 strain under in vitro and in vivo conditions. Int. J. Food Microbiol. 141:56 –164.

Bhatti, M., Veeramachaneni, A., and Shelef, L. A. (2004). Factors affecting the antilisterial effects of nisin in milk. Int. J. Fd. Microbiol. 97: 215–219.

Blanco-Padilla, A., Soto, K. M., and Hernández, I. M. (2014). Food Antimicrobials Nanocarriers. Hindawi Publishing Corporation. The Scientific World Journal. Art. ID. 837215, 11. http://dx. doi. org/10. 1155/2014/837215.

Blaskovich, M. A. T., Hansford, K. A., Butler, M. S., Jia, Z., Mark, A. E., and Cooper, M. A. (2018). Developments in Glycopeptide Antibiotics. ACS. Infect. Dis. 11:4(5):715-735. doi: 10. 1021/acsinfecdis. 7b00258.

Blazquez, A. B., and Berin, M. C. (2017). Microbiome and Food Allergy. Transl Res. 179:199-203. doi: 10. 1016/j. trsl. 2016. 09. 003. Epub 2016.

Blin, K., Medema, M. H., Kazempour, D., Fischbach, M. A., Breitling, R., Takano, E., et al. (2013). antiSMASH 2. 0 a versatile platform for genome mining of secondary metabolite producers. Nucleic acids Res. 41: W204–W212.

Bloomfield, S. F., Exner, M., Signorelli, C., Nath, K. J., and Scott, E. A. (2012). The chain of infection transmission in the home and everyday life settings, and the role of hygiene in reducing the risk of infection.   http://www. ifh-homehygiene. com/best-practice-review/chain-infection-transmission-home-and-everyday-life-settings-and-role-hygiene.

Blunt R., McOrist S., McKillen J., McNair I., Jiang T., and Mellits K. (2011). House fly vector for porcine circovirus 2b on commercial pig farms. Veterin. Microbiol. 149, (3-4): 452–455.

Bogaardt, C., van Tonder, A. J., and Brueggemann, A. (2015). Genomic analyses of *Pneumococci* reveal a wide diversity of bacteriocins–including pneumocyclicin, a novel circular bacteriocin. BMC Genomics. 16:554. doi:10. 1186/s12864-015-1729-4.

Bonadè, A., Murelli, F., Vescovo, M., and Scolari, G. (2001). Partial characterization of a bacteriocin produced by *Lactobacillus helveticus*. Lett. Appl. Microbiol. 33:153–158. doi: 10. 1046/j. 1472-765x. 2001. 00969. x.

Bonfoh, B., Wasem, A., Traore, An., Fane, A., Spillman, H., Simbe, C.F., Alfaroukh, I. O., Nicolet, J., Farah, Z., and Zinsstag, J. (2003). Microbiological quality of cow's milk taken at different intervals from the udder to the selling point in Bamako, Mali. Fd. Cntrl. 14: 495-500.

Bonner, J. J. (2004). The Biology of Food. ERIC Educational Resources Information Center.

Borchers, A., Teuber, S. S., Keen, C. L., and Gershwin, M. Eric. (2010). Food safety. Clinic. Rev. Allerg. Immunol. 39:95-141.

Bourland, C T. (1999). Food systems for space travel. Life Support Biosph. Sci. 6(1):9-12.

Bouttefroy, A., and Milliere, J. (2000). Nisin-curvaticin 13 combinations for avoiding the regrowth of bacteriocin resistant cells of bacteriocin resistant cells of *Listeria monocytogenes* ATCC 15313. Int. *J.* Fd. Microbiol. 5, 62(1-2):65-75.

Bowdish, D. M., Davidson, D. J., and Hancock, R. E. (2005). A re-evaluation of the role of host defence peptides in mammalian immunity. Curr. Protein Pept. Sci. 6:35–51. doi: 10. 2174/ 1389203053027494.

Bower, C. K., McGuire, J., and Daeschel, M. A. (1995). Suppression of *Listeria monocytogenes* colonization following adsorption of nisin onto silica surfaces. Appl. Environ. Microbiol. 61, 3: 992–997.

Boziaris, I., Humpheson, L., and Adams, M. (1998). Effect of nisin on heat injury and inactivation of *Salmonella enteritidis* PT4. Int. J. Fd. Microbiol. 43, 7–13.

Breukink, E., and de Kruijff, B. (2006). Lipid II as a target for antibiotics. Nat. Rev. Drug Discov. 5, 321–332. doi: 10. 1038/nrd2004.

Briers, Y., and Lavigne, R. (2015). Breaking barriers: expansion of the use of endolysins as novel antibacterials against Gram-negative bacteria. Future Microbiol. 10(3):377-90. doi: 10. 2217/fmb. 15. 8.

Brown, L. G., Ripley, D., Blade, H., Reimann, D., Everstine, K., Nicholas, D., Egan, J., Koktavy, N., and Quilliam, D. N. (2012). Restaurant food cooling practices. *J Fd.* Prot. 75:2172–2178.

Brown, laura green: ripley, danny: blade, henry: reimann, dave: everstine, karen: nicholas, dave: egan, jessica: koktavy, nicole: quilliam, daniela N. 2017. Restaurant Food Cooling Practices.

Brown, M. E., Antle, J. M., Backlund, P. W., Carr, E. R., Easterling, W. E., Walsh, M., Ammann, C. M., Attavanich, W., Barrett, C. B., Bellemare, M. F., Dancheck, V., Funk, C., Grace, K., Ingram, J. S. I., Jiang, H., Maletta, H., Mata, T., Murray, A., Ngugi, M., Ojima, D. S., O'Neill, B. C., and Tebaldi, C. (2015). Climate Change and Global Food Security: Food Access, Utilization, and the US Food System. NASA Astrophysics Data System (ADS).

Bryan, F. L., Teufel, P., Riaz, S., Roohi, S., Qadar, F. and Malik, Z. U. R. (1992a). Hazards and critical control points of vending operations at a railway station and bus station in Pakistan. J. Fd. Prot. 55: 534-541.

Bryan, F. L., Teufel, P., Riaz, S., Roohi, S., Qadar, F. and Malik, Z. U. R. (1992b). Hazards and critical control points of street-vending operations in a mountain resort town in Pakistan. J. Fd. Prot. 55: 701-707.

Bryan, F. L., Teufel, P., Riazj, S., Roohij, S., Quadar, F., and Malik, Z. (1992). Hazards and critical control points of street-vended chat, a regionally popular foods in Pakistan. J. Fd. Prot. 55: 708-713.

Brynjolfsson, A. (1985). Use of Irradiated Foods. NASA Technical Reports Server (NTRS).

Budu-Amoako et al., (1999). Combined effect of heat and moderate heat on destruction of

214   Food Safety

Listeria monocytogenes on Cold-Pack lobster meat. J. Fd. Prot. 62(1):46-50.

Buluswar, S. (2018). Urban Food Initiative. Science Cinema .

Bures, J., Horák, V., Fixa, B., Komárková, O., Zaydlar, K., Lonský, V., et al. (1986). Colicinogeny in colorectal cancer. Neoplasma. 33: 233–237.

Burks, A. W., Tang, M., Sicherer, S., Muraro, A., Eigenmann, P. A., Ebisawa, M, , Fiocchi, A., Chiang, W., Beyer, K., Wood, R., Hourihane, J., Jones, S. M., Lack, G., Sampson, H. A. (2012). ICON: Food allergy. J. Allergy Clin. Immunol. 129(4):906-20.

Burton, J. P., Wescombe, P. A., Macklaim, J. M., Chai, M. H., Macdonald, K., Hale, J. D., et al. (2013). Persistence of the oral probiotic *Streptococcus salivarius* M18 is dose dependent and megaplasmid transfer can augment their bacteriocin production and adhesion characteristics. PLoS ONE 8:e65991. doi: 10. 1371/journal. pone. 0065991.

Bush, W. H. (1974). Skylab food system. NASA Technical Reports Server (NTRS).

Bushra, R., Aslam, N., and Khan, A. Y. (2011). Food-Drug Interactions. Oman Med. J. 26(2):77-83. doi: 10. 5001/omj. 2011. 21.

Business Standard staff . (2017). FSSAI issues detailed guidelines on recall of food products. Press Trust of India | New Delhi . 22:15. IST.

Buzby, J. C., Frenzen, P. D., and Rasco, B. (2001). Product liability and microbial foodborne illness. Food and Rural Economics Division, Economic Research Service, U. S. Department of Agriculture. Agricultural Economic Report (AER) N0. 799.

Camargo, A. C., Woodward, J. J., Call, DR, and Nero, L. A., (2017). *Listeria monocytogenes* in Food-processing facilities, food contamination, and human listeriosis: The Brazilian Scenario. Foodborne Pathog. Dis. 14(11):623-636. doi: 10. 1089/fpd. 2016. 2274. Epub 2017.

Camargo, M. C. R., and Toledo, M. C. F. (2003). Polycyclic aromatic hydrocarbons in Brazilian vegetables and fruits. Fd. Cntrl. 14, (1): 49-53.

Campbell, J. B., Skoda, S. R., Berkebile D. R. et al., (2001). Effects of stable flies (*Diptera: Muscidae*) on weight gains of grazing yearling cattle. J. Econ. Entom. 94, (3) 780–783.

Cao-Hoang, L., Chaine, A., Grégoire, L., and Waché, Y. (2010). Potential of nisin-incorporated sodium caseinate films to control *Listeria* in artificially contaminated cheese. Fd. Microbiol. 27, 940–944.

Capps, O., and Park, J. L. (2003). Food retailing and food service. Vet. Clin. North Am. Fd. Anim. Pract. 19(2):445-61.

Cardozo, G., Barbieri, Van Dender M., Trento I., F., and Kuyae A., (2009). *Musca domestica* L. as a vector of pathogenic microorganisms in Ultra-Filtered fresh Minas cheese. Brazilian J. Fd. Technol. 12, (2) 85–91.

Carnio, M. C., Höltzel, A., Rudolf, M., Henle, T., Jung, G and Scherer, S. (2000). The macrocyclic peptide antibiotic micrococcin P1 is secreted by the food-borne bacterium *Staphylococcus equorum*WS 2733 and inhibits *Listeria monocytogenes* on Soft Cheese. Appl. Environ. Microbiol. 66: 2378–2384.

Carolina, P.N.C. (2002). Carolina Recycling Assn. The bottle bill battle [factsheet]. Recycling Assn. 2 p.

Carrard, A., Rizzuti, D., and Sokollik, C. (2015). Update on food allergy. Allergy. 70(12):1511-20.doi:10.1111/all.12780.

Carson, D. A., Barkema, H. W., Naushad, S., and DeBuck. (2017). Bacteriocins of Non-aureus *Staphylococci* isolated from Bovine Milk. J. Appl. Environ. Microbiol. 17, 83: e01015-17.

Cascales. E., Buchanan. S. K., Duché, D., et al. (2007). "Colicin Biology". Mol. Biol. Rev. DOI https://10. 1128/MMBR. 00036-06. 71, (1):158-229.

Casey, P. G. , Gardiner, G. E. , Casey, G. , Bradshaw, B. , Lawlor, P. G. , Lynch, P. B., Leonard, F. C. , Stanton , C. , Ross, R. P. , Fitzgerald, G. F. , and Hill, C. (2007). A five-strain probiotic combination reduces pathogen shedding and alleviates disease signs in pigs challenged with *salmonella enterica*serovar *typhimurium*. Appl. Environ. Microbiol.

73(6):1858–1863. doi:10.1128/AEM. 01840-06.

Castellano, P., Belfiore, C., Fadda, S., and Vignolo, G. (2008). Review of bacteriocinogenic lactic acid bacteria used as bioprotective cultures in fresh meat produced in Argentina. Meat Sci. 79:483–499.

CDC (centre for disease control and prevention. (2016). Multistate Outbreak of Listeriosis Linked to Packaged Salads Produced at Springfield, Ohio Dole Processing Facility (Final Update).

Célia, C. G. Silva., Sofia, P. M. S., and Susana, C. R. (2018). Application of Bacteriocins and Protective Cultures in Dairy Food Preservation. Front. Microbiol. https://doi. org/10. 3389/fmicb. 2018. 00594.

Célia, C. G. Silva., Sofia, P. M. S., and Susana, C. R. (2019). Application of bacteriocins and protective cultures in dairy food preservation. Front. Microbiol. 10:563 https://doi. org/ 10. 3389/fmicb. 2018. 00594.

Center, N. (2013). Cancer Facts and Figures. Goyang: Director of National Cancer Center Minister of Health and Welfare.

Chang, C., Feemster, K. A., Coffin, S and Handy, L. K. J. (2017). Treatment related complications in children hospitalized with disseminated lyme disease. Pediatric Infect. Dis . Soc. 1:6(3):e152-e154.

Chapman, B., and Gunter, C., (2018). Local food systems food safety concerns. Health Psychol. 03 25:37(3):262-270. Epub 2018.

Chauliac, M. et al. (1987). Children in the Tropics. Weaning Fds. n167-168.

Chauliac, M., et al. (1987). Weaning Foods. ERIC Educational Resources Information Center.

Chen, H., and Hoover, D. G. (2003). Bacteriocin and their food application. Comprehensive Rev. Fd Sci. Tech. 2:82-100.

Chen, J., Deng, S., and Li, J. (2013). Preparation of an novel botanic biopreservative and its efficacy in keeping quality of peeled *Penaeus vannamei*. journal

Chi. H., and Holo, H. (2018). Synergistic Antimicrobial activity between the broad spectrum bacteriocin garvicin ks and nisin, farnesol and polymyxin b against gram-positive and gram-negative bacteria. Curr. Microbiol. 75(3):272-277.

Chopra, L., Singh, G., Kautilya, K, J., and Sahoo, D. K. (2015). Sonorensin: A new bacteriocin with potential of an anti-biofilm agent and a food biopreservative. Scientific Reports. 5, Art. no. 13412.

Christine, M. Strapp., Adrienne, E. H., Shearer., and Rolf, D. J. (2003). Survey of retail alfalfa sprouts and mushrooms for the presence of *Escherichia coli* O157:h7, *Salmonella*, and *Listeria* with bax, and evaluation of this polymerase chain reaction–based system with experimentally contaminated samples. J. Fd. Protectn. 66:2, 182-187.

Chumchalová, J. and Smarda, J. (2003). Human tumor cells are selectively inhibited by colicins. Folia Microbiologica (Praha), 48: 111-115.

Cintas, L. M., Rodriguez, J. M., Fernandez, M. F., Sletten, K., Nes, I. F., Harnander, P. E., and Holo, H. (1995). Isolation and characterization of Pediocin F 50, a new bacteriocin from *Pediococcus acidilacti* with a broad inhibitory spectrum. Appl. Environ. Microbiol. 61: 2643 – 2648.

Cintas, L., Casaus, P., Fernández, M and Hernández, P. (1998). Comparative antimicrobial activity of enterocin L50, pediocin PA-1, nisin A and lactocin S against spoilage and foodborne pathogenic bacteria. Fd. Microbiol. 15: 289–298. doi: 10. 1006/fmic. 1997. 0160

Clarke, L., McQueen, J., et al. (1996). The dietary management of food allergy and food intolerance in children and adults. Aust. J. Nutr. Diet. 53 (3): 89–98. ISSN 1032-1322.

Claudia, A., Brid, B., Emanuele, Z., Ambrose, F., Aidan, C., and Arendt, E., K. (2016). Antifungal sourdough lactic acid bacteria as biopreservation tool in quinoa and rice bread. Int. J. Fd. Microbiol. 19:239:86-94. doi:10. 1016/j. ijfoodmicro. 2016. 05. 006. Epub 2016.

Claudia, A., Emanuele, Z., and Arendt, E. K. (2017). Mold spoilage of bread and its biopreservation:A review of current strategies for bread shelf life extension. 2:57(16):3528-3542. doi: 10. 1080/10408398. 2016. 1147417.

Cláudia, B. , Helena, A. , Joana, S. , and Paula, T. (2013). Role of Flies as Vectors of Foodborne Pathogens in Rural Areas. ISRN Microbiology. Article ID718780, 7. http://dx. doi. org/10. 1155/2013/718780.

Cleveland, J., Montville, T. J., Nes, I. F., and Chikindas, M. L. (2001). Bacteriocins:safe, natural antimicrobials for food preservation. Int. J. Fd. Microbiol. 71: 1–20.

CLSI–Clinical and Laboratory Standards Institute. (2007). Performance standards for antimicrobial susceptibility testing: seventeenth informational supplement. Clin. Lab. Stand. Inst. 27, (1): 98–141.

Codex Alimentarius (2012). How it all began Food and Agriculture Organization of the United Nations website.

Codex Alimentarius commission twenty-fifth session rome, italy 30 june - 5 july 2003 report of the third session of the codex ad hoc intergovernmental task force on foods derived from biotechnology yokohama, japan 4-8 march 2002.

Codex Alimentarius Commission. (2004). Report of the 20th session of the Codex committee on general principles. Joint FAO/WHO food standards programme. 2–7: 44.

Coelho, M., Silva, C., Ribeiro, S., Dapkevicius, M and Rosa, H. (2014). Control of *Listeria monocytogenes* in fresh cheese using protective lactic acid bacteria. Int. J. Fd. Microbiol. 191: 53–59.

Coles, (2003). Introduction. In: Coles R, McDowell D, Kirwan M. J., editors. Fd Packaging Technol. London , U. K. Blackwell Publishing, CRC Press. 1–31.

Collado, M. C., Isolauri, E., Salminen, S., and Sanz, Y. (2009). The impact of probiotic on gut health. Curr. Drug Metab. 10:68–78.

Coma, V., Sebti, I., Pardon, P., Deschamps, A., and Pichavant, F. H. (2001). Antimicrobial edible packaging based on cellulosic ethers, fatty acids and nisin incorporation to inhibit *Listeria innocua* and *Staphylococcus aureus*. *J. Fd. Prot.* 64: 470–475.

Combase Consortium. (2008). ComBase. Available at http, //www. combase. cc.

Concha-Meyer, A., Schöbitz, R., Brito, C., and Fuentes, R. (2011). Lactic acid bacteria in an alginate film inhibit *Listeria monocytogenes* growth on smoked salmon. Fd. Cntrl. 22: 485–489.

Container Recycling Inst. (2006a). Beverage container deposit systems in the United States: key features. Washington, D. C. Container Recycling Inst. http://www. bottlebill. org/ legislation/usa_deposit. htm. Accessed 2007 Jan 9.

Container Recycling Inst. (2006b). Bottle bill impact: litter. Washington , D. C. : Container Recycling Inst. Available from: http://www. bottlebill. org/impacts/litter. htm.

Cooley, M. B., Miller, W. G., and Mandrell, R. E. (2003). Colonization of Arabidopsis thaliana with *Salmonella enterica* and enterohemorrhagic *Escherichia coli* O157:H7 and competition by *Enterobacter asburiae*. Appl. Environ. Microbiol. 69 (8), 4915–4926, ISSN 0099-2240

Corbett, C. S., De Buck, J., Orsel, K., and Barkema, H. W. (2017). Fecal shedding and tissue infections demonstrate transmission of *Mycobacterium avium* subsp. *paratuberculosis* in group-housed dairy calves. Vet. Res. 48(1):27.

Corr, S. C., Hill, C., and Gahan, C. G. M. (2009). Understanding the mechanisms by which probiotics inhibit gastrointestinal pathogens. Adv. Fd. Nutr. Res. 56:1–15.

Cotter, P. D., Hill, C., and Ross, R. P. (2005). Bacteriocins: developing innate immunity for food. Nat. Rev. Microbiol. 3: 777–788.

Cotter, P. D., Hill, C., and Ross, R. P. (2006). What's in a name? Class distinction for bacteriocins. Nature Rev. Microbiol. 4(2). doi:10. 1038/nrmicro1273-c2.

Cotter, P. D., Ross, R. P., and Hill, C. (2013). Bacteriocins—a viable alternative to antibiotics? Nat. Rev. Microbiol. 11: 95–105. doi: 10. 1038/nrmicro2937

Covaci, A., Gerecke, A. C., Law, R. J., Voorspoels, S., Kohler, M., Heeb, N. V., et al. (2006). Hexabromocyclododecanes (HBCDs) in the environment and humans: a review. Environ. Sci. Technol. 40(12):3679–3688.

Cowden, J. M. (2000). Food poisoning notification: time for rethink. Health Bulletin. 58:328-331.

Cronquist, A. B., Mody, R. K., Atkinson, R., Besser, J., Tobin. D'Angelo., Hurd, S., Robinson, T., Nicholson, C., and Mahon, B. E. (2012). Impacts of culture-independent diagnostic practices on public health surveillance for bacterial enteric pathogens. Clin. Infect. Dis. 54, Suppl 5:S432-9. doi: 10. 1093/cid/cis267.

Cunha, B. A. (1997). Intravenous-to-oral antibiotic switch therapy. A cost-effective approach. Postgrad. Med. 101(4):111-2, 115-8, 122-3 passim.

Cutler, S. A., N. A. Cornick., Lonergan S. M., and Stahl C. H. (2007). Dietary Colicin E1 prevents post weaning diarrhea in swine. Antimicro. Agents and Chemo. (submitted).

Czerwiecki, L. (2004). Problems of food authenticity. Rocz. Panstw. Zakl. Hig. 55(1):9-19.

da Silva Malheiros, P., Daroit, D. J., and Brandelli, A. (2010). Food applications of liposome-encapsulated antimicrobial peptides. Trends Fd. Sci. Technol. 21:284–292.

da Silva Sabo, S., PÃ©rez-RodrÃ-guez, N., DomÃ-nguez, Josa M., de Souza, O., and Ricardo, P. (2017). Inhibitory substances production by *Lactobacillus plantarum* ST16Pa cultured in hydrolyzed cheese whey supplemented with soybean flour and their antimicrobial efficiency as biopreservatives on fresh chicken meat. Food Res Int. 2017 Sep:99(Pt 1):762-769. doi: 10. 1016/j. foodres. 2017. 05. 026. Epub 2017.

da Silva Sabo, S., Vitolo, M., González, J. M. D., and De Souza Oliveira, R. P. (2014). Overview of *Lactobacillus plantarum* as a promising bacteriocin producer among lactic acid bacteria. Fd. Res. Int. 64: 527–536.

DÃ-az Yubero, Miguel Ãngel. 2015. Milk and food security.

Dabour, N., Zihler, A., Kheadr, E., Lacroix, C., and Fliss, I. (2009). *In vivo* study on the effectiveness of pediocin PA-1 and *Pediococcus acidilactici* UL5 at inhibiting *Listeria monocytogenes*. Int. J. Fd. Microbiol. 133:225–233.

Dal Bello, B., Cocolin, L., Zeppa, G., Field, D., Cotter, P. D and Hill, C. (2012). Technological characterization of bacteriocin producing *Lactococcus lactis* strains employed to control *Listeria monocytogenes* in cottage cheese. Int. J. Fd. Microbiol. 153:58–65.

Das, D., and Goyal, A. (2014). Characterization of a noncytotoxic bacteriocin from probiotic *Lactobacillus plantarum* DM5 with potential as a food preservative. Fd. Funct. 5(10): 2453-62. doi: 10. 1039/c4fo00481g. Epub 2014.

Davari, B., Kalantar, E., Zahirnia, A., and Moosa-Kazemi, S. H. (2010). Frequency of resistance and susceptible bacteria isolated from houseflies. Iranian J. Arthropod-Borne Dis. 4, (2): 50–55.

Davenport, T. (2005). The Coming Commodization of Processes. Harvard business review. 83, 100-8: 149.

Davies, E., Bevis, H., and Delves-Broughton, J. (1997). The use of the bacteriocin, nisin, as a preservative in ricotta-type cheeses to control the food-borne pathogen *Listeria monocytogenes*. Lett. Appl. Microbiol. 24: 343–346.

Dawid, S., Roche, A. M., and Weiser, J. N. (2007). The *blp* Bacteriocins of *Streptococcus pneumoniae* Mediate Intraspecies Competition both *In Vitro* and *In Vivo*. Inf and Immun. 75: 443-451. doi: 10. 1128/IAI. 01775-05.

de Arauz, L. J., Jozala, A. F., Mazzola, P. G and Penna, T. C. V. (2009). Nisin biotechnological production and application: a review. Trends Fd. Sci. Technol. 20:146–154.

De Decker, A., Verbeken, S., Sioen, I., Van Lippevelde, W., Braet, C., Eiben, G., Pala, V., Reisch, L, A., and De Henauw, S. (2017). Palatable food consumption in children: interplay between (food) reward motivation and the home food environment. Eur. J. Pediat. 176(4). p. 465-474. doi:10. 1007/s00431-017-2857-4.

de Jong, A., van Hijum, S. A. F. T., Bijlsma, J. J E., Kok, J and Kuipers, O. P. (2006). BAGEL: a web-based bacteriocin genome mining tool. Nucleic Acids Res. 34: W273–W279.

de Souza Barbosa, M., Todorov, S. D., Ivanova, I., Chobert, J. -M., Haertlé, T and De Melo Franco, B. D. G. (2015). Improving safety of salami by application of bacteriocins produced by an autochthonous *Lactobacillus curvatus* isolate. Fd Microbiol. 46: 254–262.

Degnan , A. J., Yousef, A. E., and Luchansky, J. B. (1992). Use of *Pediococcus acidilactici* to control *Listeria monocytogenes* in temperature abused vacuum packaged wieners. J. Fd. Prot. 55:98-103.

DeJesus, J. M., Gelman, S. A., Viechnicki, G. B., Appugliese, D. P., Miller, A. L., Rosenblum, K. L., and Lumeng, J. C. (2018). An investigation of maternal food intake and maternal food talk as predictors of child food intake. Appetite. 1, 127:356-363. DOI: 10. 1016/j. appet. 2018. 04. 018.

Dennis O'Brien. (2019). New Food and Nutrient Data System for Researchers, Consumers Launches Today USDA-ARS?s Scientific Manuscript database. .

Dept. of Health and Human Services (U. S.). (2005). Report on carcinogens. 11thed. Research Triangle Park , N. C. National Toxicology Program, Public Health Service, HHS.

Derakhshani, H., De Buck, J., Mortier, R., Barkema, H. W., Krause, D. O and Khafipour E. (2016). The Features of Fecal and Ileal Mucosa-Associated Microbiota in Dairy Calves during Early Infection with *Mycobacterium avium* Subspecies *paratuberculosis*. Front. Microbiol. 31, 7: 426.

Deshmukh, P. V., and Thorat, P. R. (2013). Isolation and antimicrobial spectrum of new bacteriocin from *Lactobacillus rennanquilfy* WHL 3, Int. J. Pharm. Sci. Rev. Res. 22, 2: 180 – 185.

DeVoe, J. J. (2008). Addressing Food Allergies, ERIC Educational Resources Information Center.

DeVuyst, L., and Leroy, F. (2007). Bacteriocins from lactic acid bacteria: production, purification, and food applications. J. Mol. Microbiol. Biotechnol. 13: 194–199.

DeVuyst, L., Moreno, M. F., and Revets, H. (2003). Screening for enterocins and detection of hemolysin and vancomycin resistance in enterococci of different origins. Int. J. Fd. Microbiol. 84: 299–318.

Di Cagno R., et al. (2010). Quorum sensing in sourdough *Lactobacillus plantarum* DC400: induction of plantaricin A (PlnA) under co-cultivation with other lactic acid bacteria and effect of PlnA on bacterial and Caco-2 cells. Proteomics 10:2175–2190.

Di, R., Vakkalanka, M. S., Onumpai, C., Chau, H. K., White, A., Rastall, R. A., Yam, K. and Hotchkiss, A. T. (2017). Pectic oligosaccharide structure- function relationships: prebiotics, inhibitors of Escherichia coli O157:H7 adhesion and reduction of Shiga toxin cytotoxicity in HT29 cells. Food Chemistry, 227: 245-254. ISSN 0308-8146. doi: https:/ /doi. org/10. 1016/j. foodchem. 2017. 01. 100. http://centaur. reading. ac. uk/69158/.

Diez-Gonzalez, F., Callaway, T. R., Kizoulis, M. G., and Russell, J. B. (1998). Grain feeding and the dissemination of acid-resistant *Escherichia coli* from cattle. Sci. 281, 1666–1668.

Dinesh Kumar and Renu Agrawal. (2008). Preparation of probiotic fermented milk using a native isolate of *Pediococcus pentosaceous* MTCC 5151. Res. J. Biotechnol. 3 (4): 28-31.

Dirix, G., Monsieurs, P., Marchal, K., Vanderleyden, J., and Michiels, J. (2004). Screening genomes of Gram-positive bacteria for double-glycine-motif-containing peptides. Microbiol. 150: 1121–1126.

Dolan, L. C., Matulka, R. A., and Burdock, G. A. (2010). Naturally occurring food toxins. Toxins (Basel). 2(9): 2289–2332.

Domingos-Lopes, M., Stanton, C., Ross, P., Dapkevicius, M and Silva, C. (2017). Genetic diversity, safety and technological characterization of lactic acid bacteria isolated from artisanal Pico cheese. Fd. Microbiol. 63: 178–190.

dos Santos Nascimento J., dos Santos K. R., Gentilini E., Sordelli D., and de Freire Bastos Mdo, C. (2002). Phenotypic and genetic characterisation of bacteriocin-producing strains of *Staphylococcus aureus* involved in bovine mastitis. Vet. Microbiol. 1:85(2):133-44.

dos Santos Nascimento, J., Fagundes, P. C., de Paiva Brito, M. A., dos Santos, K. R., do Carmo and de Freire Bastos, M. (2005). Production of bacteriocins by coagulase-negative *Staphylococci* involved in bovine mastitis. Vet. Microbiol. 20, 106(1-2):61-71.

Doyle, M. P., Marilyn, C., Erickson, W. A., Jennifer, C. X. D., Ynes, O., Mary, A. S., and Tong, Z., (2015). The food industry's current and future role in preventing microbial foodborne illness within the United States. *Clin. Infect. Dis.* 61, 2:252–259. https://doi.org/10. 1093/cid/civ253.

Doyle, M. P., Erickson, M. C., (2008). Summer meeting 2007 —the problems with fresh produce: an overview. J. Appl. Microbiol. 105: 317–33

Doyle, M. P., Morris, J. G., Painter, J., Singh, R., Tauxe, R V., Taylor, MR and Wong, Danilo M. A. Lo Fo. (2005). Attributing illness to food. Emerg. Infect. Dis. 11(7): 993–999.

Dr. Edward Group (2013). Founder of global healing center. DrEdwardGroup. com.

Drewnowski, A., and Popkin, B. M. (1997). The nutrition transition: new trends in the global diet. Nutri. Rev. 55:31-43.

Eaton, T. J., and Gasson, M. J. (2001). Molecular screening of *Enterococcus* virulence determinants and potential for genetic exchange between food and medical isolates. Appl. Environ. Microbiol. 67: 1628–1635.

EFSA (2007). Scientific committee. introduction of a qualified presumption of safety (QPS) approach for assessment of selected microorganisms referred to EFSA1. Opinion of the Scientific Committee (Question No EFSA-Q-2005-293. EFSA J. 587: 1–16.

EFSA(2008). The maintenance of the list of QPS microorganisms intentionally added to foods or feeds. Scientific opinion of the panel on biological hazards. EFSA J. 9231–48.

Egan, K., Field, D., Rea, M. C., Ross, R. P., Hill, C and Cotter, P. D. (2016). Bacteriocins: novel solutions to age old spore-related problems?Front. Microbiol. 7:461.

Ehlermann, Dieter, A. E. (2016). Wholesomeness of irradiated food. NASA Astrophysics Data System (ADS).

Eijsink, V. G. H., Brurberg, L. Axelsson, D. B. Diep, L. S. Havarstein, H. Holo and I. F. Nes. (2002). Production of class II bacteriocins by lactic acid bacteria: an example of biological warfare and communication. Antonie van Leewenhoek. 81: 639-654.

Eijsink, V. G., Skeie, M., Middelhoven, P. H., Brurberg, M. B and Nes, I. F. (1998). Comparative studies of class IIa bacteriocins of lactic acid bacteria. Appl. Environ. Microbiol. 64: 3275–3281.

Eissa, N., Hussein, H., Kermarrec, L., Grover, J., Metz-Boutigue, M. E and Bernstein, C. N. J. E. (2017). Chromofungin Ameliorates the Progression of colitis by regulating alternatively activated macrophages. Front. Immunol. 8:1131.

El-Agamy, E. (2009). Bioactive components in camel milk. In Y. W. Park, ed. Bioactive components in milk and dairy products. 159–194. Ames, IA, USA, Wiley-Blackwell.

Environmental Protection Agency (U. S, (1990). Characterization of municipal solid waste in the United States: 1990 update. EPA530-SW-90-042. Washington, D. C. EPA.

Environmental Protection Agency (U. S.). (2005). WasteWise 2005 annual report—forging ahead. EPA530-R-05-021. Washington , D. C. EPA. p 13.

Environmental Protection Agency (U. S.). (2006a). Municipal solid waste in the United States: 2005 facts and figures. EPA530-R-06-011. Washington , D. C. EPA. p 153.

Environmental Protection Agency (U. S.). (2006b). *Plastics*. Washington, D. C. :EPA. Available from: http://www. epa. gov/epaoswer/non-hw/muncpl/plastic. htm. Accessed 2006.

Environmental Protection Agency (U. S.). (2006c). Bioreactors. Washington , D. C. : http:// www. epa. gov/epaoswer/non-hw/muncpl/landfill/bioreactors. htm. European Union. 2005. Directive2005/84/EC. http://europa. eu. int/eur-ex/lex/LexUriServ/site/en/oj/2005/l_344/ l_34420051227en00400043. pdf.

Environmental Protection Agency (U. S.). (1989). The solid waste dilemma: an agenda for action. EPA530-SW-89-019. Washington , D. C. EPA. p 70.

Environmental Protection Agency (U. S.). (1997). Landfill reclamation. EPA530-F-97-001. Washington , D. C. EPA. p 8.

Environmental Protection Agency (U. S.). (2002). Solid waste management: a local challenge with global impacts. EPA530-F-02-026. Washington , D. C. EPA. p 5.

Environmental Protection Agency (U. S.). (2004). WasteWise 2004 annual report—sustaining excellence. EPA530-R-04-035. Washington , D. C. EPA. p 13.

Ercolini, D., Ferrocino, I., La Storia, A., Mauriello, G., Gigli, S., Masi, P., et al. (2010). Development of spoilage microbiota in beef stored in nisin activated packaging. Fd. Microbiol. 27: 137–143.

Essays, UK. (November 2018). An Overview On Food Safety Management System Commerce Essay. Retrieved from https://www. ukessays. com/essays/commerce/an-overview-on-food-safety-management-system-commerce-essay. php?vref=1.

Essays, UK. (November 2018). An Overview On Food Safety Management System Commerce Essay. Retrieved from https://www. ukessays. com/essays/commerce/an-overview-on-food-safety-management-system-commerce-essay. php?vref=1 (Panagiotis, 2009) .

European Union. 2005. EU (2006): Regulation (EC) No 178/2002 of the European Parliament and of the Council of 28 January 2002. In: Official Journal of the European Communities, 1. 2. 2002. http://www. bfr. bund. de/cm/209/2002_178_en_efsa. pdf on 29th of June, 2006.

Evers, E. G., and Chardon, J. E. (2008). A swift quantitative microbiological risk assessment (sQMRA) tool. Abstract K7, Food Micro 2008, 1–4 September. Aberdeen, Scotland.

Fagundes, P. C., De Farias, F. M., Da Silva Santos, O. C., Da Paz, J. A. S., Ceotto-Vigoder, H., Alviano, D. S., et al. (2016). The four-component aureocin A70 as a promising agent for food biopreservation. Int. J. Fd Microbiol. 237: 39–46.

Fahim, H. A., Khairalla, A. S., and El-Gendy, A. O. (2016). Nanotechnology: A Valuable Strategy to Improve Bacteriocin Formulations. Fd. Microbiol. 7:1385.

Falvey, L. (1997). Food and Environmental Science. ERIC Educational Resources Information Center.

FAO [FAO] Food and Agriculture Organization of the United Nations. (1989). Prevention of post-harvest food losses: fruits, vegetables, and root crops. A training manual. Rome, Italy: FAO. Code:17, AGRIS: J11, ISBN 92-5-102766-8.

FAO and WHO. (2006). Probiotics in Food: Health and nutritional properties and guidelines for evaluation. Rome: FAO.

FAO. (1989). Radioactive fall out in soils, crops and food. http://www. fao. org/docrep/T0228E/ T0228E00. htm

FAO/WHO (2007). Joint FAO/WHO food standards programme FAO/WHO coordinating committee for Africa, seventeenth session, Rabat, Morocco. 23-26.

FAO/WHO meetings on pesticide residues. Geneva. (2019). http://www. fao. org/agriculture/ crops/thematic-sitemap/theme/pests/jmpr/en/

FAO/WHO. (2002). Principles and guidelines for incorporating microbiological risk assessment in the development of food safety standards, guidelines and related texts. Report on joint FAO/WHO consultation, 18-22 March, Kiel , Germany.

FAO/WHO. (2003). Hazard characterization for pathogens in food and water: guidelines. [FAO/WHO] Microbiological Risk Assessment Series, No. 3.

FAO/WHO. (2008). Microbiological hazards in fresh leafy vegetables and herbs- microbiological risk assessment series 14. Rome: FAO and WHO.

FAO/WHO. (2016). Risk communication applied to food safety handbook. Rome: FAO and WHO.

Faulde, M., and Spiesberger, M. (2013). Hospital infestations by the moth fly, *Clogmia albipunctata* (Diptera: Psychodinae), in Germany. J. Hospital Infect. 83, (1): 51–60.

Favaro, L., Penna, A. L. B., and Todorov, S. D. (2015). Bacteriocinogenic LAB from cheeses–application in biopreservation?Trends Fd. Sci. Technol. 41: 37–48.

FDA (2018). The U. S. Food and Drug Administration, along with the Centers for Disease Control and Prevention (CDC) and state and local partners, investigated a multistate outbreak of E. coli O157:H7 illnesses linked to romaine lettuce from the Yuma growing region.

FDA [FDA] Food and Drug Administration (U. S). (2002). PVC devices containing plasticizer DEHP [FDA public health notification]. 2002. Washington, D. C. FDA.

FDA Report on the Occurrence of Foodborne Illness Risk Factors in Selected Institutional Foodservice, Restaurant, andRetailFoodStoreFacilityTypes(2009). https://www. fda. gov/food/guidanceregulation/retailfoodprotection/foodb orneillnessriskfactorreduction/default. htm

FDA, Food Code. (2001). Department of Health and Human Services. office of inspector general janet rehnquist inspector general. retail food safety. OEI-05-00-00540.

FDA, USF and DA. (2017). Summary report on antimicrobials sold or distributed for use in food-producing animals. (released in Dec. 2018).

FDA. (2017). summary report. Antimicrobials Sold or Distributed for Use in Food-Producing Animals.

FDA. (2019). Investigation Summary): Factors Potentially Contributing to the Contamination of Romaine Lettuce Implicated in the Fall 2018 Multi-State Outbreak of *E. coli* O157:H7.

Felicio, B. A., Pinto, M. S., Oliveira, F. S., Lempk, M. W., Pires, A. C. S., and Lelis, C. A. (2015). Effects of nisin on *Staphylococcus aureus* count and physicochemical properties of Minas Frescal cheese. J. Dairy Sci. 98: 4364–4369.

Fellows, P., and Axtell, B. (2002). Packaging materials. In: Appropriate food packaging: materials and methods for small businesses. Eds. Fellows P, Axtell B, Essex , U. K. ITDG Publishing. 25–77.

Ferreira, M., and Lund, B. (1996). The effect of nisin on *Listeria monocytogenes* in culture medium and long-life cottage cheese. Lett. Appl. Microbiol. 22: 433–438.

Fleming, D. W., Cochi, S. L., MacDonald, K. L., Brondum, J., Hayes, P. S., Plikaytis, B. D., Holmes, M. B., Audurier, A., Broome, C. V., and Reingold, A. L. (1985). Pasteurized milk as a vehicle of infection in an outbreak of listeriosis. N Engl. J. Med. 312(7):404-7.

Florida Press Educational Services Inc. (FPES). (2016). Food safety matters. email ktower@press. com or jpushkin@tampabay. com.

Foegeding, P. M. and Roberts, T. (1996). Assessment of Risks Associated with Foodborne Pathogens: An Overview of a Council for Agricultural Science and Technology Report. J. Fd. Prot. 59, (13): 19-23.

Fohey, M. F., Sauer, R. L., Westover, J. B., and Rockafeller, E. F. (1978). Food packages for Space Shuttle. NASA Technical Reports Server (NTRS). Food Nanotechnology-Food Packaging Applications. USDA-ARS?s Scientific Manuscript database.

Food and Drug Administation (FDA). (1998). Guide to minimize microbial food safety hazards for fresh fruits and vegetables. http:///www. cfsan. fda. gov/<"dms/prodguid. html viewed February 4, 2008.

Food Code. (1993). Recommendations of the United States Public Health Service, Food and Drug Administration, National Technical Information Service Publication PB94-113941.

Food packets, NASA Image and Video Library, (2012). ISS032-E-019031 (15 Aug. 2012).

Food safety and foodborne illness. Retrieved (2010). World Health Organisation.

Food Safety and Inspection Service (FSIS)'s Food Safety Process Management Programs. Risk Management and Decision Processes Center The Wharton School, University of Pennsylvania, 3730 Walnut Street, Jon Huntsman Hall, Suite 500 Philadelphia, PA, 19104 USA.

Food Standards Agency Annual Report (2007/08) Presented to Parliament under Section 4 of the Food Standards Act 1999 Ordered by the House of Commons to be printed 17 July 2008 HC 805 SG/2008/26 NIA 191/07–08

Food table on ISS. NASA Image and Video Library. (2015). ISS043E091650 (04/08/2015).

Forsyth, A., Lytle, L., and Riper, D. V. (2011). Finding food. Food Nanotechnology-Food Packaging Applications. USDA-ARS?Scientific Manuscript database.

Foulquié Moreno et al. (2006). The role and application of enterococci in food and health. International J. Fd. Microbiol. 106(1):1-24.

Francis, S. L. (2014). Hybrid food preservation program improves food preservation and food safety knowledge. ERIC Educational Resources Information Center.

Francis, Sarah L. 2014. . Hybrid Food Preservation Program Improves Food Preservation and Food Safety Knowledge. ERIC Educational Resources Information Center.

Franz, C., Huch, M., Abriouel, H., Holzapfel, W., and Gálvez, A. (2011). *Enterococci* as probiotics and their implications in food safety. Int. J. Fd. Microbiol. 151(2):125–40.

Franz, E., van Diepeningen, A. D., De Vos, O. J., van Bruggen, A. H. (2005). Effects of cattle feeding regimen and soil management type on the fate of *Escherichia coli* O157:H7 and *Salmonella enterica* serovar *Typhimurium* in manure, manure-amended soil, and lettuce. App. Environ. Microbiol. 71: 6165–6174.

Frenkel, V. S., Cummings, G. A., Maillacheruvu, K. Y., and Tang, W. Z. (2017). Food-Processing Wastes. science. gov, your gateway to federal science.

Frenkel, V. S., Cummings, G. A., Maillacheruvu, K. Y., and Tang, W. Z. (2016). Food-processing Wastes. Water Environ Res. 88(10):1395-408. doi: 10. 2175/106143016X14696400495091.

Frenkel, Val., Cummings, G. A., Maillacheruvu, K. Y., and Tang, W. Z. (2015). Food Processing Wastes. Water Environ Res. 2015 Oct:87(10):1360-72. doi: 10. 2175/106143015 X14338845155868.

FSIS Compliance Guideline for Controlling Meat and Poultry Products Pending FSIS Test Results:http://www. fsis. usda. gov/wps/wcm/connect/6b7b5a65-9ad3-4f89-927d078e564aca24/Compliance_Guide_Test_Hold_020113. pdf?MOD=AJPERES

FSMA. 11th US Congress. (2011). To amend the Federal Food, Drug, and Cosmetic Act with respect to the safety of the food supply.

FSN (Food safety News). (2018). North Dakota confirms E. coli outbreak case: 26 states hit.

Ft. Lee. (1971). Prevention of Food Poisoning. ERIC Number: ED121956, Record Type: Non-Journal, Publication Date: 1971: 106.

Fuqua C. (2006). The QscR quorum-sensing regulon of *Pseudomonas aeruginosa*: an orphan claims its identity. J Bacteriol. 188:3169–3171. doi:10. 1128/JB. 188. 9. 3169-3171. 2006.

Fuska, J., Fusková, A., Smarda, J., and Mach, J. (1979). Effect of colicin E3 on leukemia cells P388 *in vitro*. Experientia. 35: 406-407.

Futter, M. (2001). Communication Essentials in Poster Development"., pub. SADC-REEP, "Resource Materials Development".

Gale, F., and D. Hu. (2009). Supply Chain Issues in China's Milk Adulteration Incident. Paper presented at International Association of Agricultural Economists' 2009 Conference. 16–22 August 2009, Beijing, China.

Galvez, A., Abriouel, H., Omar, N. Ben., and Lucas, R. (2016) Food Applications and Regulation, NASA Astrophysics Data System (ADS).

Galvez, A., Lopez, R. L., Abriouel, H., Valdivia, E., and Omar, N. B. (2008). Application of bacteriocins in the control of foodborne pathogenic and spoilage bacteria. Crit. Rev. Biotechnol. 28: 125–152.

GÃmez-Sala, B., Herranz, C., DÃ-az-Freitas, B., HernÃ¡ndez, P. E., Sala, A., and Cintas, L. M. (2016). Strategies to increase the hygienic and economic value of fresh fish: Biopreservation using lactic acid bacteria of marine origin. Information Systems Division, National Agricultural Library The National Agricultural Library is one of four national libraries of the United States, with locations in Beltsville, Maryland and Washington, D. C. http:// www. nal. usda. gov/

Ganzin, M. (1975). Food for all. Food Nutr (Roma). 1(3):2-7.

Garber, A. K., and Lustig, R. H. (2011). Is fast food addictive?

Garde, S., Avila, M., Arias, R., Gaya, P., and Nunez, M. (2011). Outgrowth inhibition of Clostridium beijerinckii spores by a bacteriocin-producing lactic culture in ovine milk cheese. Int. J. Fd. Microbiol. 150: 59–65.

Gardner, A., West, S. A., and Buckling, A. (2004). Bacteriocins, spite and virulence. Proc. Roy. Soc. Lond. Ser. B-Biol. Sci. 271: 1529–1535.

Gardner, T., McLaughlin, J. (2008). Campylobacteriosis outbreak due to consumption of raw peas —Alaska, 2008. State of Alaska Epidemiology Bulletin. 20.

Garriga, M., Aymerich, M., Costa, S., Monfort, J., and Hugas, M. (2002). Bactericidal synergism through bacteriocins and high pressure in a meat model system during storage. Fd. Microbiol. 19: 509–518.

Garsa, A. K., Kumariya, R., Kumar, A., Lather, P., Kapila, S., and Sood, S. (2014). Industrial cheese whey utilization for enhanced production of purified pediocin PA-1. LWT. Fd. Sci. Technol. 59:656–665.

Gayathri Krishnan and Renu Agrawal. (2012). Antioxidant activity and fatty acid profile of fermented milk prepared by Pediococcus pentosaceus. J. Fd. Sci. Technol. doi: 10. 1007/ s13197-012-0891-9.

GFSI Technical Committee – September. 2007.

Ghosh, D. (2014). Food safety regulations in Australia and New Zealand Food Standards. J. Sci. Fd. Agric. 94(10):1970-3. doi: 10. 1002/jsfa. 6657. Epub 2014

Gibson, G. R., Saavedra, J. M., Macfarlane, S., and Macfarlane, G. T. (1997). Probiotics and intestinal infections. In: Probiotics 2: Applications and practical Aspects. ed. R. Fuller. 10 – 39. New York: Chapman and Hall.

Gil, R., Latorre, A., and Moya, A. (2004). Bacterial endosymbionts of insects: insights from comparative genomics. Environ. Microbiol. 6, 11:1109–1122.

Gillor O., Giladi I., and Riley M. A. (2009). Persistence of colicinogenic Escherichia coli in the 976 mouse gastrointestinal tract. BMC Microbiol. 9: 165.

Gillor, O., Etzion, A., Riley, M. A. (2008). The dual role of bacteriocins as anti- and probiotics. Appl. Microbiol. Biotechnol. 81:591–606.

Giraffa, G. (1995). Enterococcal bacteriocins: their potential as anti-Listeria factors in dairy technology. Fd. Microbiol. 12: 291–299.

Giraffa, G., and Carminati, D. (1997). Control of Listeria monocytogenes in the rind of Taleggio, a surface-smear cheese, by a bacteriocin from Enterococcus faecium 7C5. Sci. Aliments 17:383–391.

Girling, P. J. (2003). Packaging of food in glass containers. In: Food packaging technology. Eds. Coles, R., McDowell, D., Kirwan, M. J. London, U. K. :Blackwell Publishing, CRC Press. 152–73.

Golan, E., Krissoff, B., Kuchler, F., Calvin, L., Nelson, K., and Price, G. (2004). Traceability in the U. S. food supply: economic theory and industry studies. Agricultural economic report nr 830. Washington , DC : Economic Research Service, U. S. Dept. Agri. p 48.

Golem, D. L., and Byrd-Bredbenner, C. (2015). Emergency food supplies in food secure households. Prehosp. Disaster Med. 2015. 30(4):359-64. doi: 10. 1017/ S1049023X15004884. Epub 2015.

Gómez-Torres, N., Avila, M., Gaya, P., and Garde, S. (2014). Prevention of late blowing defect by reuterin produced in cheese by a *Lactobacillus reuteri* adjunct. Fd. Microbiol. 42, 82–88.

Gómez-Torres, N., Garde, S., Peiroten, A., and Avila, M. (2015). Impact of *Clostridium* spp. on cheese characteristics: microbiology, color, formation of volatile compounds and off-flavors. Fd. Cntrl. 56: 186–194.

Gorris, L. G. M. (2005). Food safety objective: An integral part of food chain management. Fd. Cntrl. 16, 9: 801-809, ISSN 0956-7135.

Graham, J. P., Price, L. B., Evans, S. L., Graczyk, T. K., and Silbergeld, E. K. (2009). Antibiotic resistant *Enterococci* and *Staphylococci* isolated from flies collected near confined poultry feeding operations. Sci. Total Environ. 407, 8: 2701–2710.

Gram, L., Ravn, L., Rasch, M., Bruhn, J. B., Christensen, A. B., and Givskov, M. (2002). Food spoilage—interactions between food spoilage bacteria. Intern. J. Fd. Microbiol. 78(1): 79–97.

Gran, H. M., Wetlesen, A., Mutukumira, A. N., Narvhus, J. A. (2002). Smallholder dairy processing in Zimbabwe: hygienic practices during milking and the microbiological quality of the milk at the farm and on delivery. Fd. Cntrl. 13: 41–47.

Grande Burgos MJ, Pulido RP, López guayo MDC, Gálvez A, and Lucas R. (2014). The cyclic antibacterial peptide enterocin AS-48: isolation, mode of action, and possible food applications. Int. J. Mol. Sci. 15:22706–22727. doi:10. 3390/ijms151222706.

Grande Burgos, M. J., Romero, J. L., Pérez, P. R., Cobo, M. A., Gálvez, A., and Lucas, R. (2018). Analysis of potential risks from the bacterial communities associated with air-contact surfaces from tilapia (*Oreochromis niloticus*) fish farming. Environ. Res. 160:385-390. doi: 10. 1016/j. envres. 2017. 10. 021. Epub 2017.

Gray, J. A., Chandry, P. S., Kaur, M., Kocharunchitt, Chawalit., Bowman, J. P., Fox., and Edward, M. (2018). Novel biocontrol methods for *Listeria monocytogenes* biofilms in food production facilities. Front. Microbiol. 9:605. doi:10. 3389/fmicb. 2018. 00605. eCollection 2018.

Greger, M, M. D. (2017) Nutrition facts. org .

Greger. (2017). The Food Safety Risk of Organic versus Conventional. Written by Michael Greger M. D. FACLM.

Grief, S. N. (2016). Food Allergies. Primary care: clinics in office practice. 43, 3: 375-391.

Grinter, R., Milner, J., and Walker, D. (2012). Bacteriocins active against plant pathogenic bacteria. Biochem. Soc. Trans. 40:1498–1502. doi: 10. 1042/BST20120206.

Guandalini, S., and Newland, C. (2011). Differentiating food allergies from food intolerances. Curr. Gastroenterol. Rep. 13(5):426-34. doi: 10. 1007/s11894-011-0215-7.

Guaní-Guerra, E., . Santos-Mendoza, T., Lugo-Reyes, S. O., and L. M. Terán. (2010). "Antimicrobial peptides: general overview and clinical implications in human health and disease, " Clin. Immun. 135, 1 : 1–11.

Guidelines for food Recall-fssai. (2017). https:// www. fssai. gov. in/dam/jcr. . . a327. . . / Guidelines_Food_Recall_28_11_2017. pdf.

Gupta, M., Cox, A., Nowak, W., Anna, W., Julie. (2018). Diagnosis of Food Allergy.

Gupta, R. S., Kim, J. S., Springston, E. E., Smith, B., Pongracic, J. A., Wang, X., and Holl, J. (2009). Food allergy knowledge, attitudes, and beliefs in the United States. Ann. Allergy Asthma Immunol. 103(1):43-50. doi: 10. 1016/S1081-1206(10)60142-1.

HACCP Principles & Application Guidelines. (1997). National advisory committee on microbiological criteria for foods. Content current as of:12/19/2017.

Haddad, K., Schmelcher, M., Sabzalipoor, H., Seyed, H. E., and Moniri, R. (2018). Recombinant endolysins as potential therapeutics against antibiotic-resistant *Staphylococcus aureus*: Current status of research and novel delivery strategies. Clin. Microbiol. Rev. 31(1). pii:e00071-17.

Hafstro, et. al. (2001). A vegan diet free of gluten improves the signs and symptoms of rheumatoid arthritis: the effects on arthritis correlate with a reductionin antibodies to food antigens. Rheumatology. 40:1175–1179.

Hald, T. (2013). Advances in microbial food safety: 2. Pathogen update: Salmonella. Elsevier Inc. Chapters. p. 2. 2. ISBN 9780128089606.

Hale, C. R., Scallan, E., Cronquist, A. B., Dunn, J., Smith, K., Robinson, T., Lathrop, S., Tobin-D'Angelo, M., and Clogher, Paula. (2012). Estimates of enteric illness attributable to contact with animals and their environments in the United States. Clin. Infect. Dis. 54(S5):S472–9.

Hall, A. J., Wikswo, M. E., Manikonda, K., Roberts, V. A., Yoder, J. S., Gould, L. H. (2013). Acute gastroenteritis surveillance through the National Outbreak Reporting System, United States. Emerg. Infect. Dis. http://dx. doi. org/10. 3201/eid1908. 130482.

Halvorsen, R., Eggesb, M., and Botten, G. (1995). Reactions to food. Tidsskr Nor Laegeforen. 10:115(30):3730-3.

Hammami, R., Zouhir, A., Ben Hamida, J., Fliss, I. (2007). "BACTIBASE: a new web-accessible database for bacteriocin characterization". BMC Microbiol. 7:89. doi:10. 1186/1471-2180-7-89. PMC 2211298 . PMID 17941971.

Han, J. H. (2003). Antimicrobial food packaging. in: Novel Food Packaging Techniques. Ed. R. Ahvenainen (Cambridge: Wood head Publishing). 50–70.

Hanlin, M. B., Kalchayanand, N., Ray, P. and Ray, B. (1993). Bacteriocins of lactic acid bacteria in combination have greater antibacterial activity. J. Fd. Prot. 56: 252–255.

Hartel, Ri. W. (2013). Advances in food crystallization. Annual Rev. Fd. Sci. Technol. 4:277-292. https://doi. org/10. 1146/annurev-food-030212-182530.

Hartman, R., Brown, L. R., and Hobbs, J. (2017). Food web conceptual model. USGS Publications Warehouse.

Hauge, H. H, Mantzilas , D., Moll, G. N., Konings, W. N., Driessen, A. J. M., Eijsink, V. G. H., and Nissen, M. J. (1998). Plantaricin A is an amphiphilic alpha-helical bacteriocin-like pheromone which exerts antimicrobial and pheromone activities through different mechanisms. Biochem. 37:16026–16032.

Havenaar, R., Brink, B., and Huis in't Veld, J. H. J. (1992). Selection of strains for probiotic use, In: Probiotics, the scientific Basis, ed. R. Fuller. 209–224. London, UK: Chapman and Hall.

Havlíèek, V., Patchett, M. L., and Norris, G. E. (2011). Cysteine s-Glycosylation, a new post-translational modification found in glycopeptide bacteriocins. FEBS Letts. 585: 645–650.

Health, N. J. (2015). Partnership for Food Safety Education (USA): Laboratory of Clinical and Epidemiological Virology, Catholic University of Leuven (Belgium). Univesity of Medicine and Dentistry of New Jersey (USA).

Healthy food trends — flaxseeds. (2015). J Nutr . 145(4): 758-765.

Heikkilä, L., Reinikainen, A., Katajajuuri, J. M., Silvennoinen, K., and Hartikainen, H. (2016). Elements affecting food waste in the food service sector. Waste Management. 56: 446-453.

Heine, R. G. (2015). Gastrointestinal food allergies. Chem. Immunol. Allergy. 101:171-80. doi: 10. 1159/000371700. Epub 2015.

Heinonen, M. (2014). Food additives and healthiness. Duodecim. 130, 7:683-8.

Henderson, J., Coveney, J: and Ward, P. (2010). Who regulates food? Australians' perceptions of responsibility for food safety. Australian J. Primary Health. 16, 4 :344-51.

Heng, C. K. N., Wescombe, P. A., Burton, J. P., Jack, R. W., and Tagg, J. R. (2007). The diversity of bacteriocins in Gram-positive bacteria. In: Bacteriocins: Ecology and Evolution. 1st ed., Riley, M. A. & Chavan, M. A., Eds. Springer, Hildberg. 45-83.

Hernandez, J. (2000). Food safety: to keep food safe, stay out of the danger zone. Restaurant Hospitality. 84, 6: 104-110.

Heron, M., and Anderson, R. N. (2016). Changes in the leading cause of death: Recent patterns in heart disease and cancer mortality. NCHS data brief, no 254. Hyattsville, MD: National Center for Health Statistics.

Hillman, J., Dzuback, A., and Andrews, S. (1987). Colonization of the human oral cavity by a *Streptococcus mutans* mutant producing increased bacteriocin. J. Dent. Res. 66:1092–1094.

Hillman, J., et al. (1998). Genetic and biochemical analysis of mutacin 1140, a lantibiotic from *Streptococcus mutans*. Infect. Immun. 66: 2743–2749.

Hintz, T., Matthews, K. K., and Di, R. (2015). The use of plant antimicrobial compounds for food preservation. Bio med. Res. Int. 246264.

Hogan and Hartson. (2007). "A Guide to Federal Food Labeling Requirements for Meat, Poultry, and Egg Products, " prepared for USDA Food Safety and Inspection Service.

Hotchkiss, J. J. (1997). Food-packaging interactions influencing quality and safety. Fd. Addit. Contam. 14. (6-7): 601–7.

Howden, B. P. (2005). Recognition and management of infections caused by vancomycin-intermediate *Staphylococcus aureus* (VISA) and heterogenous VISA (hVISA). Intern. Med. J. Dec:35 Suppl 2:S136-40.

http://100kgenome. vetmed. ucdavis. edu/.

http://100kgenome. vetmed. ucdavis. edu/.

http://dx. doi. org/10. 1016/S1466-8564(02)00012-7.

http://dx. doi. org/10. 1155/2013/718780

http://focustaiwan. tw/news/aipl/201608300011. aspx

http://www. cdc. gov/amd/.

http://www. emd. saccounty. net/EH/FoodProtect-RetailFood/Pages/CalCode. aspx

http://www. fda. gov/Food/FoodScienceResearch/WholeGenomeSequencingProgramWGS/ucm363134. htm.

http://www. genomicepidemiology. org.

http://www. globalmicrobialidentifier. org : https://doi. org/10. 21775/978191019037

https://health. howstuffworks. com

Huffman, M. M. (2015). Food and environmental allergies. Primary care: Clinics in office practice. 42, 1:113-28. doi: 10. 1016/j. pop. 2014. 09. 010.

Hunt, R. G., Sellers, V. R., Franklin, W. E., Nelson, J. M., Rathje, W. L., Hughes, W. W., and Wilson, D. C. (1990). Estimates of the volume of MSW and selected components in trash cans and land fills. Report prepared by the Garbage Project and Franklins Associates Ltd. for the Council for Solid Waste Solutions, Tucson, Ariz.

Hussein, H. S., Sakuma, T. (2005). Prevalence of shiga toxin-producing Escherichia coli in dairy cattle and their products. J. Dairy Sc. 88: 450–465.

Ibarguren, C., Céliz, G., Díaz, A. S., Bertuzzi, M. A., Daz, M., and Audisio, M. C. (2015). Gelatine based films added with bacteriocins and a flavonoid ester active against food-borne pathogens. Innov. Fd. Sci. Emerg. Technol. 28: 66–72.

IFS(2014). IFS Standard for auditing quality and food safety of food products Version 6. Berlin: IFS.

Information Center. 15, 1: 34-41. https://doi. org/10. 1111/1541-4329. 12079.

Ingrid, H. V., Eijsink G. H., Brurberg, M. B. (2008). Bacteriocins from plant pathogenic bacteria. FEMSMicrobiol. Letts. 280, 1:1–7. https://doi. org/10. 1111/j. 1574-6968. 2007. 01010.

x

Institute of Food Technologists. (1988). Migration of toxicants, flavors, and odoractive substances from flexible packaging materials to food [IFT scientific status summary]. Risch S. J, author. Food Tech. 42, 7:95–102.

Institute of Food Technologists. (1991). Effective management of food packaging: from production to disposal [IFT scientific status summary]. Marsh KS, author. Food Tech 45, (5):225–34. 318.

Institute of Food Technologists. (1997). Edible and biodegradable polymer films: challenges and opportunities [IFT scientific status summary]. Krochta JM, DeMulder-Johnston C, authors. Food Tech 51, 2: 61–74.

International Organization Standards. (ISO). *(2017)*. "ISO 22000 Revision". International Organization Standards ISO.

Isadore, R., and Howard, K., (2010). Roles for third parties in implementing USDA,

ISO 11290-2(1998). Microbiology of food and animal feeding stuffs-Horizontal method for the detection and enumeration of *Listeria monocytogenes*—Part 2: Enumeration method.

ISO 16649-2 (2001). Microbiology of food and animal feeding stuffs -Horizontal method for the enumeration of beta-glucuronidase-positive *Escherichia coli*—Part 2:Colony-count technique at 44°C using 5-bromo-4-chloro-3-indolyl beta-D-glucuronide.

ISO 22000:2005. Food safety management systems — Requirements for any organization in the food chain. ICS :67. 020. Processes in the food industry. 03. 100. 70. Management systems.

ISO 9000:(2000). Quality management systems — Fundamentals and vocabulary.

Ito, Y., Kawai, Y., Arakawa, K., Honme , Y., Sasaki, T., and Saito, T. (2009). Conjugative plasmid from *Lactobacillus gasseri* LA39 that carries genes for production of and immunity to the circular bacteriocin gassericin A. Appl. Environ. Microbiol. 75, 6340–6351. 10. 1128/AEM. 00195-09.

Iweala, O. I., and Choudhary, S. K. (2018). Food Allergy. Curr Gastroenterol Rep. 5, 20(5) : 17. doi: 10. 1007/s11894-018-0624-y.

Jackson, P. and Viehoff, V. (2016). Reframing convenience food. Appetite, 98. 1-11. ISSN 0195-6663.

Jamuna, M., and Jeevaratnam, K. (2004). Isolation and partial characterization of bacteriocins from *Pediococcus* species. Appl. Microbiol. Biotechnol. 65, 433–439.

Jaouani, I., Abbassi, M., Ribeiro, S., Khemiri, M., Mansouri, R., Messadi, L., et al. (2015). Safety and technological properties of bacteriocinogenic enterococci isolates from Tunisia. J. Appl. Microbiol. 119, 1089–1100.

Jarvenpaa, S. L., and Todd, P. A. (1997). Consumer Reactions to Electronic Shopping on the World Wide Web. International Journal of Electronic Commerce, 1:59-88.

Jay, J. M. (1995). Modern Food Microbiology. Fd Sci. Text series.

Jay, J. M. (1996a). Intrinsic and extrinsic parameters of foods that affect microbial growth. Pages 38–66 in Modern Food Microbiology. 5th ed. Chapman and Hall, New York, NY.

Jay, J. M. (1996b). Foodborne listeriosis. Pages 478–506 in Modern Food Microbiology. 5th ed. Chapman and Hall, New York, NY.

Jedermann, R., Nicometo, M., Uysal, I., and Lang, W. (2014). Reducing food losses by intelligent food logistics. Philos Trans A Math Phys Eng Sci. 372(2017): 20130302. doi: 10. 1098/rsta. 2013. 0302.

Joerger, R. D. (2003). Alternatives to antibiotics: Bacteriocins, antimicrobial peptides and bacteriophages. Poultry Sci. 82, 4:640-647.

John, D. F. Hale, Philip A. Wescombe, John R. Tagg and Nicholas C. K. Heng. (2016). Streptococcal Bacteriocin-producing Strains as Oral Probiotic Agents. Caister Academic Press, U. K. 103-126. DOI: https://doi. org/10. 21775/9781910190371. 06.

Jones, K. E., Patel, N. G., Levy, M. A. et al., (2008). Global trends in emerging infectious diseases. Nature. 451, No. 7181: 990–993,

Jones, R. (1998). Food Safety, Farm to Fork. ERIC Educational Resources Information Center.

Joo, N. E., Ritchie, K., Kamarajan, P., Miao, D., and Kapila, Y. I. (2012). Nisin, an apoptogenic bacteriocin and food preservative, attenuates HNSCC tumorigenesis via CHAC1. Cancer Med. 1: 295–305.

Joon, C., In Sun, J., Jun, H. C., Kyoung, H. J., Eun, J. C., Soon, H. L., and In Gyun, H. (2012). Prevalence and characterization of foodborne bacteria from meat products in Korea. Fd. Sci. Biotechnol. 21, 5:1257–1261.

Jordan, D. J. (1998). Food Concerns. Research Notes. ERIC Educational Resources Information Center.

Jordi Bart, J. A. M., Boutaga, K., Heeswijk Caroline, M. E. van., Knapen, F, van ., and Lipman Len, J. A. (2001). Sensitivity of Shiga toxin-producing *Escherichia coli* (STEC) strains for colicins under dierent experimental conditions. FEMS Microbiol. Lett. 204 : 329-334.

Jordi K., Grebel E. K., and Ammon K. (2006). A&A, 460: 339.

Joseph, L. C. (2000). Food Quest for Health. ERIC Educational Resources Information Center.

Josephson, E. S. (1977). Food irradiation and sterilization. NASA Astrophysics Data System (ADS). Radiation Physics and Chemistry. 18, 1-2: 223-239.

Josephson, E. S. (1983). Radappertization of meat, poultry, finfish, shellfish, and special diets. In Preservation of Food by Ionizing Radiation. In: (E. S. Josephson and M. S. Peterson, eds.) Chap. 8, 3: 231-251, CRC Press, Boca Raton, Florida.

Józefiak, D., Kieroñczyk, B., Juœkiewicz, J., Zduñczyk, Z., Rawski, M., D³ugosz, J., Sip, A., and Højberg, O. (2013). Dietary nisin modulates the gastrointestinal microbial ecology and enhances growth performance of the broiler chickens. PloS one. 8. e85347.

Juneja, V. K., Dwivedi, H. P., and Yan, X. (2012). Novel natural food antimicrobials. Annu. Rev. Fd. Sci. Technol. 3:381-403. doi: 10. 1146/annurev-food-022811-101241.

Justyna, A., Wojdyla, G. P., and Colin, K. (2016). Nuclease Colicins: Mode of Action, Immunity and Mechanism of Import into *Escherichia coli*. Caister Academic Press, U. K 35-64. DOI: https://doi. org/10. 21775/9781910190371. 03.

Kabuki, T., Saito, T., Kawai, Y., Uemura, J., and Itoh, T. (1997). Production, purification and characterization of reutericin 6, a bacteriocin with lytic activity produced by *Lactobacillus reuteri* LA6". Internat. J. Fd. Microbiol. 34 (2): 145–56. doi:10. 1016/s0168-1605(96)01180-4. PMID 9039561. Retrieved 2015-01-19.

Käferstein, F. K., Motarjemi, Y., and Bettcher, D. W. (1997). Foodborne disease control: a transnational challenge. Emerging Infectious Diseases. 31: 503–510.

Kalchayanand, N., Hanlin, M., and Ray, B. (1992). Sublethal injury makes Gram-negative and resistant Gram-positive bacteria sensitive to the bacteriocins, pediocin AcH and nisin. Lett. Appl. Microbiol. 15: 239–243.

Kawai, Y., Kemperman, R., Kok, J., and Saito, T. (2004). The circular bacteriocins gassericin A and circularin A. Current Protein Peptide Sci. 5 (5): 393–8. doi:10. 2174/1389203043379549. PMID 15544534. Retrieved 2015-01-19.

Kellie O'Dare Wilson. (2016). Place matters: Mitigating obesity with the person-in-environment approach. Social Work in Health Care. 55:3, 214-230.

Kerr, W. L., and Reid, D. S. (1993). Thermodynamics and Frozen Foods. ERIC Educational Resources Information Center.

Khan, H., Flint, S., and Yu, P. L. (2010). Enterocins in food preservation. Int. J. Fd. Microbiol. 141: 1–10.

Khan, I., and Oh, D. H. (2016). Integration of nisin into nanoparticles for application in foods. Innov. Fd. Sci. Emerg. Technol. 34:376–384.

Kibler, K. M., Reinhart, D., Hawkins, C., Motlagh, A. M., and Wright, J. (2018). Food waste and the food-energy-water nexus: A review of food waste management alternatives. 74: 52-62. doi: 10. 1016/*j*. wasman. 2018. 01. 014. Epub *2018.*

Kicza , A., Pureka, C., Proctor, D., Riley, M., and Dorit, R. The Phenotypic and Genotypic Landscape of Colicin Resistance in: The Bacteriocins: Current Knowledge and Future Prospects (Edited by: Robert L. Dorit, Sandra M. Roy and Margaret A. Riley). Caister Academic Press, U. K. (2016) *141-154*. DOI: https://doi. org/10. 21775/9781910190371. 08

King, G., Walsh, J. E., and Martin, S. (2003). Spectroscopic study of food and food toxins. NASA Astrophysics Data System (ADS).

Kirwan, M. J. (2003). Paper and paperboard packaging. In: Coles R, McDowell D, Kirwan M. J, editors. *Food packaging technology*. London , U. K. : Blackwell Publishing, CRC Press. 241–81.

Kirwan, M. J., Strawbridge, J. W. (2003). Plastics in food packaging. In: Coles, R., McDowell, D., Kirwan, M. J., editors. Food packaging technology. London, U.K.: Blackwell Publishing, CRC Press. 174–240.

Kiser, J. V. L., and Zannes, M. (2004). The 2004 IWSA directory of waste-to-energy plants. Washington , D. C. : Integrated Waste Services Association. 31.

Klaenhammer, T. R. (1988). Bacteriocins of lactic acid bacteria. Biochimie. 70:337–349.

Klein, S. S., and DeWaal, C. S. (2008). Center for Science in the Public Interest, https://cspinet. org/sites/default/files/attachment/ddreport.

Kodani, S., Lodato, M. A., Durrant, M. C., Picart, F., and Willey, J. M. (2005) SapT, a lanthionine-containing peptide involved in aerial hyphae formation in the streptomycetes. Mol. Microbiol. 58: 1368–1380.

Koivisto, H. U K. (1999). Factors influencing children's food choice. Annals of Medicine 31:S1. 26-32.

Kolata, G. (1982). Food affects human behavior. ERIC Educational Resources Information Center.

Kondrotiene, K., Kasnauskyte, N., Serniene, L., Gölz, G., Alter, T., Kaskoniene, V., et al. (2018). Characterization and application of newly isolated nisin producing *Lactococcus lactis* strains for control of *Listeria monocytogenes* growth in fresh cheese. LWT Fd. Sci. Technol. 87: 507–514.

Kong, K. L., and Epstein, L. H. (2016). Food reinforcement during infancy. Prev. Med. 92: 100-105. doi: 10. 1016/j. ypmed. 2016. 06. 031. Epub 2016.

Koppel, K. (2014). Sensory analysis of pet foods. J. Sci. Fd. Agri. 94, 11: 2148-2153.

Krochta, J. M., and DeMulder-Johnston, C. (1997). [IFT] Institute of Food Technologists. Edible and biodegradable polymer films: challenges and opportunities [IFT scientific status summary]. Fd. Tech. 51, 2: 61–74.

Kuiper, H. A., Noteborn, H. P. J. M., and Peijnenburg, A. A. C. M. (1999). Adequacy of methods for testing the safety of genetically modified foods. Lancet. 354: 1315 – 1316. DOI: 10. 1016/s0140-6736(99)00341-4.

Kuipers, O. P., Beerthuyzen, M. M., de Ruyter, P. G., Luesink, E. J., and de Vos, W. M. (1995). Autoregulation of nisin biosynthesis in *Lactococcus lactis* by signal transduction. J. Biol. Chem. 270: 27299–27304.

Kykkidou, S., Pournis, N., Kostoula, O. K., and Savvaidis, I. N. (2007). Effects of treatment with nisin on the microbial flora and sensory properties of a Greek soft acid-curd cheese stored aerobically at 4oC. Int. Dairy J. 17: 1254–1258.

Lagos, R., Tello, M., Mercado, G., García, V., and Monasterio, O. (2009). Antibacterial and antitumorigenic properties of microcin E492, a pore-forming bacteriocin. Current pharmaceutical biotechnol. 10, 1: 74–85. PMID 19149591.

Lakshmi. V., et al., (2012). Food adulteration. IJSIT. 1, 2: 106-113.

Lancaster, K. (1971). Consumer Demand: A New Approach. New York: Columbia University Press.

Lancaster, L. E., Wintermeyer, W., and Rodnina, M. V. (2007). Colicins and their potential in cancer treatment. Blood Cells Mol. Dis. 38: 15–18 10. 1016/j. bcmd. 2006. 10. 006.

Land, M., Hauser, L., Jun, S. R., et al. (2015). Insights from 20 years of bacterial genome sequencing. Funct Integr Genomics. 15(2):141–161. doi:10. 1007/s10142-015-0433-4

Lau, O. W., and Wong, S. K. (2000). Contamination in food from packaging materials. J. Chrom. A 882. (1–2):255–70.

Lauková, A., and Czikková, S. (1999). The use of enterocin CCM 4231 in soy milk to control the growth of Listeria monocytogenes and Staphylococcus aureus. J. Appl. Microbiol. 87: 182–182.

Lauková, A., Czikková, S., and Burdová, O. (1999a). Anti-staphylococcal effect of enterocin in Sunar and yogurt. Folia Microbiol. 44: 707–711.

Lauková, A., Czikková, S., Dobránsky, T., and Burdová, O. (1999b). Inhibition of Listeria monocytogenes and Staphylococcus aureus by enterocin CCM 4231 in milk products. Fd. Microbiol. 16: 93–99.

Lauková, A., Vlaemynck, G., and Czikkova, S. (2001). Effect of enterocin CCM 4231 on Listeria monocytogenes in Saint-Paulin cheese. Folia Microbiol. 46: 157–160.

Le Blay, G., Lacroix, C., Zihler, A., and Fliss, I. (2007). In vitro inhibition activity of nisin A, nisin Z, pediocin PA-1 and antibiotics against common intestinal bacteria. Lett. Appl. Microbiol. 45: 252–257.

Lebel, A., Noreau, D., Tremblay, L., Oberl, Ã., Girard-Gadreau, M., Duguay, M., and Block, J. P. (2016). Identifying rural food deserts: Methodological considerations for food environment interventions. PubMed.

Lee, J. H., et al. (2008). Comparative genomic analysis of the gut bacterium Bifidobacterium longum reveals loci susceptible to deletion during pure culture growth. BMC Genomics 9:247.

Lee, J. H., Karamychev, V. N., Kozyavkin, S. A., Mills, D., Pavlov, A. R., Pavlova, N. V., Polouchine, N. N., Richardson, P. M., Shakhova, V. V., Slesarev, A. I., Weimer, B., and O'Sullivan, D. J. (2008). Comparative genomic analysis of the gut bacterium Bifidobacterium longum reveals loci susceptible to deletion during pure culture growth. BMC Genomics. 9: 247. doi: 10. 1186/1471-2164-9-247.

Lee, N. K., and Paik, H. D. (2016). Status, antimicrobial mechanism and regulation of natural preservatives in livestock food systems. Korean J. Fd. Sci. Anim. Resour. 36, 4: 547–557.

Lee, N. K., Jung, B. S., Na, D. S., Yu, H. H., Kim, J. S., and Paik, H. D. (2016). The impact of antimicrobial effect of chestnut inner shell extracts against Campylobacter jejuni in chicken meat. LWT-Fd. Sci. Technol. 65: 746-750.

Lemieux, C. (1980). Convenience Food. "ERIC Educational Resources Information Center.

Leroy, F., and De Vuyst, L. (2004). Lactic acid bacteria as functional starter cultures for the food fermentation industry. Trends Fd. Sci. Technol. 15: 67–78.

Leroy, F., and De Vuyst, L. (2010). Bacteriocins of lactic acid bacteria to combat undesirable bacteria in dairy products. Australian Journal of Dairy Technology 65(3):143-9

Leyva, S. M., Mounier, J., Valence, F., Coton, M., Thierry, A., and Coton, E. (2017). Antifungal Microbial Agents for Food Biopreservation-A Review. Microorganisms. 5. 37. 10. 3390/microorganisms5030037.

Liliam, K. H., Erica, C. S., Welida, F. C., Fernando, S. D. F., Marta, V., Krystyna, D., Victor, N. Krylovc., and Victor, M. B. (2018). Biotechnological applications of bacteriophages:State of the art. Microbiological Res. 212–213.

Lin, L. C., Chang, S. C., Ge, M. C., Liu, T. P., and Lu, J. J. (2018). Novel single-nucleotide variations associated with vancomycin resistance in vancomycin-intermediate Staphylococcus aureus. Infect Drug Resist. 11:113-123.

Linares, J. F., Gustafsson, I., Baquero, F., and Martinez, J. L. (2006). Antibiotics as intermicrobial signaling agents instead of weapons. Proc. Natl. Acad. Sci. USA. 103:19484–19489.

Lindeman, A. K. (2002). Functional Foods for Women's Health. ERIC Educational Resources Information Center.

Localharvest. org

Lopez, C. E. , L. L. Ramos., S. S. Ramadàn., and L. C. Bulacio. (2003). Presence of aflatoxin M1 in milk for human consumption in Argentina. Fd. Cntrl. 14, 1:31-34.

Lopez-Rubio, A., Almenar, E., Hernandez-Munoz, P., Lagaron, J. M., Catala, R., and Gavara, R. (2004). Overview of active polymer-based packaging technologies for food application. Fd. Rev. Int. 20, 4: 357–87.

Lopman, B. A., Reacher, M. H., van Duijnhoven, Y., Hanon, F. X., Brown, D., and Koopmans, M. (2003). Viral gastroenteritis outbreaks in Europe, 1995-2000. Emerging Infect. Dis. 9: 90-96.

Lorena, Z., Immaculada, V., Lucía, P., and María, B. I. (2018). Evaluation of postharvest calcium treatment and biopreservation with *Lactobacillus rhamnosus* GG on the quality of fresh-cut 'Conference' pears: Evaluation of the effect of calcium and probiotic bacteria on quality of fresh-cut pears. J. Sci. Fd. Agricul. 98:13.

Louise, E. L., Omar, N., Amarat, H., Simonne, C., and M. Bruh. (2012). Consumer perceptions on food safety in Asian and Mexican restaurants. Fd. Cntrl. 26: 531e538.

Løvseth, A., Loncarevic, S., and Berdal K. G. (2004). Modified multiplex PCR method for detection of pyrogenic exotoxin genes in Staphylococcal isolates. J. Clin. Microbiol. 42, 8: 3869–3872.

Lucy, H. D., Paul, D. C., Colin, H., and Paul, R. (2006). Review. Bacteriocins: Biological tools for bio-preservation and shelf-life extension. Internat. Dairy J. 16, 9:1058-1071.

Lynch, M., Tauxe, R., and Hedberg, C. (2009). The growing burden of foodborne outbreaks due to contaminated fresh produce: risks and opportunities. Epidemiol. Infect. 137:307-315.

Maarten, G. K., Ghequire., and René, De Mot. (2014). Ribosomally encoded antibacterial proteins and peptides from *Pseudomonas*. FEMS Microbiol. Rev. 38, 4 :523–68.

MaÃtre, S., Maniu, C. M., Buss, G., Maillard, M. H., Spertini, F., and Ribi, C. (2014). Food allergy or food intolerance. pubmed central. science. gov.

Machaidze, G., and Seelig, J. (2003). Specific binding of cinnamycin (Ro 09-0198) to phosphatidylethanolamine. Comparison between micellar and membrane environments. Biochemistry. 42:12570-12576.

Macovei, L., and Zurek L. (2007). Influx of Enterococci and associated antibiotic resistance and virulence genes from ready-to-eat food to the human digestive tract. Appl. Environ. Microbiol. 73, 21: 6740–6747.

Macovei, L., and Zurek, L. (2006). Ecology of antibiotic resistance genes: characterization of Enterococci from houseflies collected in food settings. Appl. Environ. Microbiol. 72, 6: 4028–4035.

Mahboubi, M., (2019). Therapeutic Potential of Zataria multiflora Boiss in treatment of irritable bowel syndrome(IBS). J. DietSuppl. 16, 1:119-128. doi: 10. 1080/19390211. 2017. 1409852. Epub 2018

Maisnier-Patin, S., Deschamps, N., Tatini, S., and Richard, J. (1992). Inhibition of *Listeria monocytogenes* in Camembert cheese made with a nisin-producing starter. Le Lait 72: 249–263.

Majeed, H., Gillor, O., Kerr, B., Riley, M. A. (2011). Competitive interactions in *Escherichia coli* populations: the role of bacteriocins. ISME J. 5:71–81.

Makras, L., Triantafyllou, V., Fayol-Messaoudi, D., Adriany, T., Zoumpopoulou, G., Tsakalidou, E., Servin, A., and De Vuyst, L. (2006). Kinetic analysis of the antibacterial activity of probiotic *Lactobacilli* towards *Salmonella enterica* serovar *Typhimurium* reveals a role for lactic acid and other inhibitory compounds. Res Microbiol. 157.

Malheiros, P. S., Cuccovia, I. M., and Franco, B. (2016). Inhibition of *Listeria monocytogenes in vitro* and in goat milk by liposomal nanovesicles containing bacteriocins produced by *Lactobacillus sakei* subsp sakei 2a. Fd. Cntrl. 63: 158–164.

Mandal, P. K. , Biswas, A. K. , Choi, K., and Pal, U. K. (2011). Methods for Rapid Detection of Foodborne Pathogens:An Overview. Amer. J. Fd. Technol. 6 (2): 87-102.

Manning, L., and Baines, R. N. (2004). Effective management of food safety and quality. British Fd. J. 106. 598-606. 10. 1108/00070700410553594.

Maria, G. B., Rubén, P. P., María, d. C. L. A., and Antonio, G. (2014). The Cyclic Antibacterial Peptide Enterocin AS-48:Isolation, mode of action, and possible food applications. Intern. J. Mol. Sci. 15(12): 22706-22727.

Marques, J. D., Funck, G. D., Dannenberg, G. D., Cruxen, C. E. D., El Halal, S. L. M., Dias, A. R. G., et al., (2017). Bacteriocin-like substances of *Lactobacillus curvatus* P99: characterization and application in biodegradable films for control of *Listeria monocytogenes* in cheese. Fd. Microbiol. 63: 159–163.

Marsh, K. S. (1991). [IFT] Institute of Food Technologists. Effective management of food packaging:from production to disposal [IFT scientific status summary]. Fd. Tech 45(5):225–34.

Marsh, K. S. (1993). Packaging as a supplier of health, filler of landfills, and excuse for trade barriers [abstract]. In: International symposium on packaging, economic development and the environment. Chicago Ill. Herndon, Va . : Inst. of Packaging Professionals. 23–36.

Martel, J. L., Chaslus-Dancla, E., Coudert, M., Poumarat. R., and Lafont, J. P. (1995). Survey of antimicrobial resistance in bacterial isolates from diseased cattle in France. Microbial Drug Resistance. 1, 3:273-283.

Martin , B. , Mukesh, D., et al., (2018). Out in the cold:Identification of genomic regions associated with cold tolerance in the biocontrol fungus *Clonostachys rosea* through genome-wide association mapping. Frontiers Microbiol . 9:2844. www. frontiersin. Org

Martin K, Havens E, Boyle K, Matthews G, Schilling E, et al., (2012). If you stock it, will they buy it? Healthy food availability and customer purchasing behavior within corner stores in Hartford, CT. Public Health Nutrition 15: 1973–1978.

Martin, S. (1971). Food quality evaluation-a learning technique. J. Nutri. Edu. and Bev. 3, 2: 70-72.

Martindale, W. (2017). The potential of food preservation to reduce food waste. Proceedings of the nutrition society. 76, 1:28-33.

Martínez-Cuesta, M. C., Bengoechea, J., Bustos, I., Rodríguez, B., Requena, T., and Peláez, C. (2010). Control of late blowing in cheese by adding lacticin 3147-producing *Lactococcus lactis* IFPL 3593 to the starter. Int. Dairy J. 20:18–24.

Martínez-Cuesta, M. C., Bengoechea, J., Bustos, I., Rodríguez, B., Requena, T., and Peláez, C. (2010). Control of late blowing in cheese by adding lacticin 3147-producing *Lactococcus lactis* IFPL 3593 to the starter. Int. Dairy J. 20: 18–24.

Martins, J. T., Cerqueira, M. A., Souza, B. W., Carmo Avides, M. D., and Vicente, A. A. (2010). Shelf life extension of ricotta cheese using coatings of galactomannans from nonconventional sources incorporating nisin against *Listeria monocytogenes*. J. Agric. Fd. Chem. 58:1884–1891.

McAuliffe, O., Hill, C., and Ross, R. (1999). Inhibition of *Listeria monocytogenes* in cottage cheese manufactured with a lacticin 3147-producing starter culture. J. Appl. Microbiol. 86: 251–256.

McAuliffe, O., Ryan, M. P., Ross, R. P., Hill, C., Breeuwer, P., and Abee, T. (1998). Lacticin 3147, a broad-spectrum bacteriocin which selectively dissipates the membrane potential. Appl. Environ. Microbiol. 64: 439–445.

McBride, A. E. (2010). Food porn. *Gastronomica:The J. Crit. Fd. Stud.* 10 , 1:38-46. doi: 10. 1525/gfc. 2010. 10. 1. 38

McDonough, W., and Braungart, M. (2002). Cradle to cradle: remaking the way we make things. New York : North Point Press. 212 .

McKown, C. (2000). Containers. In: Coatings on glass - technology roadmap workshop. Sandia National Laboratories report. 8–10.

McLeod, A., Zagorec, M., Champomier-VergÃ¨s, M. C., Naterstad, K., and Axelsson, L. (2010). Proteomic approach identifies differentially expressed proteins caused by the change of carbon source. Primary metabolism in *Lactobacillus sakei* food isolates by proteomic analysis. 10:120. doi: 10. 1186/1471-2180-10-120.

McMeekin, T. A., and Ross, T. (2002). Predictive microbiology, providing a knowledge-based framework for change management. Intern. J. Fd. Microbiol. 78: 133–153.

McMeekin, T. A., Baranyi, J., Bowman, J., Dalgaard, P., Kirk, M., Ross, T., Schmid, S., Zwietering, M. H. (2006). Information systems in food safety management. International J. Fd. Microbiol. 112:181–194.

McMeekin, T. A., Mellefont, L. A., and Ross, T. (2007). Predictive microbiology, past, present and future. In: Brul, S., van Gerwen, S., Zwietering, M. (Eds.), Modelling microorganisms in food. Woodhead, Cambridge (UK). 7–21.

Mehrdad Tajkarimi. (2007). Materials from Maha Hajmeer. HACCP , GMPs, SSOPs. PHR250 5/9/07, 7p.

Melo, J., Andrew, P., and Faleiro, M. (2015). *Listeria monocytogenes* in cheese and the dairy environment remains a food safety challenge: the role of stress responses. Fd. Res. Int. 67: 75–90.

Melton-Celsa, A. R. (2014). Shiga Toxin (Stx)Classification, Structure, and Function. *Microbiol. Spectrum.* 2(2).

Metlitskaya, A., Kazakov, T., Kommer, A., Pavlova, O., Praetorius-Ibba, M., Ibba, M., et al. (2006). Aspartyl-tRNA synthetase is the target of peptide nucleotide antibiotic Microcin C. *J. Biol. Chem.* 281: 18033–18042. 10. 1074/jbc. M513174200.

Meyer, J. N. , Oppegård, C. , Rogne, P. , Haugen, H. S. and Kristiansen, P. E. (2010). Structure and Mode-of-Action of the Two-Peptide (Class-IIb) Bacteriocins. Probiotics Antimicrob Proteins. 2(1): 52–60. Published online 2009 Nov 3. doi: 10. 1007/s12602-009-9021-z.

Michael, G. M. D. (2017). The food safety risk of organic versus conventional. In health, NutritionFacts. org

Michael, P., Doyle, M. C., Erickson, W. Alali., Jennifer, C., Xiangyu, D., Ynes, O., Mary, A., and Smith, T. Z. (2015). The food industry's current and future role in preventing microbial food borne illness within the United States. Clin. Infect. Dis. 61, (2): 252–259.

Michel-Briand, Y., and Baysse, C. (2002). The pyocins of *Pseudomonas aeruginosa*. Biochimie. 84 (5–6): 499–510. doi:10. 1016/s0300-9084(02)01422-0. PMID 12423794.

MikÅ¡-Krajnik, M., Yuk, H. G., Kumar, A., Yang, Y., Zheng, Q., Kim, M., Ghate, V., Yuan, W., and Pang, X. (2015). Ensuring food security through enhancing microbiological food safety. NASA Astrophysics Data System (ADS).

Miles, D. C., and Briston, J. H. (1965). Polymer technology. New York : Chemical Publishing Co. Inc. p. 444.

Miller, C. (1993). Waste product profile: steel cans. Waste Age. 69.

Millette, M., Cornut, G., Claude, S. F. O., Archambault, D., and Onique, L. (2008). Capacity of human nisin- and pediocin-producing lactic acid bacteria to reduce intestinal colonization by vancomycin-resistant enterococci. Appl. Environ. Microbiol. 1997–2003. 74, 7.

Mills, S., Serrano, L., Griffin, C., O'connor, P. M., Schaad, G., Bruining, C., et al. (2011). Inhibitory activity of *Lactobacillus plantarum* LMG P-26358 against *Listeria innocua* when used as an adjunct starter in the manufacture of cheese. Microbial Cell Factories. 10: S7.

Ming, X., G. H. Weber, J. W. Ayres, and W. E. Sandine. (1997). Bacteriocins applied to food packaging materials to inhibit *Listeria monocytogenes* on meats. J. Fd. Sci. 62:413–415.

Montesinos, E. (2007). Antimicrobial peptides and plant disease control. FEMS Microbiol. Letts. 270, 1: 1-11.

Moreno, M. F., Sarantinopoulos, P., Tsakalidou, E., and De Vuyst, L. (2006). The role and application of enterococci in food and health. Int. J. Fd. Microbiol. 15, 106(1):1-24. Epub 2005.

Morgan, S. M., Hickey, R., Ross, R. P., and Hill, C. (2000). Efficient method for the detection of microbially produced antibacterial substances from food systems. J. Appl. Microbiol. 89, 1: 56-62.

Morgan, S., Galvin, M., Ross, R., and Hill, C. (2001). Evaluation of a spray-dried lacticin 3147 powder for the control of *Listeria monocytogenes* and *Bacillus cereus* in a range of food systems. Lett. Appl. Microbiol. 33: 387–391.

Morgan, S., O'sullivan, L., Ross, R., and Hill, C. (2002). The design of a three strain starter system for Cheddar cheese manufacture exploiting bacteriocin-induced starter lysis. Int. Dairy J. 12: 985–993.

Morgan, S., Ross, R., and Hill, C. (1997). Increasing starter cell lysis in Cheddar cheese using a bacteriocin-producing adjunct. J. Dairy Sci. 80: 1–10.

Mørtvedt, C. I., Nissen-Meyer, J., Sletten, K., and Nes, I. F. (1991). Purification and amino acid sequence of lactocin S, a bacteriocin produced by *Lactobacillus sake* L45. Appl. Environ. Microbiol. 57, 6 :1829–1834. PMC 183476. PMID 1872611.

Mosupye, F. M., and von Holy, A. (1999). Microbiological quality and safety of ready-to-eat street-vended foods in Johannesburg, South Africa. J. Fd. Prot. 62: 1278–1284.

Motiwala, A. S., Li, L., Kapur, V., and Sreevatsan, S. (2006). Current understanding of the genetic diversity of *Mycobacterium avium subsp. paratuberculosis*. Microbes Infect. 8, 5:1406-18. Epub 2006.

Muhialdin, Belal, J., Hassan, Z, M., and Sadon, S, Kh., (2011). Antifungal activity of *Lactobacillus fermentum* Te007, *Pediococcus pentosaceus* Te010, *Lactobacillus pentosus* G004, and *L. paracasei* D5 on selected foods. J. Fd. Sci. 76, 7 :M493-9. doi: 10. 1111/j. 1750-3841. 2011. 02292. x. Epub 2011.

Mulet-Powell et al., (1998). Interactions between pairs of bacteriocins from lactic bacteria. J. Fd. Prot. 61, 9:1210-2.

Müller, I., Lurz, R., and Geider, K. (2012). Tasmancin and lysogenic bacteriophages induced from *Erwinia tasmaniensis* strains. Microbiol. Res. 167, 7:381–387. doi:10. 1016/j. micres. 2012. 01. 005.

Muncke, J., Peterson Myers, J., Scheringer, M., and Porta, M. (2014). Food packaging and migration of food contact materials: will epidemiologists rise to the neotoxic challenge? Epidemiology and Community Health. Intl. J. Polymer Sci. DOI: 10. 1136/jech-2013-202593. Article ID 302029, 11 pages. http://dx. doi. org/10. 1155/2012/302029.

Muñoz, A., Ananou, S., Gálvez, A., Martínez-Bueno, M., Rodríguez, A., Maqueda, M., et al. (2007). Inhibition of *Staphylococcus aureus* in dairy products by enterocin AS-48 produced *in situ* and *ex situ*: bactericidal synergism with heat. Int. Dairy J. 17, 760–769.

Murkovic, M. (2004). Acrylamide in Austrian foods. J Biochem BiophysMethods. 29 , 61(1-2):161-7.

Nandane, A., Tapre, A., Ranveer., and Dr. Rahul. (2007). applications of bacteriocins as bio-preservative in foods:a review. ADIT J of Engg. 4: 50-55.

Narayanan, A., and Ramana, K. V. (2013). Synergized antimicrobial activity of eugenol incorporated polyhydroxybutyrate films against food spoilage microorganisms in conjunction with pediocin. Appl. Biochem. Biotechnol. 170: 1379–1388.

Nardis, C., Mastromarino, P., and Mosca, L. (2013). Vaginal microbiota and viral sexually transmitted diseases. Annali di Igiene. 25, 5: 443–56.

Narsaiah, K., Wilson, R. A., Gokul, K., Mandge, H., Jha, S., Bhadwal, S., et al. (2015). Effect of bacteriocin-incorporated alginate coating on shelf-life of minimally processed papaya (*Carica papaya* L.). Postharvest Biol. Technol. 100: 212–218.

NASA Image and Video Library. (2017). Food Production Info Sharing.

NASA Technical Reports Server (NTRS). (1981). Food packing optimization.

NASA Technical Reports Server (NTRS). (1992). Food Service System.

NASA Technical Reports Server (NTRS). (1997). Food Processing Control.

Natalie, R., Sabrina, K., and Beatrice, G. A. (2013). Mood food chocolate and depressive symptoms in a cross-sectional analysis. archives internal medicine. 170, 8: 699-703.

National Food Processors Association's Microbiology and Food Safety Committee. (1992). HACCP and total quality management-Winning concepts for the 90's. J. Food Prot. 55:459-462.

Natrajan, N., and Sheldon, B. W. (2000). Inhibition of *Salmonella* on poultry skin using protein- and polysaccharide-based films containing a nisin formulation. J. Fd. Protec. 63, 9:1268-72.

Neilson, A. (1985). Psychobiology and food perception. NASA Technical Reports Server (NTRS).

Netz, D. J. A., Sahl, N., Oliveira , S., and Bastos M. C. F. (2001). Molecular characterisation of aureocin A70, a multiple-peptide bacteriocin isolated from *Staphylococcus aureus*. J. Mol. Biol. 311: 939–949.

Netz, D. J., Pohl, Beck-Sickinger, A. G., Selmer, Pierik., and Sahl, H. G. (2002). Biochemical characterisation and genetic analysis of aureocin A53, a new, atypical bacteriocin from *Staphylococcus aureus*. J. Mol. Biol. 319: 745–756.

Netz, D. J., Sahl, H. G., Marcelino, R., dos Santos, Nascimento J., de Oliveira, S. S, Soares, M. B., do Carmo., and de Freire, Bastos, M. J. (2001). Molecular characterisation of aureocin A70, a multi-peptide bacteriocin isolated from *Staphylococcus aureus*. J. Mol. Biol. 31, 311(5) : 939-49.

Newslow, D. L. (2002). Effective HACCP plan: not just a fairytale. Food Quality. 28-37.

Nguyen, T. , Wilcock, A. and Aung, M. (2004). Food safety and quality systems in Canada. International Journal of Quality & Reliability Management. 21, 6:655-671. https://doi.org/10. 1108/02656710410542052

Nielsen, S. S. (2016). Introduction to food analysis. NASA Astrophysics Data System (ADS).

Nishie, M., Nagao, J., and Sonomoto, K. (2012). Antibacterial peptides "bacteriocins": an overview of their diverse characteristics and applications. Biocontrol Sci. 17, 1:1-16.

Nissen, H., Holo, H., Axelsson, L., and Blom, H. (2001). Characterization and growth of *Bacillus* spp. in heat-treated cream with and without nisin. J. Appl. Microbiol. 90: 530–534.

Nissen, H., Holo, H., Axelsson, L., and Blom, H. (2001). Characterization and growth of *Bacillus* spp. in heat-treated cream with and without nisin. J. Appl. Microbiol. 90: 530–534.

Nissen-Meyer, J., Rogne, P., Oppegård, C., Haugen, HS., and Kristiansen, P. E. (2013). Structure-function relationships of the non-lanthionine-containing peptide (class II) bacteriocins produced by gram-positive bacteria. Curr Pharm Biotechnol. 10: 19–37.

Nithya, V., Prakash, M., and Halami, P, M. (2018). Utilization of industrial waste for the production of bio-preservative from *Bacillus licheniformis* Me1 and Its Application in Milk and Milk-Based Food Products. Probiotics Antimicrob Proteins. 10(2):228-235. doi: 10. 1007/s12602-017-9319-1.

Novembre, E., de Martino, M., and Vierucci, A. (1988). Foods and respiratory allergy. J Allergy Clin. Immunol. 81(5 Pt 2):1059-65.

Nowotarska, S. W., Nowotarski, K., Grant, I. R., Elliott, C. T., Friedman, M., and Situ, C. (2017). Mechanisms of antimicrobial action of cinnamon and oregano oils, cinnamaldehyde, carvacrol, 2, 5-dihydroxybenzaldehyde, and 2-hydroxy-5-methoxybenzaldehyde against *Mycobacterium avium subsp. paratuberculosis* (Map). Foods. 6, 9: pii: E72.

Nunez, M., Rodriguez, J., Garcia, E., Gaya, P., and Medina, M. (1997). Inhibition of *Listeria monocytogenes* by enterocin 4 during the manufacture and ripening of Manchego cheese. J. Appl. Microbiol. 83: 671–677.

O'Connor, P. M., Ross, R. P., Hill, C., and Cotter, P. D. (2015). Antimicrobial antagonists against food pathogens: a bacteriocin perspective. Curr. Opin. Fd. Sci. 2: 51–57.

O'Driscoll B, Gahan CG, Hill C. (1996). Adaptive acid tolerance response in *Listeria monocytogenes*: isolation of an acid-tolerant mutant which demonstrates increased virulence. Appl. Environ. Microbiol. 62:1693–1698.

O'Sullivan, L., Ryan, M. P., Ross, R. P., and Hill, C. (2003). Generation of food-grade lactococcal starters which produce the lantibiotics lacticin 3147 and lacticin 481. Appl. Environ. Microbiol. 69:3681–3685.

Oexle, N., Barnes, T. L., Blake, C. E., Bell, B. A., and Liese, A. D. (2015). Neighborhood fast food availability and fast food consumption. Appetite. 92:227-32. doi: 10. 1016/j. appet. 2015. 05. 030. Epub 2015 May 27.

Ogunremi, O.R., Sanni, A.I. and Agrawal, R. 2015. Hypolipidemic and antioxidant effects of functional cereal-mix produced with probiotic yeast in rats fed high cholesterol diet. *Journal of Functional Foods*, 17: 742-748. Elsevier.

Ogunremi, O.R., Sanni, A.I. and Agrawal, R. 2015. Starter-culture and probiotic potentials of yeasts isolated from some cereal-based Nigerian traditional fermented food products. *Journal of Applied Microbiology*, 119:797-808. Wiley.

Oliveira, M., Abadias, M., ColÃ¡s-MedÃ , P., Usall, J., and ViÃas, I. (2015). Biopreservative methods to control the growth of foodborne pathogens on fresh-cut lettuce. Int. J. Fd. Microbiol. 2:214:4-11. doi: 10. 1016/j. ijfoodmicro. 2015. 07. 015. Epub 2015.

Olsen, A. R. (1998). Regulatory action criteria for filth and other extraneous materials. III. Review of flies and foodborne enteric disease. Regulatory Toxicol. Pharmacol. 28, 3:199–211.

Olszyna-Marzys, A E. 1991. Radioactivity and food. Bulletin of the Pan American Health Organization (ýPAHO). . 25, ý1: ýhttps://apps. who. int/iris/handle/10665/306330.

Ordinance and Code Regulating Eating and Drinking Establishments, Recommended by the United States Public Health Service. (1943). FSA, Public Health Bulletin No. 280 (Republished in 1955), DHEW, PHS Publication No. 37.

Orunremi, O.R. Sanni, A.I. and Agrawal, R. 2015. Development of cereal-based functional food using cereal-mix substrate fermented with probiotic strain-PIchia kudriazevii OG 32. *Food Science and Nutrition*, 3(6): 486-494. Wiley.

Oshima, S., Hirano, A., Kamikado, H., Nishimura, J., Kawai, Y., and Saito, T. (2014). Nisin A extends the shelf life of high-fat chilled dairy dessert, a milk-based pudding. J. Appl. Microbiol. 116:1218–1228.

Ott, D. B. (1988). Trends in Food Packaging. ERIC Educational Resources Information Center. Intern. J. Res. Innov. Soc. Sci. (IJRISS). ISSN No. 2454-6186.

Oumer, A., Garde, S., Gaya, P., Medina, M., and Nunez, M. (2001). The effects of cultivating lactic starter cultures with bacteriocin-producing lactic acid bacteria. J. Fd. Prot. 64: 81–86.

Ovchinnikov, K. V., Chi, H., Mehmeti, I., Holo, H., Nes , I. F., and Diep, D. B. (2016). Novel Group of Leaderless Multipeptide Bacteriocins from Gram-Positive Bacteria. Appl. Environ. Microbiol. 82, 17: 5216-24.

Padgett, T., Han, I. Y., and Dawson, P. L. (1998). Incorporation of food-grade antimicrobial compounds into biodegradable packaging films. J. Fd. Protect. 61, 10:1330–1335.

Pag, U., and Sahl, H. G. (2002). Multiple activities in lantibiotics–models for the design of novel antibiotics. Curr. Pharm Des. 8: 815–833. doi: 10. 2174/1381612023395439.

Page, B., Edwards, M., and May, N. (2003). Metal cans. In:Coles, R., McDowell, D., Kirwan, M. J. Eds. Food packaging technology. London, U. K. :Blackwell Publishing, CRC Press. 121–51.

Pandey, N., Malik, R. K., Kaushik, J. K., and Singroha, G. (2013). Gassericin A: a circular bacteriocin produced by lactic acid bacteria *Lactobacillus gasseri*. World J. Microbiol. Biotechnol. 29 , 11: 1977–87. doi:10. 1007/s11274-013-1368-3. PMID 23712477.

Parada, J. L., Caron, C. R., Bianchi, A. P. M., and Carlos, R. S. (2007). Bacteriocins from Lactic Acid Bacteria: Purification, Properties and use as Biopreservatives. Brazilian Archives Biol. Technol. 50, 3 : 521-542 . ISSN 1516-8913, Printed in Brazil.

Parente , E., Maria, A. G., Annamaria, R., and Francesco, C. (1998). The combined effect of nisin, leucocin F10, pH, NaCl and EDTA on the survival of *Listeria monocytogenes* in broth. Intern. J. Fd. Microbiol. 40, (1-2):65-75.

Parfitt, J., Barthel, M., and Macnaughton, S., (2010). Food waste within food supply chains: quantification and potential for change to 2050. *in* Philosophical Transactions of The Royal Society B. Biol. Sci. 365, 1554:3065-81.

Parks W. M., Bottrill A. R., Pierrat O. A., Durrant M. C., Maxwell A. (2007). The action of the bacterial toxin, microcin B17, on DNA gyrase. *Biochimie*. 89: 500–507. 10. 1016/j. biochi. 2006. 12. 005.

Patten, E. V., Alcorn, M., Watkins, T., Cole, K., Paez, P. (2017). Observed Food Safety Practices in the Summer Food Service Program. ERIC Educational Resources Information Center.

Pedersen, G. Dan. (2015). Development, validation and implementation of an *in vitro* model for the study of metabolic and immune function in normal and inflamed human colonic epithelium. Med J. 62(1):B4973.

Pedro, R. L., Juan, J. R. H., Daniel, V. S., and Marta, L. C. (2018). Review. Current knowledge on *Listeria monocytogenes* biofilms in food-related environments: incidence, resistance to biocides, ecology and biocontrol. Foods. 7, 85:1-19.

Peleg, M. (1993). Fractals and foods. Crit. Rev. Fd. Sci. Nutr. 33, 2:149-65.

Pellegrini, M., Chaves-Lapez, C., Mazzarrino, G., DaAmato, S., Lo Sterzo, C. (2018). Characterization of essential oils obtained from abruzzo autochthonous plants: antioxidant and antimicrobial activities assessment for food application. Foods. 7(2). pii:E19. doi: 10. 3390/foods7020019.

Perchonok, M., and Russo, D. M. (2001). Space Food Systems Laboratory. NASA Technical Reports Server (NTRS).

Perez, R. H., Zendo, T., and Sonomoto, K. (2014). Novel bacteriocins from lactic acid bacteria (LAB): various structures and applications. Microb. Cell Fact. 13: S3.

Piard, J. C., Muriana, P., Desmazeaud, M., and Klaenhammer, T. (1992). Purification and partial characterization of lacticin 481, a lanthionine containing bacteriocin produced by *Lactococcus lactis* subsp. lactis CNRZ 481. Appl. Environ. Microbiol. 58: 279–284.

Pimentel-Filho, N. D. J., Mantovani, H. C., Carvalho, A. F., Dias, R. S., and Vanetti, M. C. D. (2014). Efficacy of bovicin HC5 and nisin combination against *Listeria monocytogenes* and *Staphylococcus aureus* in fresh cheese. Int. J. Fd Sci. Technol. 49: 416–422.

Pingitore, E. V., Todorov, S. D., Sesma, F., and De Melo Franco, B. D. G. (2012). Application of bacteriocinogenic *Enterococcus mundtii* CRL35 and *Enterococcus faecium* ST88Ch in the control of *Listeria monocytogenes* in fresh Minas cheese. Fd. Microbiol. 32:38–47.

Popov, I. G. (1975). Food and water supply. NASA Technical Reports Server (NTRS).

Powell, D. J. (2011). Enhancing food safety culture to reduce rates of foodborne illness. Fd. Cntrl. 22: 817–822.

Prasert, S. (2012). Innovative Integrated Management System (IIMS) for Sustainable Food Industry. Nang Yan Business Journal. 1. 1, (Paper, 2 - 09):137.

Prudêncio, C. V., Dos Santos, M. T., and Vanetti, M. C. D. (2015). Strategies for the use of bacteriocins in Gram-negative bacteria: relevance in food microbiology. J. Fd. Sci. Technol. 52: 5408–5417.

Pucci, M. J., Vedamuthu, E. R., Kunka, B. S., and Vandenbergh, P. A. (1988). Inhibition of *Listeria monocytogenes* by using bacteriocin PA-1 produced by *Pediococcus acidilactici* PAC 1. 0. Appl. Environ. Microbiol. 54:2349–2353.

Pulido, Rubén ., Grande Burgos, Maria ., Gálvez, Antonio., and López, Rosario. (2015). Application of bacteriophages in post-harvest control of human pathogenic and food spoiling bacteria. Critical Reviews in Biotechnology. 36. 10. 3109/07388551. 2015. 1049935.

Quintavalla, S., and Vicini, L. (2002). Antimicrobial food packaging in meat industry. Meat Sci. 62, 373–380.

Rabbi, M. F., Eissa, N., Munyaka, P. M., Kermarrec, L., Elgazzar, O., Khafipour, E., Bernstein, C. N., and Ghia, J. E. (2017). Reactivation of intestinal inflammation is suppressed by catestatin in a murine model of colitis via m1 macrophages and not the gut microbiota. Front Immunol. 8:985.

Rai, M., Pandit, R., Gaikwad, S., and Kövics, G. (2016). Antimicrobial peptides as natural bio-preservative to enhance the shelf-life of food. J. Fd. Sci. Technol. 53(9):3381-3394. DOI: 10. 1007/s13197-016-2318-5.

Ramos, Ó. L., Pereira, J., Silva, S. I., Fernandes, J. C., Franco, M., Lopes-Da-Silva, J., et al. (2012). Evaluation of antimicrobial edible coatings from a whey protein isolate base to improve the shelf life of cheese. J. Dairy Sci. 95: 6282–6292.

Ramsaran, H., Chen, J., Brunke, B., Hill, A., and Griffiths, M. (1998). Survivial of bioluminescent *Listeria monocytogenes* and *Escherichia coli* 0157: H7 in soft cheeses. J. Dairy Sci. 81:1810–1817.

Ramu, R., Shirahatti, P. S., Devi, A. T., and Prasad, A. (2015). Bacteriocins and their applications in food preservation. Crit. Rev. Fd. Sci. Nutr. doi: 10. 1080/10408398. 2015. 1020918 [Epub ahead of print].

Rana, M. (2009). Effect of water, sanitation and hygiene intervention in reducing self reported waterborne diseases in rural Bangladesh, Research and Evaluation Division, BRAC Centre, 75 Mohakhali, Dhaka 1212, Bangladesh, www. brac. net/research.

Randell, A. W., and Whitehead, A. J. (1997). Codex Alimentarius:food quality and safety standards for international trade. Revue scientifique et technique (International Office of Epizootics) 16, 2:313-21.

Rao, M. S., Chander, R., and Sharma, A. (2005). Development of shelf stable intermediate moisture meat products using active edible chitosan coating and irradiation. J. Fd. Sci. 70: M325-331.

Rappole, C. L., and Louvier, S. A. (1985). Food Service Management NASA Technical Reports Server (NTRS).

Rasch, M., and Knochel, S. (1998). Variations in tolerance of *Listeria monocytogenes* to nisin, pediocin PA-! and bavaricin A. Lett. ppl. Microbiol. 27: 275-278.

Rather, I. A., Koh, W. Y. , Paek, W. K. , and Lim, J. (2017). The sources of chemical contaminants in food and their health implications. Front Pharmacol. 8: 830.

Rathje, W. L., Reilly, M. D., and Hughes, W. W. (1985). Household garbage and the role of packaging—the United States/Mexico City household refuse comparison. Tucson , Ariz. Solid Waste Council of the Paper Industry Publishing. p 116.

Rayman, K., Aris, B., and Hurst, A. (1981). Nisin: a possible alternative or adjunct to nitrite in the preservation of meats. Appl. Environ. Microbiol. 41:375-80.

Rayman, K., Malik, N., and Hurst, A. (1983). Failure of nisin to inhibit outgrowth of *Clostridium botulinum* in a model cured meat system. Appl. Environ. Microbiol. 46: 1450–1452.

Raymond Communications. (2005). U. S. recycling laws: state by state guide. College Park, Md. Raymond Communications Inc. Publishing. p 168.

Raymond Communications. (2006). Recycling laws international Vol. 12. [electronic subscription]. College Park , Md. : Raymond Communications Inc. Publishing.

Reale, A., Di Renzo, T., Rossi, F., Zotta, T., Iacumin, L., Preziuso, M., Parente, E., Sorrentino, E., and Coppola, R. (2015). Tolerance of *Lactobacillus casei, L. paracasei* and *L. rhamnosus* strains to stress factors encountered in food processing and in the gastro-intestinal tract. LWT Fd. Sci. Technol. 60: 721–728.

Rekhif, N., Atrih, A., and Lefebvre, G. (1994). Selection and properties of spontaneous mutants of *Listeria monocytogenes* ATCC 15313 resistant to different bacteriocins produced by lactic acid bacteria strains. Curr. Microbiol. 28 : 237-241.

Renu Agrawal and Shylaja Dharmesh. (2011). An antishigella protein from *Pediococcus pentosaceous* MTCC 5151. Turkish J. Biol. 36: 177-185.

Renu Agrawal., Vanaja, G., Villetchka, Gotcheva., and Angel, A. (2010). Differences in biochemical and electrophoretic properties in native *Lactobacillus plantarum* and probiotic culture strain *Lactobacillus plantarum* cfr on adaptation to GIT conditions with functional properties. *Res. J. Biotechnol.* 5, 2: 67-73.

Renye, J., Somkuti, G., Garabal, J., and Du, L. (2011). Heterologous production of pediocin for the control of *Listeria monocytogenes* in dairy foods. Fd. Cntrl. 22: 1887–1892.

Resa, C. P. O., Gerschenson, L. N., and Jagus, R. J. (2014). Natamycin and nisin supported on starch edible films for controlling mixed culture growth on model systems and Port Salut cheese. Fd. Cntrl. 44, 146–151.

Reuters. (2010). Industry Has Sway Over Food Safety System: US Study. Archived from the original on 14 September 2010.

Rezaei Javan, R., van Tonder, A. J., King, J. P., Harrold, C. L., and Brueggemann, A. B. (2018). Genome Sequencing Reveals a Large and Diverse Repertoire of Antimicrobial Peptides. Front Microbiol. 9:2012. doi: 10. 3389/fmicb. 2018. 02012. eCollection 2018.

Ribeiro, S. C., O'Connor, P. M., Ross, R. P., Stanton, C., and Silva, C. C. (2016). An anti-listerial *Lactococcus lactis* strain isolated from Azorean Pico cheese produces lacticin 481. Int. Dairy J. 63: 18–28.

Ribeiro, S. C., Ross, R. P., Stanton, C., and Silva, C. C. (2017). Characterization and application of antilisterial enterocins on model fresh cheese. J. Fd. Prot. 80: 1303–1316.

Rice, S., McAllister, E. J., and Dhurandhar, N. V. (2007). Fast food: friendly? Int J Obes (Lond). 31(6):884-6. Epub 2007

Riley , W. W. (2015). Advances in food technology and nutritional sciences. Food security: a long term issue. ISSN 2377-8350. Adv. Fd. Technol. Nutr. Sci. Open J. SE(1): Se3-Se4. doi: 10. 17140/AFTNSOJ-SE-1-e002.

Riley, M. A., and Wertz, J. E. (2002a). Bacteriocins: evolution, ecology, and application. Annu. Rev. Microbiol. 56:117–137. 10. 1146/annurev. micro. 56. 012302. 161024.

Riley, M. A., and Wertz, J. E. (2002b). Bacteriocin diversity: ecological and evolutionary perspectives. Biochimie 84:357–364.

Ring, J., Brockow, K., and Behrendt, H. (2001). Adverse reactions to foods. J. Chromatogr B. Biomed. Sci. Appl. 25:756(1-2):3-10.

Risch, S. J. (1988). [IFT] Institute of Food Technologists. Migration of toxicants, flavors, and odor-active substances from flexible packaging materials to food [IFT scientific status summary]. Fd. Tech. 42, 7:95–102.

Roberts, C. M., and Hoover, D. G., (1996). Sensitivity of *Bacillus coagulans* spores to combination of high hydrostatic pressure, heat, acidity and nisin. J. Appl. Bacteriol. 81: 363-368.

Roberts, T. (2007). WTP Estimates of the Societal Costs of U. S. Food-Borne Illness. *Amer. J. Agri. Econ.* 89, 5:1183–1188, https://doi. org/10. 1111/j. 1467-8276. 2007. 01081. x

Rochon, K., Lysyk, T. J., and Selinger, L. B. (2005). Retention of *Escherichia coli* by house fly and stable fly (Diptera: Muscidae) during pupal metamorphosis and eclosion. J. Med. Ent. 42, 3: 397–403.

Rodgers, S. (2016). Minimally Processed Functional Foods: Technological and Operational Pathways. J. Fd. Sci. 81, 10: R2309-R2319. doi:10. 1111/1750-3841. 13422. Epub 2016.

Rodney, H. P. , Takeshi, Z., and Kenji, S. (2014). Novel bacteriocins from lactic acid bacteria (LAB): various structures and applications. Microb. Cell Fact. 13(Suppl 1): S3.

Rodrígez, E., Arqués, J. L., Gaya, P., Nunez, M., and Medina, M. (2001). Control of *Listeria monocytogenes* by bacteriocins and monitoring of bacteriocin-producing lactic acid bacteria by colony hybridization in semi-hard raw milk cheese. J. Dairy Res. 68: 131–137.

Rodriguez, E., Arques, J. L., Nunez, M., Gaya, P., and Medina, M. (2005). Combined effect of high-pressure treatments and bacteriocin-producing lactic acid bacteria on inactivation of *Escherichia coli* O157: H7 in raw-milk cheese. Appl. Environ. Microbiol. 71: 3399–3404.

Rodríguez, E., Arqués, J. L., Rodríguez, R., Nuñez, M. and Medina, M. (2003). Reuterin production by *Lactobacilli* isolated from pig faeces and evaluation of probiotic traits. Lett. Appl. Microbiol. 37: 259-263.

Rodriguez-Pardo, D., Pigrau, C., Campany, D., Diaz-Brito, V., Morata, L., de Diego, I. C., Sorlí , L., Iftimie, S., Pérez-Vidal, R., García-Pardo, G., Larrainzar-Coghen, T., and Almirante, B. (2016). Effectiveness of sequential intravenous-to-oral antibiotic switch therapy in hospitalized patients with gram-positive infection: the sequence cohort study. Eur J. Clin. Microbiol. Infect. Dis. 35, 8:1269-76.

Rogers, L. A. (1928). The inhibiting effect of *Streptococcus lactis* on *Lactobacillus bulgaricus*. J. Bacteriol. 16: 321–325.

Rosalie, M. B. (2005). Scientific Manuscript database. USDA- ARS?s. Electrotechnologies for processing foods. "What's in the Foods You Eat—Search Tool. "

Ross, R. P., Galvin, M., Mcauliffe, O., Morgan, S. M., Ryan, M. P., Twomey, D. P., et al. (1999). Developing applications for lactococcal bacteriocins. Antonie Van Leeuwenhoek 76: 337–346.

Ross, R. P., Morgan, S., and Hill, C. (2002). Preservation and fermentation: past, present and future. Int. J. Fd. Microbiol. 79: 3–16.

Ross, R. P., Stanton, C., Hill, C., Fitzgerald, G. F., and Coffey, A. (2000). Novel cultures for cheese improvement. Trends Fd. Sci. Technol. 11:96–104.

Ruef, C. (2004). Epidemiology and clinical impact of glycopeptide resistance in *Staphylococcus aureus*. Infection. 32, 6:315-27.

Ryan, M. P., Rea, M. C., Hill, C., and Ross, R. P. (1996). An application in cheddar cheese manufacture for a strain of *Lactococcus lactis* producing a novel broad-spectrum bacteriocin, lacticin 3147. Appl. Environ. Microbiol. 62: 612–619.

SABS ISO 9000: 2000. Quality Management Systems ñ Fundamentals and Vocabulary. Pretoria, South Africa.

Sacharow, S., and Griffin Jr., R. C. (1980). The evolution of food packaging. In: Sacharow S, Griffin Jr. RC, Eds. Principles of food packaging. 2nd ed. Westport, Conn. AVI Publishing Co. Inc. 1–61.

Sacramento. (2004). In:California food processing industry technology roadmap. prepared by the Food Industry Advisory Committee & California Institute of Food and Agricultural Research University of California, Davis for the California Energy Commission Sacramento, CA July 2004.

Sahingil, D., Ýsleogl, H., Yildirim, Z., Akcelik, M., and Yildirim, M. (2011). Characterization of lactococcin BZ produced by *Lactococcus lactis* subsp. lactis BZ isolated from boza. Turk. J. Biol. 35: 21–33.

Salgado, P. R., Ortiz, C. M., Musso, Y. S., Di Giorgio, L., and Mauri, A. N. (2015). Edible films and coatings containing bioactives. Curr. Opin. Fd. Sci. 5: 86–92.

Salleh, NA., Rusul, G., Hassan, Z., Reezal A., Isa, SH., Nishibuchi M., and Radu S. (2003). Incidence of *Salmonella* spp. in raw vegetables in Selangor Malaysia. Fd. Cntrl. 14:475-479.

Samadi, N., Sharifan, A., Emam-Djomeh, Z., Sormaghi, M. H. Salehi. 2012-01-01. Biopreservation of hamburgers by essential oil of Zataria multiflora.

Sánchez-González, L., Saavedra, J. I. Q., and Chiralt, A. (2013). Physical properties and antilisterial activity of bioactive edible films containing *Lactobacillus plantarum*. Fd. Hydrocoll. 33: 92–98.

Sánchez-Hidalgo, M., Montalbán-López, M., Cebrián, R., Valdivia, E., Martínez-Bueno, M., and Maqueda, M. (2011). AS-48 bacteriocin: close to perfection. Cell. Mol. Life Sci. 68: 2845–2857.

Sandra, M. Roy., Margaret, A., Riley., and Joseph H . Crabb. (2016). Treating Bovine Mastitis with Nisin: A Model for the Use of Protein Antimicrobials in Veterinary Medicine. Prions: Current Progress in Advanced Research(2nd, Edition). 127-140. doi: https://doi. org/10. 21775/9781910190371. 07

Sathe, S. K., and Sharma, G. M. (2009). Effects of food processing on food allergens. Mol. Nutr. Fd. Res. 53, 8: 970-8. doi: 10. 1002/mnfr. 200800194.

Sawabe, K., Hoshino, K., Isawa, H. et al. (2006). Detection and isolation of highly pathogenic H5N1 avian influenza A viruses from blow flies collected in the vicinity of an infected poultry farm in Kyoto, Japan, 2004. Amer. J. Trop. Med. Hyg. 75, 2: 327–332.

Scallan, E., Griffin, P. M., Angulo, F. J., et al. (2011). Foodborne illness acquired in the United States—unspecified agents. Emerg. Infect. Dis. 17: 16–22.

Scallan, E., Hoekstra, R. M., Angulo, F. J., Tauxe, R. V., Widdowson, M. A., Roy, S. L., Jones, J. L., and Griffin., P. M. (2011). Foodborne illness acquired in the United States-major pathogens. Emerg. Infect. Dis. 1:7-15. doi: 10. 3201/eid1701. P11101.

Scallan, Elaine & Hoekstra., Mike & J Angulo., Frederick & V Tauxe., Robert & Widdowson., Marc-Alain & Roy, Sharon & L Jones., Jeffery & M Griffin, Patricia. (2011). Foodborne Illness Acquired in the United States—Major Pathogens. Emerging infectious diseases. 17:7-15. 10. 3201/eid1701. 091101p1.

Scannell, A. G., Hill, C., Ross, R., Marx, S., Hartmeier, W., and Arendt, E. K. (2000b). Development of bioactive food packaging materials using immobilised bacteriocins Lacticin 3147 and Nisaplin. Int. J. Fd. Microbiol. 60: 241–249.

Scannell, A. G., Hill, C., Ross, R., Marx, S., Hartmeier, W., and Arendt, E. (2000a). Continuous production of lacticin 3147 and nisin using cells immobilized in calcium alginate. J. Appl. Microbiol. 89: 573–579.

Scharff, R. L., (2010). health related-costs-from-foodborne-illness-in-the-united-states. http:// www. pewtrusts. org/en/research-and-analysis/reports/0001/01/01/healthrelated-costs-from-foodborne-illness-in-the-united-states.

Schillinger, U., Geisen, R., and Holzapfel, W. H. (1996):Potential of antagonistic microorganisms and bacteriocins for the biological preservation of foods. Trends Fd. Sci. Technol. 7, 5:158-164.

Schlyter, J. H., Glass, K. A., Loeffelholz, J., Degnan, A. J., Luchansky, J. B. (1993). The effects of diacetate with nitrite, lactate, or pediocin on the viability of *Listeria monocytogenes* in turkey slurries. Int. J. Fd. Microbiol. 19: 271–281.

Schmidt, L. E., and Dalhoff, Kim. (2002). Food-drug interactions. Drugs. 62, 10:1481-502. doi: 10. 2165/00003495-200262100-00005

Schneider, K. R., Silverberg, R., Chang, A., Goodrich, S., Renée, M. (2015). Preventing Foodborne Illness: *Clostridium botulinum. edis. ifas. ufl. edu*

School Business Affairs. (1983). Upgrading Food Service Operations. ERIC Educational Resources Information Center.

Schuenzel, K. M., and Harrison, M. A. (2002). Microbial antagonists of foodborne pathogens on fresh, minimally processed vegetables. J. Fd. Prot. 65, 12:1909-15.

Schwenk, N. E. (1991). Food Trends. ERIC Educational Resources Information Center.

Semeraro, E. F., Giuffrida, S., Cottone, G., and Cupane, A. (2017). Biopreservation of Myoglobin in Crowded Environment:A Comparison between Gelatin and Trehalose Matrixes. J. Phys. Chem. B. 121, 37:8731-8741. doi: 10. 1021/acs. jpcb. 7b07266. Epub 2017.

Servin, A. L. (2004). Antagonistic activities of *Lactobacilli* and *Bifidobacteria* against microbial pathogens. FEMS Microbiol. Rev. 28: 405– 440.

Settanni, L., and Corsetti, A. (2008). Application of bacteriocins in vegetable food biopreservation. Int. J. Fd. Microbiol. 121: 123 138.

SGS, (2013). Global Food Safety Conference 2013. Moderator: Rob Parrish, Vice President Global Food, SGS.

Shane, S. M., Montrose, M. S., and Harrington, K. S., (1985). Transmission of *Campylobacter jejuni* by the housefly (*Musca domestica*). Avian Diseases. 29, 2: 384–391

Shea, K. M. (2003). Pediatric exposure and potential toxicity of phthalate plasticizers. *Pediatrics.* 111, 6:1467–74.

Shekhar, S. R. et al. (1992). Food-Borne Disease Outbreaks in the Twin Cities of Hyderabad and Secunderabad (India) during 1984 to 1989. In: Proceedings of the 3rd World Conference of Food-Borne Infections and Intoxications. Berlin.

Shobharani, P., and Renu Agrawal. (2007). Therapeutic importance of volatile compounds produced by *Leuconostoc paramesenteroides*. Turkish J. Biol. 31:35-40.

Shobharani, P., and Renu Agrawal. (2008). Effect on cellular membrane fatty acids in the stressed cells of *Leuconostoc mesenteroides* sub sp. Dextranicum. A native probiotic lactic acid bacterium. Fd. Biotechnol. 22: 47-63.

Shobharani, P., and Renu Agrawal. (2010). Enhancement of cell stability and viability of probiotic *Leuconostoc mesenteroides* MTCC 5209 on freeze drying to be used in food formulation. Intern. J. Fd. Sci. Nutr. 60 , 56: 70-83.

Shobharani, P., and Renu Agrawal. (2010). Interception of quorum sensing signal molecule by furanone to enhance shelf life of fermented milk. Fd. Cntrl. 21, 1: 61-69.

Shobharani, P., and Renu Agrawal. (2011). Isolation and characterization of lactic acid bacteria from cheddar cheese as a potent probiotic strain. Ind. J. Microbiol. 51, 3: 251–258.

Shobharani, P., Ramesh, B., and Renu Agrawal. (2006). Strain improvement by mutagenesis and optimum conditions for culture parameters by response surface methodology for lactose tolerance in a novel native culture isolate *Leuconostoc mesenteroides* sub sp. Res. J. Biotechnol. 1, 2: 5-11.

Sicherer, S. H., Lack, G., Jones, S. M. (2014). Food allergy management. In: Adkinson NF Jr, Bochner BS, Burks AW, et al., (eds.) *Middleton's Allergy: Principles and Practice.* 8th ed. Philadelphia, PA: Elsevier Saunders, chap: 84.

Siegris M, Stampfli N, Kastenholz H, Keller C. (2008). Perceived risks and perceived benefits of different nanotechnology foods and nanotechnology food packaging. Appetite. 51, 2:283–290.

Silveira, A. C., Conesa, A., Aguayo, E., and Artes, F. (2008). Alternative sanitizers to chlorine for use on fresh cut "Galia" (*Cucumis melo* var. *catalupensis*) melon. J. Fd. Sci. 73, 9: M405– M411.

Simha, B. V., Sood, S., Kumariya, R., and Garsa, A. K. (2012). Simple and rapid purification of pediocin PA-1 from *Pediococcus pentosaceous* NCDC 273 suitable for industrial application. Microbiol. Res. 167: 544–549.

Sinclair. (1906). Food safety:A historical Look. Madeline Drexler, Copyright © 2011 Schuster Institute for Investigative Journalism at Brandeis University, Waltham, MA.

Singh, M. (2014). Mood, food, and obesity. Front Psychol. 1, 5:925. doi: 10. 3389/fpsyg. 2014. 00925. eCollection 2014.

Singh, V. Pal. (2018). Recent approaches in food bio-preservation - a review. Open Vet. J. 8, 1:104-111.

Siragusa, G. R. et al. (1998). The incidence of *Escherichia coli* on beef carcasses and its association with aerobic mesophilic plate count categories during the slaughter process. J. Fd. Prot. 61:1269-1274. PMid:9798140.

Siragusa, G. R., Cutter, C. N., and Willett, J. L. . (1999). Incorporation of bacteriocin in plastic retains activity and inhibits surface growth of bacteria on meat. Fd. Microbiol. 16: 229-235.

Sirmons, T. A., Cooper, M. R., and Douglas, G. L. (2016). food fortification stability study. NASA Technical Reports Server (NTRS).

Sirmons, T. A., Cooper, M. R., and Douglas, G. L. (2017). Food Fortification Stability Study. NASA Technical Reports Server (NTRS).

Smaoui, S., Hsouna, A. B., Lahmar, A., Ennouri, K., Mtibaa-Chakchouk, A., Sellem, I., Najah, S., Bouaziz, M., and Mellouli, L. (2016). Bio-preservative effect of the essential oil of the endemic *Mentha piperita* used alone and in combination with BacTN635 in stored minced beef meat. Meat Sci. 117: 196-204. doi: 10. 1016/j. meatsci. 2016. 03. 006. Epub 2016 Mar 10.

Smith, C., and White. P. (2000). Life cycle assessment of packaging. In: Levy GM. Ed. Packaging, policy, and the environment. Gaithersburg. Md. Aspen. 178-204.

Smith, K. D. (1977). ERIC Educational Resources Information Center.

Smith, M. C., Jr., Heidelbaugh, N. D., Rambaut, P. C., Rapp, R. M., Wheeler, H. O., Huber, C. S., and Bourland, C. T. (1975). Apollo food technology, NASA Technical Reports Server (NTRS).

Sobal, J., and LaChaussee, J. (1993). Social science perspectives in nutrition. Appetite, 21: 304.

Sobal, J., Bisogni, C. A. (2009). Constructing food choice decisions. Ann Behav Med. 38 Suppl 1:S37-46. doi: 10. 1007/s12160-009-9124-5.

Sobrino, C. C., Perianes, C. A., Puebla, J. L., Barrios, V., and Arilla, F. E. (2014). Peptides and food intake. Front. Endocrinol. (Lausanne) 5: 58.

Sobrino-López, A., and Martín-Belloso, O. (2008). Use of nisin and other bacteriocins for preservation of dairy products. Int. DairyJ. 18:329–343.

Soroka, W. (1999). Paper and paperboard. In: Emblem A, Emblem H, editors. *Fundamentals of packaging technology.* 2nd ed. Herndon , Va. : Inst. of Packaging Professionals. 95–112.

Space Food NASA Technical Reports Server. (NTRS). (1994).

Sparo, M. D., Confalonieri, A., Urbizu, L., Ceci, M., and Bruni, S. F. SÁ¡nchez. (2013). Bio-preservation of ground beef meat by *Enterococcus faecalis* CECT7121. Braz. J . Microbiol. 44, 1: 43-9. doi: 10. 1590/S1517-83822013005000003. eCollection 2013.

Speer, F. (1975). Multiple food allergy. Ann Allergy. 34: 71.

Spencer, M. (1974). Food additives. Food Safety. Science. gov Websites. veggies?Federal Pesticide Regulation Pesticides and Human Health Regulating Organic Food Production fruit and veggies? Federal Pesticide Regulation Pesticides and HumanHealth Regulating Organic Food:Environment Human Health Animal Health Safe Use Practices Food Safety Environment Air Water Soil Wildlife.

Staden, U., Rolinck-Werninghaus, C., Brewe, F., Wahn, U., Niggemann, B., and Beyer, K. (2007). Specific oral tolerance induction in food allergy in children: efficacy and clinical patterns of reaction. Allergy. 62: 1261–1269.

Stahl, C. H., T. R. Callaway., Lincoln L. M., Lonergan S. M., and Genovese K. J. . (2004). Inhibitory activities of colicins against *Escherichia coli* strains responsible for postweaning diarrhea and edema disease in swine. Antimicrob. Agents Chemother. 48:3119-21.

Stanton, R. A. (2015). Food Retailers and Obesity. Curr. Obes. Rep. 4, 1:54-9. doi: 10. 1007/s13679-014-0137-4.

Stern, et al. (2006). Stern Review on the Economics of Climate Change. ISBN: 0-521-70080-9). http://www. cambridge. org/9780521700801.

Stern, N. J., Svetoch, E. A., Eruslanov, B. V., Perelygin, V. V., Mitsevich, E. V., Mitsevich, I. P., et al. (2006). Isolation of a *Lactobacillus salivarius* strain and purification of its bacteriocin, which is inhibitory to *Campylobacter jejuni* in the chicken gastrointestinal system. Antimicrob. Agents Chemother. 50: 3111–3116 10. 1128/AAC. 00259-06

Stewart, M., Brown, J. B., Donner, A., McWhinney, I. R., Oates, J., Weston, W. W., Jordan, J. (2000). The impact of patient-centered care on outcomes. J. Fam. Pract. 49, 9:796-804.

Stilwell, E. J., Canty, R. C., Kopf, P. W., Montrone, A. M. (1991). Packaging for the environment: a partnership for progress. New York: American Management Assn. (Amacom) Publishing. 262 p.

Suda, S., Cotter, P. D., Hill, C., and Paul, R. R. (2012). Lacticin 3147-biosynthesis, molecular analysis, immunity, bioengineering and applications. Curr. Protein Pept. Sci. 13: 193–204.

Sudershan R. V. , Naveen Kumar, R. , Kashinath, L. , Bhaskar, V., and Polasa, K. (2014). Foodborne Infections and Intoxications in Hyderabad, India. Epidemiol. Res. Intern. 942961, 5. http://dx. doi. org/10. 1155/2014/942961

Sudha, N., Shobharani, P., and Renu Agrawal. (2006). Studies on the stability and viability of a local probiotic isolate *Pediococcus pentosaceous* (MTCC 5151) under induced gastrointestinal tract conditions. *J. Fd. Sci. Technol.* 43, 6: 677-678.

Sundh, I., and Melin, P. (2011). Safety and regulation of yeasts used for biocontrol or biopreservation in the food or feed chain. Antonie Van Leeuwenhoek. 99, 1:113-9. doi: 10. 1007/s10482-010-9528-z. Epub 2010.

Suphan, B. (2016). The use of pyocins in treating *Pseudomonas aeruginosa* infections. The Bacteriocins: Current knowledge and future prospects (Eds: Robert, L. Dorit, Sandra M. Roy and Margaret A. Riley). Caister Academic Press, UK. 81-102. DOI: https://doi. org/10. 21775/9781910190371. 05

Survival Kit - Food Kit NASA Image and Video Library1962-10-01S62-08743 (1962) - Food kit used by Mercury astronauts.

Surya Chandra Rao., Rachappa, S. B., and Renu Agrawal. (2007). Role of oxygen scavengers in improving the stability and viability of probiotic lactic acid bacteria. Res. J. Biotechnol. 2, 1: 26-32.

Swaminathan, M. S., and Bhavani, R. V. (2013). Food production & availability - Essential prerequisites for sustainable food security. Ind. J Med. Res. 138, 3: 383-391.

Sylvie, R. (2016). Microcins and Other Bacteriocins: Bridging the Gaps Between Killing Stategies, EcologyandApplications. CaisterAcademicPress, U. K. 11-34. https://doi. org/10. 21775/9781910190371. 02.

Synge, B. A., 2000. Verocytotoxin-producing Escherichia coli, a veterinary view. J. Appl. Microbio. 88: 31S–37.

SzÃpfalusi, Z. (2012). Food allergy in childhood.

Tasara, T., and Stephan, R. (2006). Review cold stress tolerance of *Listeria monocytogenes*: A review of molecular adaptive mechanisms and food safety implications. J. Fd. Prot. 69:1473–1484.

Tauxe, R. V., Doyle, M. P., Kuchenmüller, T. et al. (2010). Evolving public health approaches to the global challenge of foodborne infections. Int. J. Fd. Microbiol. 139:S16–28.

Taylor A. E., and Taylor, J. (2004). Using qualitative psychology to investigate HACCP implementation barriers. International journal of environmental health research. 14:53-63. 10. 1080/09603120310001633877.

Taylor, S. L. (1987). Allergic and sensitivity reactions to food components. In "Nutritional Toxicology, " 2, ed. J. N. Hathcock, 173-198. Academic Press, N.Y.

Texas Food Establishment Rules. Texas DSHS website: Texas Department of State Health Services. (2015). Texas Department of State Health Services Division for Regulatory Services Environmental and Consumer Safety Section Policy, Standards, and Quality Assurance Unit Public Sanitation and Retail Food Safety Group . p. 6.

Tharanathan, R. N. (2003). Biodegradable films and composite coatings: past, present and future. Trends Fd. Sci. Tech. 14, 3:71–8.

The ASQ the Quarterly Quality Report. (2007). "Managing the Customer Experience: A Measurement-Based Approach" by Morris Wilburn. ASQ Quality Press, Milwaukee (2006).

Therdtatha, P., Tandumrongpong, C., Pilasombut, K., Matsusaki, H., Keawsompong, S., and Nitisinprasert, S. (2016). Characterization of antimicrobial substance from *Lactobacillus salivarius* KL-D4 and its application as biopreservative for creamy filling. Springer Plus 5:1060.

Thurmond, B, A., Gillan, D. J., Perchonok, M, G., Marcus, B. A., and Bourland, C. T. (1986). Space Station Food System NASA Technical Reports Server (NTRS).

Tivadar, B., and Luthar, B. (2005). Food, ethics and aesthetics. Appetite, 44, 2:215–233.

Todorov, S. D., Vaz-Velho, M., Franco, B. D. G. M., Holzapfel, W. H. (2013). Partial characterization of bacteriocins produced by three strains of *Lactobacillus sakei*, isolated from salpicão, a fermented meat product from North-West of Portugal. Fd. Cntrl. 30: 111-121.

Tordesillas, L., Berin, M. C., and Sampson, H. A. (2017). Immunology of Food Allergy. Immunity. 18:47(1):32-50. doi: 10. 1016/j. immuni. 2017. 07. 004.

Torres, A. G., Jeter, C., Langley, W., and Matthysse, A. G. (2005). Differential binding of *Escherichia coli* O157:H7 to alfalfa, human epithelial cells, and plastic is mediated by a variety of surface structures. Appl. Environ. Microbiol. 71: 8008–8015.

Twomey, D., Ross, R. P., Ryan, M., Meaney, B., and Hill, C. (2002). Lantibiotics produced by lactic acid bacteria: structure, function and applications. Antonie van Leeuwenhoek. 82:165–185.

U. S. CDC (2009). Division of food borne, water borne, and Environmental diseases.

Ukuku, D. O. et al. (2005). Use of hydrogen peroxide in combination with nisin, sodium lactate and citric acid for reducing transfer of bacterial pathogens from whole melon surfaces to fresh-cut pieces. Inter. J. Fd. Microbiol. 104:225-233.

Urban, J. E., and Broce, A. (1998). Flies and their bacterial loads in greyhound dog kennels in Kansas. Curr. Microbiol. 36, 3:164–170,

Urquía-Fernández, N. (2013). Food security in Mexico. Salud Pública de México. 56: s92–s98.

US food and Drug Administration. (2018). www. ucsusa. org/2018survey. US Pharmacopeia. (2010). Adulteration of milk products and pet food with melamine underscores weaknesses of traditional methods, experts say. " ScienceDaily. ScienceDaily, 27 August. <www. sciencedaily. com/releases/2010/08/100826162007. htm>. USFDA. (2017). Summary report on antimicrobials sold or distributed for use in food- producing animals. December (2018).

USDA-ARS?s Scientific Manuscript database . Pectin in foods.

USFDA/FSIS [U. S. Food and Drug Administration/Food Safety and Inspection Agency (USDA)]. 2001. Draft Assessment of the relative risk to public health from foodborne Listeria monocytogenes among selected categories of ready-to-eat foods. Center for Food Safety and Applied Nutrition (FDA) and Food Safety Inspection Service (USDA) (Available at: www. foodsafety. gov/~dms/lmrisk. html). [Report published September 2003 as: Quantitative assessment of the relative risk to public health from foodborne Listeria monocytogenes among selected categories of ready-to-eat foods. Available at: www. foodsafety. gov/~dms/lmr2-toc. html].

Valdés, A., Ramos, M., Beltrán, A., Jiménez, A., and Garrigós, M. C. (2017). State of the art of antimicrobial edible coatings for food packaging applications. Coatings. 7:56.

Valenta, R., Hochwallner, H., Linhart, B., and Pahr, S. (2015). Food Allergies:The Basics. Gastroenterology. 148:1120-31. e4.

Valentina, S. (2012). Review Article. Food Packaging Permeability Behaviour:A Report. Hindawi Publishing Corporation Intern. J. Polymer Sci. ID302029, 11pages, doi:10. 1155/2012/302029.

Van der Spiegel, M., Luning, P. A., Ziggers, G. W., & Jongen, W. M. F. (2005). Development of the instrument IMAQE-food to measure effectiveness of food quality management. International Journal of Quality and Reliability Management, 22, 3: 234e255.

Van Heel, A. J., De Jong, A., Montalban-Lopez, M., Kok, J., and Kuipers, O. P. (2013). BAGEL3: automated identification of genes encoding bacteriocins and (non-) bactericidal posttranslationally modified peptides. Nucleic Acids Res. 41: W448–W453.

Van Willige, R. W. G., Linssen, J. P. H., Meinders, M. B. J., van der Steger, H. J., Voragen, A. G. J. (2002). Influence of flavor absorption on oxygen permeation through LDPE, PP, PC and PET plastics food packaging. Fd. Addit. Contam. 19, 3:303–13.

Vanaja, G., Villetchka, G., Angel, A., and Renu Agrawal. (2010). Volatile profile of compounds with therapeutic importance produced by probiotic culture strain of *Lb. plantarum* LB cfr. *J. Fd. Sci. Technol.* 6948, 1:110-113.

Vandera, E., Lianou, A., Kakouri, A., Feng, J., Koukkou, A. I., and Samelis, J. (2017). Enhanced control of *Listeria monocytogenes* by *Enterococcus faecium* KE82, a multiple enterocin–producing strain, in different milk environments. J. Fd. Prot. 80: 74–85.

Vanitha, R., and Renu Agrawal. (2012). Purification of protein from probiotic *Leuconostoc mesenteroides* active against *V. cholerae*. Res. J. Biotechnol. 7, 1: 38-42.

Vazquez-Guillamet, C., and Kollef, M. H. (2014). Treatment of Gram-positive infections in critically ill patients. BMC Infect. Dis. 14:92.

Veldhoen M., V. Brucklacher-Waldert. (2012). Dietary influences on intestinal immunity. Nat. Rev. Immunol. 12: 696–708.

Verhoeckx, K. C. M., Vissers, Y. M., Baumert, J. L., Faludi, R., Feys, M., Flanagan, S., Herouet-Guicheney, C., Holzhauser, T., Shimojo, R., van der Bolt, N., Wichers, H., Kimber, I. (2015). Food processing and allergenicity. Fd. Chem. Toxicol. 80:223-240. doi: 10. 1016/j. fct. 2015. 03. 005. Epub 2015.

Verma, S. K., Sood, S. K., Saini, R. K., and Saini, N. (2017). Pediocin PA-1 containing fermented cheese whey reduces total viable count of raw buffalo (*Bubalis bubalus*) milk. LWT Fd. Sci. Technol. 83: 193–200.

VidyaLaxme, B., Rovetto, A., Grau, Roberto., Agrawal, Renu. (2012). Synergistic effects of probiotic Leuconostoc mesenteroides and *Bacillus subtilis* in malted ragi (*Eleucine corocana*) food for antagonistic activity against *V. Cholerae* and other beneficial properties. J. Fd. Sci. Technol. 51: 10. 1007/s13197-012-0834-5.

Vincent P. A., and Morero R. D. (2009). The structure and biological aspects of peptide antibiotic microcin J25. *Curr. Med. Chem.* 16 : 538–549. 10. 2174/092986709787458461

Vincent, I. H., Eijsink, G. H., and Brurberg, M. B. (2008). Bacteriocins from plant pathogenic bacteria. FEMS Microbiol. Lett. 280, 1:1–7.

vom Saal, F. S., and Hughes, C. (2005). An extensive new literature concerning low-dose effects of bisphenol A shows the need for a new risk assessment. Environ Health Perspect. 113, 8: 926–33.

Wagner, H. S., Ahlstrom, B., Redden, J. P., Vickers, Z., and Mann, T. (2014). The myth of comfort food. Health Psychol. 33, 12:1552-7. doi: 10. 1037/hea0000068. Epub 2014.

Walker, K. (2005). Cafeteria Food. Research Brief. ERIC Educational Resources Information Center.

Wang, C. Yi., Huang, H. Wen., Hsu, C. P., and Yang, B. B. (2016). Recent advances in food processing using high hydrostatic pressure technology. Crit. Rev. *Fd.* Sci. Nutr. 56, 4:527-40. doi: 10. 1080/10408398. 2012. 745479.

Warnock, F. (2007). Final report on community-based intervention study of food safety practices in rural community households of Cambodia. Retrieved October 2018, from World Health Organization. http://www. who. int/foodsafety/consumer/ Cambodia_Dec07. pdf.

Watthanakulpanich, D. (2015). Food safety-related aspects of parasites in foods. J. Nutr. Sci. Vitaminol. (Tokyo). 61 Suppl:S96-7. doi: 10. 3177/jnsv. 61. S96.

Wayah, S. B., and Philip, K. (2018). Pentocin mq1: a novel, broad-spectrum, pore-forming bacteriocin from *lactobacillus pentosus* cs2 with quorum sensing regulatory mechanism and biopreservative potential. Front. Microbiol. 9:564. doi: 10. 3389/fmicb. 2018. 00564.

Weber, P. (1988). Food Business Entrepreneurship. NASA Astrophysics Data System (ADS).

Wender, E. H. (1977). Food Additives and Hyperkinesis. ERIC Educational Resources Information Center.

Werfel, T. (2016). Food allergy in adulthood. Bundesgesundheitsblatt Gesundheitsforschung Gesundheitsschutz. 59, 6:737-44. doi: 10. 1007/s00103-016-2360-5.

Wescombe, P. A., Upton, M., Dierksen, K. P., Ragland, N. L., Sivabalan, S., Wirawan, R. E., Inglis, M. A., Moore, C. J., Walker, G. V., Chilcott, C. N., Jenkinson, H. F., Tagg, J. R. (2006). Production of the lantibiotic salivaricin A and its variants by oral streptococci and use of a specific induction assay to detect their presence in human saliva. Appl. Environ. Microbiol. 72, 2: 1459–66. doi:10. 1128/aem. 72. 2. 1459-1466. 2006. PMC 1392966/ . PMID 16461700.

West, K. P., Stewart, C. P., Caballero, B., and Black, R. E. (2012) Nutrition: in Merson MH, Black RE, Mills AJ (eds): Global Health: Diseases, Programs, Systems, and Policies, Ed. 3. Burlington, Jones and Bartlett Learning. 271-304.

WHO (World Health Organization) (1999). Strategies for implementing HACCP in small and/or less developed businesses. Report on at WHO Consultation. WHO/SDE/PHE/FOS/99. 7. World Health Organization, Geneva, Switzerland.

WHO, 2019: Five keys to safer food Manual.

WHO. (2002). Reducing risks, Promoting Healthy Life. ISBN 9241562072(NLM Classification: WA 540. 1) ISSN 1020-3311.

WHO. (2004). Surveillance Programme for Control of Food-borne Infections and Toxications in Europe 8th report 1999-2000, Country reports: Turkey.

WHO. (2015). WHO HQ Food Safety: Regional Information World Health Day 2015: Food Safety.

WHO. (2017a). WHO strategic communications framework for effective communication. Retrieved October 2018, from World Health Organization: http://www. who. int/ mediacentre/communication-framework. pdf?ua=1.

WHO. (2017b). Communicating risk in public health emergencies:A WHO guideline for emergency risk communication (ERC) policy and practice. Geneva:World Health Organization.

WHO. (2017c). Did you know that superbugs can be found in food? Geneva: World Health Organization.

WHO. Food Safety and Foodborne Illness. (2002). (accessed May 3, 2006). Available: http://www . who. int/mediacentre /factsheets/fs237/en/

Wight, R. A., and Killham, J. (2014). Food mapping: a psychogeographical method for raising food consciousness. ERIC Educational Resources Information Center.

Wijnands, L. M. (2008). *Bacillus cereus* associated foodborne disease. Quantitative aspects of exposure assessment and hazard characterization. PhD thesis, WageningenUniversity.

Wild, L G: Lehrer, S B. (2001). Immunotherapy for food allergy. Curr Allergy Rep. 1, 1:48-53.

Willet, W. C., and Stampfer, M. J. (2003). Food Crystalization and Eggs. USDA-ARS?s Scientific Manuscript database. Food crystalization and eggs. Deana R. Jones, Ph. D. USDA Agricultural Research Service Egg Safety and Quality Research Unit Athens, Georgia, USA Deana. Rebuilding the Food Pyramid. ERIC Educational Resources Information Center.

Wilsey, D., and Dover, S. (2014). Personal Food System Mapping, ERIC Educational Resources Information Center. *Journal of Extension,* v52 n6 Article 6IAW5.

Winter, C. K. (2016). Presenting Food Science Effectively. ERIC Educational Resources.

Wood, W. (1990). Marketing School Food Services. ERIC Educational Resources Information Center.

Woods, E. J., and Thirumala, S. (2011). Packaging Considerations for Biopreservation. Transfus Med. Hemother. 38:149–156. https://doi. org/10. 1159/000326083.

Woods, E. J., Bagchi, A., Goebel, W. S., Vilivalam, V. D., and Vilivalam, V. D. (2010). Container system for enabling commercial production of cryopreserved cell therapy products. Regen Med. 5, 4: 659-67. doi: 10. 2217/rme. 10. 41.

Woraprayote, W., Kingcha, Y., Amonphanpokin, P., Kruenate, J., Zendo, T., Sonomoto, K.,

World Economic Forum. (2009). World Economic Forum Geneva, Switzerland. The Global Competitiveness Report 2009–2010 . by :Klaus Schwab.

World Food India. (2017). Food Regulatory Environment Knowledge Partner National Event Partner Inspiring Trust, Assuring Safe and Nutritious Food. FSSAI.

World Health Day. (2015). WHO HQ Food Safety: Regional Information World Health Day. 2015: Food Safety. FAO source: www. fao. org/animal-health/en/(as at 28 Feb 2015).

World Health Organization (WHO). (2003). Information concerning this publication can be obtained from:World Health Report World Health Organization, 1211 Geneva 27, Switzerland.

World Health Organization. (2018). Whole genome sequencing for foodborne disease surveillance: landscape paper ISBN 978-92-4-151386-9.

World Health Organization. (WHO). (2019). Five keys to safer food Manual. NLM classification WA 695. ISBN 9241594632. ISBN 9789241594639.

World Health Report. (2000). Information concerning this publication can be obtained from: World Health Report World Health Organization 1211 Geneva 27, Switzerland.

Wright, W., and Nault, K. (2013). Growing Youth Food Citizens, ERIC Educational Resources Information Center.

Wüthrich, B . (2009). Food allergy, food intolerance or functional disorder? Praxis (Bern 1994). 98, 7:375-87. doi: 10. 1024/1661-8157. 98. 7. 375.

www. cdc. gov/norovirus. (2017). Norovirus: Facts for Food Workers. CS273773-A. Revised January.

www. foodrisk. org

Xavier, B. (2007). Inadequate housing and health:an overview. Int. J. Environment and Pollution. 30: 3-4.

Xiao-Qing, Q. and Margaret, A. Riley. (2016). Capturing the Power of Van der Waals Zone in the Creation of a Novel Family of Bacteriocin-based Antibiotics. 65-80. DOI: https://doi. org/10. 21775/9781910190371. 04. Prions:Current Progress in Advanced Research (2nd

Edition) Caister Academic Press, U. K. 65-80. Edited by: Robert L. Dorit, Sandra M. Roy and Margaret A. Riley. ISBN 978-1-910190-37-1.

Yamauchi, Y., Ishii, S., Toyoda, S., and Ahiko, K. (1996). Process for the manufacture of fermented milk. U. S. Patent 5:527, 505. Washington, DC: U. S. Patent and Trademark Office.

Yang, Y., Luo, Y., Millner, P., Turner, E., Feng, H. (2012). Assessment of *Escherichia coli* O157:h7 transference from soil to iceberg lettuce via a contaminated field coring harvesting knife. Intern. J. Fd. Microbiol. 153: 345–350.

Yildirim, Z., Öncül, N., Yildirim, M., and Karabiyikli, [a]. (2016). Application of lactococcin BZ and enterocin KP against *Listeria monocytogenes* in milk as biopreservation agents. Acta Aliment. 45, 486–492.

Zacharof, M., and Lovitt, R. (2012). Bacteriocins produced by lactic acid bacteria a review article. APCBEE Procedia 2: 50–56.

Zapico, P., Medina, M., Gaya, P., and Nuñez, M. (1998). Synergistic effect of nisin and the lactoperoxidase system on *Listeria monocytogenes* in skim milk. Int. J. Fd. Microbiol. 40:35–42.

Zbrun, M. V., Frizzo, L. S., Soto, L. P., Rosmini, M. R., Sequeira, G. J., Astesana, D. M., Blajman, J. E., Rossler, E., Berisvil, A., Romero Scharpen, A., and Signorini, M. L. (2016). Poultry blood from slaughterhouses: development of a biopreservation system to improve microbiological quality prior to transforming blood into by-products.

Zendo, T. (2013). Screening and characterization of novel bacteriocins from lactic acid bacteria. Biosci. Biotechnol. Biochem. 77, 5:893-9. Epub 2013.

Zhang, K., Sparling, J., Chow, B. L. et al., (2004). New quadriplex PCR assay for detection of methicillin and mupirocin resistance and simultaneous discrimination of *Staphylococcus aureus* from coagulase-negative *Staphylococci*. J. Clin. Microbiol. 42, 11: 4947–4955.

Zhao, L., Wang, S., Liu, F., Dong, P., Huang, W., Xiong, L., et al. (2013). Comparing the effects of high hydrostatic pressure and thermal pasteurization combined with nisin on the quality of cucumber juice drinks. Innov. Fd. Sci. Emerg. Technol. 17: 27–36.

Zhong, J., and Wang, X. (2019). Evaluation Technologies for Food Quality. 1st Edition. Paperback ISBN: 9780128142172. Imprint: Woodhead Publishing. 2019. 936.

Zottola EA Sasahara KC (1994) Microbial biofilms n the food processing industry – should they be a concern?Int J Food Microbiol 23: 125–148.

Zudaire, L., Vias, I., Plaza, L., Iglesias, M. B., Abadias, M., and Aguil. Ã. Aguayo, I. (2018). Evaluation of post harvest calcium treatment and biopreservation with *Lactobacillus rhamnosus* GG on the quality of fresh-cut Conference pears. J Sci Food Agric. 98, 13:4978-4987. doi: 10. 1002/jsfa. 9031.

Zulfiqar, A. M., Afsheen, A., Shagufta, N., Anila, S., Mohammad, N, K., Seema, I. Khan. (2017). Prevalence of *Staphylococci* in Commercially Processed Food Products in Karachi-Pakistan. J. Microbiol. Infect. Dis. 7 (2): 83-87. doi: 10. 5799/ jmid. 328789.